FORENSIC SCIENCE
Fundamentals & Investigations

Anthony J. Bertino

Canandaigua Academy High School
Canandaigua, New York

Patricia Nolan Bertino

Contributing Author
Scotia-Glenville High School
Scotia, New York

SOUTH-WESTERN
CENGAGE Learning

Australia • Brazil • Japan • Korea • Mexico • Singapore • Spain • United Kingdom • United States

SOUTH-WESTERN
CENGAGE Learning™

Forensic Science: Fundamentals and Investigations, 1st edition
Anthony J. Bertino

VP/Editorial Director: Jack W. Calhoun

VP/Editor-in-Chief: Karen Schmohe

Senior Developmental Editor: Dave Lafferty

Senior Marketing Manager: Courtney Schulz

Content Project Manager: Jennifer A. Ziegler

Consulting Editor: Mike Tracy

Production Manager: Patricia Matthews Boies

Technology Project Editor: Bryan England

Web Site Project Manager: Ed Stubenrauch

Manufacturing Coordinator: Kevin Kluck

Production Service: Lachina Publishing Services

Art Director: Bethany Casey

Internal Designer: Joe Devine, Red Hangar Design, LLC

Cover Designer: Joe Devine, Red Hangar Design, LLC

Cover Image: ©istockphoto.com/Jeff Metzger

Photography Permissions Manager: Deanna Ettinger

Photography Researcher: Susan Van Etten Lawson

For product information and technology assistance, contact us at
Cengage Learning Academic Resource Center, 1-800-423-0563

For permission to use material from this text or product, submit all requests online at **www.cengage.com/permissions** Further permissions questions can be emailed to **permissionrequest@cengage.com**

ExamView® and ExamView Pro® are registered trademarks of FSCreations, Inc.

© 2008 Cengage Learning. All Rights Reserved.

Student Edition ISBN 13: 978-0-538-44586-3
Student Edition ISBN 10: 0-538-44586-6

South-Western Cengage Learning
5191 Natorp Boulevard
Mason, OH 45040
USA

Cengage Learning products are represented in Canada by Nelson Education, Ltd.

For your course and learning solutions, visit **school.cengage.com**

Printed in the United States of America
4 5 6 7 11 10 09 08

Dedication

To Arthur Nolan for his interest and invaluable technical support for this project.

Preface

WELCOME TO
FORENSIC SCIENCE: FUNDAMENTALS AND INVESTIGATIONS

Finally, a textbook that provides the science behind forensics, as well as labs and activities appropriate for high school students! *Forensic Science: Fundamentals and Investigations* is student *and* teacher friendly. Teachers can conduct a full-year's study of forensics or select topics that can be incorporated into a half-year course. As another option, teachers can use the textbook to motivate students in all science classes by using forensics to teach basic science concepts. *Forensic Science: Fundamentals and Investigations* integrates science, mathematics, and writing skills by using real-life applications and case studies, and providing complete flexibility for any science program. *Forensic Science: Fundamentals and Investigations* is the new standard in high school forensic science . . . case closed!

GETTING STARTED

Forensic Science: Fundamentals and Investigations explains the science used in forensic science techniques. It provides a chapter-by-chapter description of specific types of evidence and the techniques to collect and analyze the evidence. As students progress through the course, they refine the techniques and apply them to other areas of study. The topics covered in the seventeen chapters include crime-scene investigation; the collection, handling, and examination of trace evidence such as hair, fibers, soil, pollen, and glass; fingerprint, blood and blood spatter examination; DNA, drug, handwriting, and tool mark analysis; impressions; ballistics; forensic anthropology; and the determination of the cause and time of death.

One of the strengths of the textbook is student motivation. Areas of study are introduced in scenarios taken from headlines and popular media. These features engage students as they describe the historical development of forensic science techniques. Inexpensive, easy-to-use labs provide students with opportunities for successful laboratory experiences as well as an appreciation of the true nature of forensic science problem-solving techniques. Suggestions for research projects extend and enrich student learning and interest.

CHAPTER FEATURES

Each chapter of *Forensic Science: Fundamentals and Investigations* begins with a true-life story, student objectives, key vocabulary, and a Key to Science Topics. The Key to Science Topics identifies biology, earth science, chemistry, physics, psychology, or mathematics concepts integrated into chapter topics.

Special features include *Did You Know* margin notes that provide additional interesting facts and information, and *Digging Deeper*, additional research topics that refer students to the free online InfoTrac® database from Gale Publishing called the Forensic Science eCollection Database.

At the end of each chapter, a *Chapter Summary* reviews the main points of the chapter. A series of short *Case Studies* offer high-interest topics for critical thinking, writing, and class discussion. *Careers in Forensics* describes occupations related to forensic science. A *Chapter Review* contains both objective and short-answer questions to assess student understanding. A chapter bibliography lists the research sources.

Each chapter has *Activities* that provide hands-on experiences with forensic science techniques. Each activity has clear, step-by-step directions for students of all reading levels. For teachers, they offer easy, quick preparation and minimal expense for materials. Each activity includes objectives, materials, safety precautions, procedures, and other learning tools.

FOR THE TEACHER

The *Wrap-around Teacher's Edition* (ISBN 0-538-44633-1) contains teaching strategies and tips to engage students. It provides clarification of science content and forensic science procedures, ideas to help stimulate students, evaluation opportunities, additional questions, and suggestions for further exploration and research.

An *Instructor's Resource CD-ROM* (ISBN 0-538-44632-3) is available to teachers who adopt a classroom set of *Forensic Science: Fundamentals and Investigations*. The CD contains additional activities, PowerPoint Presentations, student activity forms, rubrics, content blueprints, and enrichment materials, as well as comprehensive teaching objectives for each chapter.

Teachers can purchase a flexible, easy-to-use *ExamView® Electronic Test Bank and Test Generation Software CD-ROM* (ISBN 0-538-44631-5) that contains objective questions that cover textbook content. The test bank includes questions for each chapter and a final exam. The ExamView software enables teachers to modify questions from the test bank or include teacher-written questions to create customized tests.

A supplementary *eBook CD-ROM* (ISBN 0-538-44629-3) is also sold separately. The eBook is a digitized version of *Forensic Science: Fundamentals and Investigations*. It offers the same rich photographs, illustrations, other graphics, and easy-to-read fonts as the printed text! Students view the PDF files on their computers.

Cengage maintains a *Web site* to support this text. Both students and teachers using *Forensic Science: Fundamentals and Investigations* may access the Web site at **school.cengage.com/forensicscience**. The site provides teacher resources and information about related products. Student resources on the site include forms, additional projects, and links to related sites. In addition, a link is provided to the *Gale Forensic Science eCollection Databas*e which allows free online research in various journals and the Gale Virtual Reference Library.

ABOUT THE AUTHORS

Anthony (Bud) Bertino has taught science for forty years. He has served as biology teacher and science supervisor at Canandaigua Academy. His awards include Outstanding Biology Teacher (NY, NABT), Woodrow Wilson Fellowship, Tandy Scholars' Award, and Outstanding Teaching Award from the University of Rochester. He is co-author of "Where's the CAT," "The Cookie Jar Mystery," and several other published activities. He presently serves as an AP biology consultant for the college board and as a clinical supervisor for the University of Albany (NY), Department of Education.

Patricia Nolan Bertino has taught science for thirty-five years. Her awards include Outstanding Biology Teacher (NY, NABT), Woodrow Wilson Fellowship, and the Tandy Scholars' Award. She has served as a scientific consultant for Video Discovery, Neo Sci, and several publishers. Patricia developed curricula and taught high school forensic science and biology at Scotia-Glenville High School for several years.

The Bertinos live near Schenectady, New York and conduct summer workshops in forensic science, criminal justice topics, and advanced placement biology. They are yearly co-presenters at both the National Association of Biology Teachers Conferences (NABT) and the National Science Teachers Conferences (NSTA).

Brief Contents

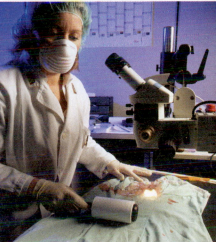

Contents

Chapter 1
Observation Skills 2

Chapter 2
Crime-Scene Investigation and Evidence Collection 20

Chapter 3
The Study of Hair 48

Chapter 4
A Study of Fibers and Textiles   76

Chapter 5
Pollen and Spore Examination   106

Chapter 6
Fingerprints   132

Chapter 7
DNA Fingerprinting 158

Chapter 8
Blood and Blood Spatter 194

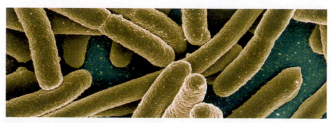

Chapter 12
Soil Examination 338

Chapter 13
Forensic Anthropology: What We Learn from Bones 360

Chapter 14
Glass Evidence 394

Chapter 15
Casts and Impressions   430

Chapter 16
Tool Marks   468

Chapter 17
Ballistics 490

Glossary 514

Appendices 523

Index 531

How to Use This Book

CHAPTER 1

Observation Skills

WAS SOMEONE STEALING THE TREES?

An officer with the Department of Natural Resources was called to a farm where a landowner had discovered missing trees. The trees were black walnut, a valuable wood used to make expensive furniture. The officer found six stumps where once there were living trees. The limbs and branches were left behind. Scattered around the woods were 20 empty beer cans.

The officer examined the area and found tracks left by a truck leading across a neighbor's field; the perpetrator of the theft had then cut through the boundary fence. By following the tracks, the officer found where the truck had slid sideways and scraped against a tree, leaving a small smear of paint. These pieces of evidence were photographed and sampled.

The landowner remembered having seen similar tire marks leading into another wooded area two miles up the road. The officer investigated these marks and found several more black walnut stumps and more empty beer cans. The officer documented numerous forms of evidence—a paint sample from the truck, tire tread impressions, and one fingerprint lifted from a beer can. The thefts stopped, and the case was considered unsolved.

Two years later, a man was caught stealing black walnut trees a couple of counties away, and his truck was impounded. The officer compared the original paint sample to matching paint from the truck. A receipt in the truck from a veneer mill (veneer is the thin layer of high-value wood put on the surface of low-quality woods to be used in furniture) suggested that the man had been selling logs for some time.

The paint on his truck was consistent with paint found at the crime scene, and his fingerprints matched the fingerprint found on the beer can at the scene. Based on the evidence, he was convicted, fined, and so six years. An observant investigation to collect sufficient evidence for a man guilty of stealing the trees.

An investigator examines paint evidence

Scenario *introduces a real-life story that sets the stage for the topics covered in the chapter.*

The Key to Science Topics *identifies the topics in a chapter that involve biology, earth science, chemistry, physics, psychology, or mathematics concepts.*

OBJECTIVES

By the end of this chapter, you will be able to:

✓ Define *observation* and describe what changes occur in the brain.

✓ Describe examples of factors influencing eyewitness accounts of events.

✓ Compare the reliability of eyewitness testimony to what actually happened.

✓ Relate observation skills to their use in forensic science.

✓ Define forensic science.

✓ Practice and improve your own observation skills.

TOPICAL SCIENCES KEY

VOCABULARY

analytical skills the ability to identify a concept or problem, to isolate its component parts, to organize information for decision making, to establish criteria for evaluation, and to draw appropriate conclusions

deductive reasoning deriving the consequences from the facts using a series of logical steps

eyewitness a person who has seen someone or something and can communicate these facts

fact a statement or assertion of information that can be verified

forensic relating to the application of scientific knowledge to legal questions

logical the process of forming conclusions drawn from assumptions and known facts

observation what a person perceives using his or her senses

opinion personal belief founded on judgment rather than on direct experience or knowledge

perception interpreting information received from the senses

Chapter Navigation tool *shows which chapter you are in and allows easier searching for information.*

Objectives *list the main learning objectives for the chapter.*

Vocabulary *lists key terms in the chapter and definitions.*

SOIL PROFILES

Soils form in horizons, or layers, that are more or less parallel to Earth's surface. The soil in each horizon has characteristic properties that differ from those in other horizons, as shown in Figure 12-2. Soil in a given area will have a unique profile or sequence of layers. Soil horizons within the profile are labeled with an uppercase letter, named as follows:

The uppermost horizon is called the O horizon. It is made up mostly of decaying organic matter, sometimes referred to as **humus**.

Beneath the O is the A horizon. The soil here is dark in color. The A horizon is also called topsoil. Topsoil is a mixture of humus and mineral particles. Seeds sprout and plant roots grow in the A horizon.

Next is the E horizon. The soil in the E horizon is light in color. It is made up mostly of sand and silt. Water dripping through the soil in this layer carries away most of the minerals and clay originally present. This process is called **leaching**.

Figure 12-2. A typical soil profile with horizons labeled.

The B horizon lies beneath the E horizon. Another name for this layer is the subsoil. The subsoil contains clay and mineral deposits that have leached out from layers above it as water drips through from the horizons above.

The C horizon is next. This layer is made of partially broken-up rock. Plant roots do not grow in this layer. Also, very little humus is found in this layer. If there is a solid rock layer underneath all of the other horizons, it is called the R horizon.

CHEMISTRY OF THE SOIL

The materials that make up a type of soil determine that soil's chemical properties. An important chemical property of soil is whether it is acidic or basic (alkaline). Chemists use a special scale, called the pH scale, to indicate how acidic or basic a substance is. The pH scale ranges from 0 to 14 (Figure 12-3). Anything with a pH of less than 7 is acidic. Substances with a pH greater than 7 are basic. A substance with a pH of 7, such as pure water, is considered neutral.

Figure 12-3. The pH scale.

	strong acid		weak acid			neutral			weak base			strong base	
pH value =	1	2	3	4	5	6	7	8	9	10	11	12	13	14

Examples	pH value (approximate)	Examples	pH value (approximate)
Acidic substances		Basic substances	
Battery acid	1	Baking soda, sea water	8.5
Lemon juice	2.5	Milk of Magnesia	10.5
Orange juice	3.0	Detergents	10.0
Vinegar	3.5	Ammonia water	11.0
Breads, pasta	5.0	Bleaches, oven cleaner	12.0
Rain (not acid)	5.5	Lye (drain cleaner)	13.5
Milk	6.5		

342   Soil Examination

Figures *provide visual support for key concepts, including photographs and diagrams; charts and tables organize key information.*

Did You Know *provides margin notes with extra information and facts.*

Science icons *identify specific science disciplines — biology, earth science, chemistry, physics, psychology, and mathematics — in the chapter.*

Digging Deeper *offers additional research topics with reference to the free online InfoTrac® database from Gale Publishing called the* **Forensic Science eCollection database**.

---

forensic toxicologists to popularize these new methods were physicians Mathieu Orfila (1787–1853) and Robert Christison (1797–1882).

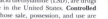

Did You Know?

If two people use the exact same amount of a drug and are tested, the person with darker hair will retain more drug in his or her hair than the lighter haired person.

MURDER BY POISON

Although poisoning is popular in murder mysteries and detective stories, in reality, it is not a common form of murder. Less than one-half of 1 percent of all homicides result from poisoning. Throughout history, some notable individuals have died from poisoning: Nazi leaders Heinrich Himmler and Hermann Goering ingested cyanide capsules in 1945; Jonestown cult members consumed cyanide-laced punch in 1978, killing approximately 900 people; Bulgarian dissident Georgi Markov was killed by ricin in 1978; and most recently, Russian ex-spy Alexander Litvinenko was exposed to radiation in 2006. Today, the commonly used poisons include arsenic, cyanide, and strychnine, as well as an assortment of industrial chemicals that were created for other uses, such as fertilizers.

Testing for a vast array of possible toxins can be a challenge to the toxicologist. Toxicologists must distinguish between acute poisoning and chronic poisoning. *Acute poisoning* is caused by a high dose over a short period of time, such as cyanide ingestion or inhalation, which immediately produces symptoms. *Chronic poisoning* is caused by lower doses over long periods of time, which produces symptoms gradually. Mercury and lead poisoning are examples of chronic poisoning in which symptoms develop as the metal concentrations slowly rise and accumulate to toxic levels in the victims' bodies over a long period of exposure.

ACCIDENTAL DRUG OVERDOSES

Accidental deaths from drug overdoses are more common than deaths from poisoning. The deaths of comedians John Belushi and Chris Farley, actor River Phoenix, and musicians Steve Clark, Janis Joplin, Jim Morrison, and Jimi Hendrix were all linked to lethal drug combinations or overdoses.

DRUGS AND CRIME

Illegal drugs, such as heroin and lysergic acid diethylamide (LSD), are drugs with no currently accepted medical use in the United States. **Controlled substances** are defined as legal drugs whose sale, possession, and use are restricted because of the effect of the drugs and the potential for abuse. These drugs are medications, such as certain **narcotics**, depressants, and stimulants, that physicians prescribe for various conditions.

Arrests for drug abuse violations have increased steadily since the early 1990s. Drug abuse violations topped the list of the seven leading arrest offenses in

Digging Deeper
with Forensic Science e-Collection

Do research on modern detection methods and the techniques of forensic toxicology or drug-testing work, such as the different chromatography and spectrometry methods. Go to the Gale Forensic Sciences eCollection on school.cengage.com/forensicscience and research the various methods. Determine which methods are more appropriate for the major types of controlled substances. Relate your findings to the chemical properties of the major controlled substances. Cite any limitations or concerns in using any of these methods for drug testing.

Drug Identification and Toxicology   259

End of Chapter

Summary *recaps the key points of the chapter, linked to the learning objectives.*

Case Studies *provide stories based on topics for critical thinking.*

Careers *highlight an occupation in forensics related to chapter content.*

Web site references *call attention to the additional resources on school.cengage.com/ forensicscience.*

End-of-Chapter Review *provides true or false, multiple choice, and short answer questions for student evaluation.*

Bibliography *lists the sources for additional research.*

SUMMARY

- Ballistics is the study of bullets and firearms, which are weapons capable of firing a projectile using a confined explosive, such as gunpowder.
- Modern firearms are divided into two basic types—long guns and handguns—that require only one or two hands, respectively, for accurate firing
- Handguns can be further classified as revolvers or semiautomatic firearms, depending on the feeding mechanism.
- Bullets fired from a firearm show patterns of lands and grooves that match the rifling pattern in the barrel of the firearm.
- A cartridge consists of primer powder, gunpowder, a bullet (a pointed projectile), and the casing material that holds them all together.
- The caliber of a cartridge is a measure of its diameter and is identified along with the name of the manufacturer on the headstamp.
- In addition to examining lands and grooves on a bullet, investigators can examine firing pin marks, breechblock marks, and extractor and ejector marks on a spent cartridge casing to match evidence at a crime scene with a specific firearm.
- Gunshot residues found on victims, shooters, or nearby objects can help investigators recreate a crime scene.
- Investigators often use national databases to match crime-scene evidence to registered weapons.
- Using at least two reference points, an investigator can recreate a bullet's trajectory and determine where a shooter was located during a crime.
- Two major forces are acting on a bullet once it is fired: the forward force of the gunshot and the downward force of gravity.
- Examination of the wounds on a body can determine where a bullet entered and exited the victim.

CASE STUDIES

Sacco and Vanzetti (1920)

Three different types of shell casings were found at the scene of a payroll holdup. Two security guards were killed. When suspects Sacco and Vanzetti were arrested, they were both in possession of loaded guns of the same caliber and ammunition from the same three manufacturers as the casings found at the crime scene. Both were anarchists who openly advocated the violent overthrow of the government.

The trial opened in 1921. More than 140 witnesses were called to testify. The one fact that seemed incontrovertible was that the bullet that killed one of the security guards was so ancient in its manufacture that no similar ammunition could be found to test-fire Sacco's weapon except those equally ancient cartridges found in his pocket upon arrest.

CAREERS IN FORENSICS

Neutron Activation Analyst

A neutron activation analyst applies techniques that can be used in many different fields, such as medicine, geology, engineering, and forensics. Given a sample of trace evidence, a nuclear activation analyst bombards the sample with neutrons produced by a nuclear reactor. Some atoms in the sample may absorb one of these neutrons and become radioactive.

The radioactive sample gives off radiation of different wavelengths, which the analyst measures to determine the elements contained within the sample. Each chemical element gives off radiation of a specific wavelength. The amount of radiation at a certain wavelength can indicate the amount of the element in the sample. Instrumental neutron activation analysis is a very sensitive technique that can detect chemicals that other methods cannot. In addition to being very sensitive, neutron activity analysis requires a very small sample, does not destroy the sample, and is very inexpensive.

Although neutron activation analysis is a powerful technique that would be useful in helping to solve many crimes, most laboratories cannot perform this type of analysis, because it requires a nuclear reactor. One laboratory that does have the facilities to perform neutron activation analysis is the Radiation Center at Oregon State University. Police departments have

called on analysts there to help solve crimes by analyzing trace evidence, such as glass fragments.

Neutron activation analysis does not work as well on paint chips and human hair, because these types of samples are porous and can absorb contaminants from the environment. Besides glass evidence, this analysis also works well on bullets. A neutron activation analyst can use this technique to match a bullet found at a crime scene to a bullet found in a suspect's possession, even when the police have not found a gun connected to the crime. The material used to make bullets contains varying amounts of several elements, such as arsenic, copper, silver, tin, mercury, or gold. The exact chemical composition of a bullet made in one batch is different from that of a bullet made in a different batch. According to one analyst, there is only one out of 10,000 chance that a bullet from one batch will match a bullet from a different batch of bullets.

An analyst from the World's Center State University was able to determine chemical composition of bullets found at the scene of a murder and link the bullets found in a suspect's home. These findings helped convict the suspect on three counts of aggravated murder. The suspect was sentenced to life in prison without parole.

Learn More About It
To learn more about a career as a neutron activation analyst, go to school.cengage.com/forensicscience.

CHAPTER 17 REVIEW

True or False

1. The lands and grooves of a barrel's rifling improve the accuracy of a bullet.
2. The caliber of a cartridge is always measured in one-hundredths of an inch.
3. Firing pin marks are found on the back of the bullet.
4. Lands and grooves help match a crime-scene bullet with its shell casing.
5. The amount of gunshot residue on a victim is usually proportional to the distance between the victim and the shooter.

Multiple Choice

6. Shotguns are examples of
   a. handguns
   b. long guns
   c. revolvers
   d. semiautomatic weapons

7. Which of the following best describes the trajectory of a projectile?
   a. the height of the shooter
   b. the path of the flight of a bullet
   c. the housing for the bullet's gunpowder
   d. the pattern of lands and grooves on the projectile

8. Which of the following is not part of a cartridge?
   a. barrel
   b. bullet
   c. gunpowder
   d. primer powder

9. The caliber of a bullet is related to its
   a. diameter
   b. length
   c. speed
   d. weight

10. Semiautomatic pistols store cartridges in a
    a. magazine (clip)
    b. cylinder
    c. firing pin
    d. muzzle

Bibliography

Books and Journals
David, A. J., and L. Exline. *Current Methods in Forensic Gunshot Residue Analysis*. Boca Raton, FL: CRC Press, 2000.
Di Maio, Vincent J. *Gunshot Wounds: Practical Aspects of Firearms, Ballistics, and Forensic Techniques.* Boca Raton, FL: CRC Press, 1999.
Evans, Colin. *The Casebook of Forensic Detection: How Science Solved 100 of the World's Most Baffling Crimes.* New York: Berkley Trade, 2007.
Heard, Brian J. *Handbook on Firearms and Ballistics: Examining and Interpreting Forensic Evidence.* Boca Raton, FL: CRC Press, 1997.
Pejsa, Arthur J. *Modern Practical Ballistics*. Minneapolis, MN: Kenwood Publishers, 2001.
Platt, Richard. *Crime Scene*. Englewood Cliffs, NJ: Prentice Hall, 2006.

Websites
Gale Forensic Science eCollection. www.thomsonedu.com/school/forensicscience.
Eng, Paul. "Uncovering Convincing Evidence." www.abc.news.com, Oct. 30, 2003.
http://en.wikipedia.org/wiki/Bullet
http://www.freemrsid.com/Bullets/bullet1.htm
http://library.med.utah.edu/WebPath/TUTORIAL/GUNS/GUNBLST.html
www.fitearms.JD.com
http://www.bls.gov/oco/ocos115.htm

ACTIVITY 13-3
THE ROMANOVS AND DNA: AN INTERNET ACTIVITY

Objectives:

By the end of this activity, you will be able to:
1. Examine how the Romanov family was identified from their remains.
2. Describe what the remains revealed about the family's fate.
3. Describe how DNA technology was used to help identify the skeletal remains of the Romanov family.

Safety Precautions:

None

Time Required to Complete Activity:

Part A: 40 minutes
Part B: 30 minutes
Part C: 60 minutes

Materials:

access to the Internet
pen and paper

Part A:
Procedure:

1. Open the Internet and type in the Internet site: http://www.dnai.org/index.htm.
2. Refer to the menu bar at the far left of the page and double-click on the fifth heading: Applications.
3. Use the scroll bar to the right to pull the screen down. At the bottom of the screen, double-click on the second module entitled "Recovering the Romanovs."
4. Move the scroll bar up to the top of the screen to double-click on the Romanov family.
5. Move the scroll bar down to the bottom of the screen. You will see 10 circles. Click on each numbered circle and answer the questions as you progress through the material. Note the number of the circle corresponds to the numbered questions below.

Questions:

1. How long did the Romanov family rule Russia?
2. Alexandria was born in what country?
3. Refer to the pedigree of Tsar Nicholas II's family.
   a. How many daughters did he have?
   b. How many sons did he have?
   c. Who were the two youngest children?

Activities *provide hands-on experience with forensic concepts and procedures with clear, step-by-step directions.*

Objectives *describe the key outcomes from the activity.*

Safety *identifies recommended precautions.*

Materials *list what is needed to complete the activity.*

Questions *ask students to think critically about the results of the activity.*

Procedures *itemize the steps to follow to perform the activity*

Questions:

1. Which measurement and relationship most accurately reflected your height?
2. Was this the same measurement that most people of your gender found to most accurately estimate their actual height? Explain.
3. Which measurement and relationship most accurately reflected your partner's height?
4. Which measurement was the least accurate in estimating your height?
5. Explain why using the Canons of Proportions on teenagers to estimate height would provide less accurate data than using the canons of proportions on adults.
6. Describe a crime scene that could use the Canons of Proportions to help estimate the height of a person.

Part B:
Procedure:

1. The distance from your elbow to armpit is roughly the length of your humerus. Record the humerus length and actual length from everyone in your class and complete Data Table 3.
2. Graph the length of the humerus (x axis) vs. height (y axis). Be sure to include on your graph the following:
   · Appropriate title for graph
   · Set up an appropriate scale on each axis
   · Label units (cm) on each of the x and y axes
   · Circle each data point

Data Table 3: Comparison of Humerus to Actual Height

Name	Length of Humerus (cm)	Actual Height (cm)
1		
2		
3		
4		
5		
6		
7		
8		
9		
10		
11		
12		
13		
14		
15		
16		
17		
18		
19		
20		

Technology

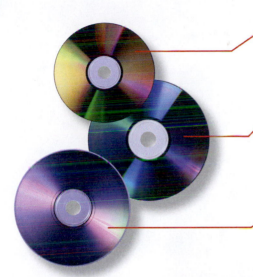

The Instructor's Resource CD-ROM (IRCD) contains tests, PowerPoint presentations, forms for activities, lesson plans, activity teacher notes, and much more!

The ExamView CD-ROM offers a test bank of objective questions for each chapter and a final exam. The ExamView software enables instructors to modify questions from the test bank or add teacher-written questions to create customized tests.

The E-book is a digital version of the textbook in PDF format that offers the same rich photos, graphics, and easy-to-read fonts as the printed text.

The Product Web site includes resources for both the student and teacher, including a link to the **Gale Forensic Science eCollection** database for online research.

The web site lists resources for each chapter and includes web links, glossary terms, flashcards, crossword puzzles, lab forms, and Interactivities emphasizing special topics.

Reviewers

Lori Allen
Guthrie High School
Guthrie, OK

Sherry Annee
Brebeuf Jesuit
Prepatory School
Indianapolis, IN

Joe Bomba
Ellet High School
Akron, OH

William Breuer
Highland High School
Erial, NJ

Rhonda M. Brown
East Ridge High School
Clermont, FL

Kacia Cain
East High School
Des Moines, IA

Dorothy Canote
D. H Hickman High
School
Columbia, MO

Becky Collins
Granada High School
Livermore, CA

Tammy Cox
Provine High School
Jackson, MS

Diameda Croker
Wiregrass Ranch
High School
Wesley Chapel, FL

Karen Lynn Cruse
Summit Country
Day School
Cincinnati, OH

Carl Davidse
Camrose Composite
High School
Camrose, AL

Kathy Dunaway
Lloyd High School
Erlanger, KY

Cheryl Ellis
The Agnes Irwin School
Rosemont, PA

JoAnn Farber-Keck
Palm Harbor
University
High School
Palm Harbor, FL

Randa Flinn
Northeast High
Oakland Park

Jim Foresman
Pittsburg High School
Pittsburg, KS

Myra Frank
Marjory Stoneman
Douglas High School
Parkland, FL

Kenneth Fusco
Plainville High School
Plainville, CT

Trevor Gallant
Kennebecasis Valley
High School
Rothesay, New
Brunswick, Canada

Lori Haines
Sierra Vista
High School
Las Vegas, NV

Mary Halsall
Mariemont High
School
Mariemont, OH

Jammy Hemphill
Forest Hill High
Jackson, MS

David Hooker
Savannah Country
Day School
Savannah, GA

Sophia Hu
McKinley High School
Honolulu, HI

Bettina Hughes
Drake High School
San Anselmo, CA

Kevin Jones
Cary Academy
Cary, NC

Mari Knutson
Lynden High School
Lynden, WA

Gloria Latta
Wheaton Warrenville
South
Wheaton, IL

Madeline Loftin
Wingfield High School
Jackson, MS

Kathleen Luczynski
Downers Grove South
High School
Downers Grove, IL

Christine McOmber
Lake Zurich High School
Lake Zurich, IL

Sharon Olson
Wheaton Warrenville
South High School
Wheaton, IL

Rob Richard
West Windsor-Plainsboro
Regional Schools
Plainsboro, NJ

Danielle Ristow
Rancho High School
Las Vegas, NV

Carol Robertson
Fulton High School
Fulton, MO

Mary Robinson
Rio Rancho
High School
Rio Rancho, NM

Jeff Roth-Vinson
Cottage Grove
High School
Cottage Grove, OR

Mary Shane
Advanced
Technologies
Academy
Las Vegas, NV

John Sharkey
Parma High School
Parma, ID

Rhead Smart
Vantage Point
High School
Thornton, CO

Sheila Smith
Jackson Public Schools
Jackson, MS

Donald C. Snyder, Jr.
South Philadelphia
High School
Philadelphia, PA

Larry Spears
John A. Logan
Community College
Carterville, IL

Christi Tamayo
Lake Highland
Prepatory School
Orlando, FL

Zach Tracy
Cleveland Heights
High School
Cleveland, OH

Sue Tramnell
John A. Logan College
Carterville, IL

Gina Vallari
Advanced Tech.
Academy
Las Vegas, NV

Paola Ventura
Lake Highland Prep
School
Orlando, FL

Nicole Whitaker
Sierra Vista High School
Las Vegas, NV

Marcia Wiger
Totino Grace High School
Fridley, MN

Leesa Wingo
South Anchorage
High School
Anchorage, AK

Pam Zeigler
Kenston High School
Chagrin Falls, OH

CHAPTER 1

Observation Skills

WAS SOMEONE STEALING THE TREES?

An officer with the Department of Natural Resources was called to a farm where a landowner had discovered missing trees. The trees were black walnut, a valuable wood used to make expensive furniture. The officer found six stumps where once there were living trees. The limbs and branches were left behind. Scattered around the woods were 20 empty beer cans.

The officer examined the area and found tracks left by a truck leading across a neighbor's field; the perpetrator of the theft had then cut through the boundary fence. By following the tracks, the officer found where the truck had slid sideways and scraped against a tree, leaving a small smear of paint. These pieces of evidence were photographed and sampled.

The landowner remembered having seen similar tire marks leading into another wooded area two miles up the road. The officer investigated these marks and found several more black walnut stumps and more empty beer cans. The officer documented numerous forms of evidence—a paint sample from the truck, tire tread impressions, and one fingerprint lifted from a beer can. The thefts stopped, and the case was considered unsolved.

Two years later, a man was caught stealing black walnut trees a couple of counties away, and his truck was impounded. The officer compared the original paint sample to match-

An investigator examines paint evidence.

©Mauro Fermariello/Photo Researchers, Inc.

ing paint from the truck. A receipt in the truck from a veneer mill (veneer is the thin layer of high-value wood put on the surface of low-quality woods to be used in furniture) suggested that the man had been selling logs for some time.

The paint on his truck was consistent with paint found at the crime scene, and his fingerprints matched the fingerprint found on the beer can at the scene. Based on the evidence, he was convicted, fined, and sent to prison for six years. An observant investigator was able to collect sufficient evidence for a jury to find the man guilty of stealing the trees.

OBJECTIVES

By the end of this chapter, you will be able to:

✔ Define *observation* and describe what changes occur in the brain.

✔ Describe examples of factors influencing eyewitness accounts of events.

✔ Compare the reliability of eyewitness testimony to what actually happened.

✔ Relate observation skills to their use in forensic science.

✔ Define *forensic science*.

✔ Practice and improve your own observation skills.

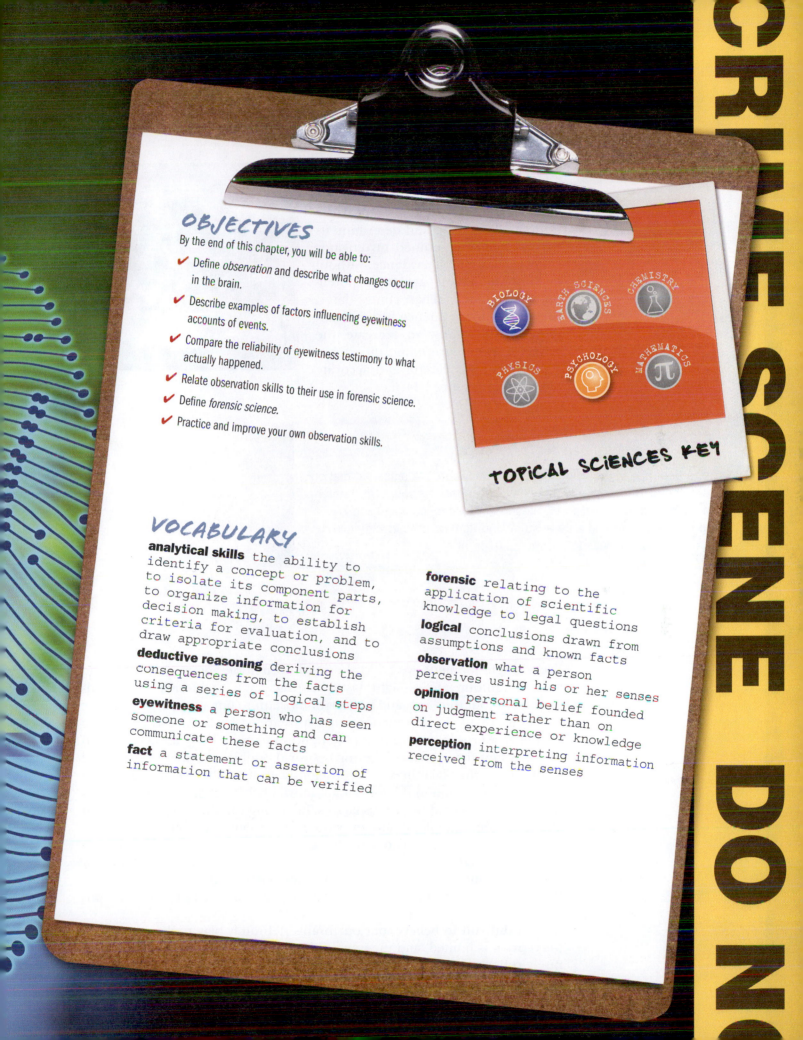

BIOLOGY   EARTH SCIENCES   CHEMISTRY

PHYSICS   PSYCHOLOGY   MATHEMATICS

TOPICAL SCIENCES KEY

VOCABULARY

analytical skills the ability to identify a concept or problem, to isolate its component parts, to organize information for decision making, to establish criteria for evaluation, and to draw appropriate conclusions

deductive reasoning deriving the consequences from the facts using a series of logical steps

eyewitness a person who has seen someone or something and can communicate these facts

fact a statement or assertion of information that can be verified

forensic relating to the application of scientific knowledge to legal questions

logical conclusions drawn from assumptions and known facts

observation what a person perceives using his or her senses

opinion personal belief founded on judgment rather than on direct experience or knowledge

perception interpreting information received from the senses

INTRODUCTION

One of the most important tools of the **forensic** investigator is the ability to observe, interpret, and report **observations** clearly. Whether observing at a crime scene or examining collected evidence in the laboratory, the forensic examiner must be able to identify the evidence, record it, and determine its significance. The trained investigator collects all available evidence, without making judgments about its potential importance. That comes later. Knowing which evidence is significant requires the ability to recreate the series of events preceding the crime. The first step is careful and accurate observation (Figure 1-1).

Figure 1-1. A crime scene is often laid out in a grid to ensure that all evidence is found.

©AP Photo/Charles Bennett

Digging Deeper
with Forensic Science e-Collection

Can a beer can in the woods lead to a conviction? A smashed dial on a safe betray the suspect? They have, and now it's your turn. Search the Gale Forensic Science eCollection on school.cengage.com/forensicscience to find a case study and demonstrate in writing how good observation skills led to the solution of a crime.

WHAT IS OBSERVATION?

Every single moment, we are gathering information about what is around us, through our senses—sight, taste, hearing, smell, and touch. We do this largely without thinking, and it is very important to our survival. Why are we not aware of all the information our senses are gathering at any time? The simple answer is that we cannot pay attention to everything at once. Instead of a constant flow of data cluttering up our thoughts, our brains select what information they take in; we unconsciously apply a filter (Figure 1-2). We simply pay attention to things that are more likely to be important. What is important is decided by various factors, including whether the environment changes. For example, if you are sitting in a room and everything is still, you are unlikely to be filled with thoughts about the color of the sofa, the shade of the light, or the size and shape of the walls. But if a cat walks in, or you hear a loud bang, you will perceive these changes in your environment. Paying attention to the details of your surroundings requires a conscious effort.

It is difficult to believe, but our brains definitely play tricks on us. Our **perception** is limited, and the way we view our surroundings may not accurately reflect what is really there. Perception is faulty; it is not always accurate, and it does not always reflect reality. For example, our brains will fill in

Figure 1-2. *How information is processed in the brain.*

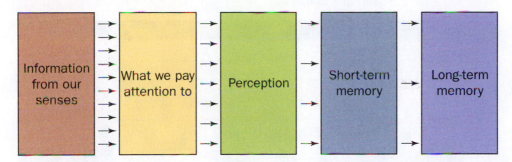

information that is not really there. If we are reading a sentence and a word is missing, we will often not notice the omission but instead predict the word that should be there and read the sentence as though it is complete.

Our brains will also apply knowledge we already have about our surroundings to new situations. In experiments with food coloring, a creamy pink dessert is perceived to be strawberry-flavored even though it tastes of vanilla. Our minds have learned to associate pink with strawberries and apply that knowledge to new situations—even when it is wrong. An interesting aspect of our perception is that we believe what we see and hear, even though our ability to be accurate is flawed. People will stick to what they think they saw, even after they have been shown that it is impossible.

If you are feeling like your brain is rather defective, do not worry: the brain, while faulty, is still good at providing us with the information we need to survive. Filtering information, filling in gaps, and applying previous knowledge to new situations are all useful traits, even if they do interfere sometimes. Understanding our limitations helps us improve our observation skills, which is extremely important in forensic science. Criminal investigations depend on the observation skills of all parties involved—the police investigators, the forensic scientists, and the witnesses.

Did You Know?

China, 1248: The suspect in a stabbing death confessed after flies were used to determine which knife in the village had blood on it. All of the knives of the village were collected, and when the flies all landed on just one knife, the man confessed.

OBSERVATIONS BY WITNESSES

One key component of any crime investigation is the observations made by witnesses. Not surprisingly, the perceptions of witnesses can be faulty, even though a witness may be utterly convinced of what he or she saw. Have you ever noticed that you can walk along the street or ride in a car and be totally unaware of your surroundings? You may be deeply involved in a serious conversation or hear some disturbing news and lose track of events happening around you. Your focus and concentration may make an accurate accounting of events difficult.

Our emotional state influences our ability to see and hear what is happening around us. If people are very upset, happy, or depressed, they are more likely not to notice their surroundings. Anxiety also plays a big part in what we see and what we can remember. Our fear at a stressful time may interfere with an accurate memory. Victims of bank robberies often relate conflicting descriptions of the circumstances surrounding the robbery. Their descriptions of the criminals committing the robbery often do not match (Figure 1-3).

Figure 1-3. *The success of a police lineup depends on an eyewitness's ability to recognize a person seen at a crime.*

©MGM/Corbis

In an unusual situation, however, our ability to observe is often heightened. For example, most people can recall exactly where they were when they first heard of the attack on the World Trade Center towers in Manhattan on September 11, 2001.

Other factors affecting our observational skills include:

• Whether you are alone or with a group of people

• The number of people and/or animals in the area

• What type of activity is going on around you

• How much activity is occurring around you

All of these factors influence the accuracy of a witness's observations.

EYEWITNESS ACCOUNTS

What we perceive about a person depends, in part, on his or her mannerisms and gestures. How a person looks, walks, stands, and uses hand gestures all contribute to our picture of his or her appearance. Think about your family members. How would you describe them? What makes them unique? We also form images of familiar places. Our homes, school, and other places we might often visit (e.g., a favorite store or restaurant) are burned into our memories and easy to recognize and remember.

Eyewitness accounts of crime-scene events vary considerably from one person to another. What you observe depends on your level of interest, stress, concentration, and the amount and kind of distraction that may be present. Our prejudices, personal beliefs, and motives also affect what we see. Memory fades with time, and our brains tend to fill in details that we feel are appropriate but may not be accurate. These factors can decrease an eyewitness's reliability in reporting a crime. The testimony of an eyewitness can be very powerful in persuading the jury one way or another; knowing the shortcomings of eyewitness testimony is necessary to ensure that justice is carried out appropriately.

THE INNOCENCE PROJECT

The Innocence Project at the Benjamin N. Cardozo School of Law at Yeshiva University in New York was created by Barry C. Scheck and Peter J. Neufeld in 1992. Its purpose was to reexamine post-conviction cases (individuals convicted and in prison) using DNA evidence to provide conclusive proof of guilt or innocence (Figure 1-4). After evaluating more than 200 wrongful convictions in the United States, the Innocence Project found that faulty eyewitness identification contributed up to 87 percent of those wrongful convictions. Eyewitness errors included mistakes in describing the age, and facial distinctiveness of the suspect. These mistakes resulted from disguised appearances, brief sightings of the perpetrator, cross-gender and cross-racial bias, and changes in the viewing environment (from crime scene to police lineup).

Figure 1-4. *Gary Dotson was the first individual shown to be innocent by the Innocence Project.*

©AP Photo/Seth Perlman

When evaluating eyewitness testimony, the investigator must discriminate between **fact** and **opinion**. What did the witness actually see? Often what we think we see and what really happened may differ. The act of someone fleeing from the site of a shooting might imply guilt but could also be an innocent bystander running away in fear of being shot. Witnesses have to be carefully examined to describe what they saw (eyewitness evidence), not what they thought happened (opinion).

On completion of witness examination, the examiner tries to piece together the events (facts) preceding the crime into a **logical** pattern. The next step is to determine if this pattern of events is verified by the evidence and reinforced by the witness testimony.

HOW TO BE A GOOD OBSERVER

We can apply what we know about how the brain processes information to improve our observation skills. Here are some basic tips:

1. *We know that we are not naturally inclined to pay attention to all of the details of our surroundings.* To be a good observer, we must make a conscious effort to examine our environment systematically. For example, if you are at a crime scene, you could start at one corner of the room and run your eyes slowly over every space, looking at everything you see. Likewise, when examining a piece of evidence on a microscope slide, look systematically at every part of the evidence.

2. *We know that we are naturally inclined to filter out unimportant information.* However, at a crime scene, we do not know what may turn out to be important. In this situation, we can consciously decide to observe everything, no matter how small or how familiar, no matter what our emotions or previous experiences. So we train ourselves to turn off our filters, and instead act more like data-gathering robots.

3. *We know that we are naturally inclined to interpret what we see, to look for patterns, and make connections.* To some degree, this inclination can lead to us

jumping to conclusions. While observing, we need to be careful that we concentrate first and foremost on gathering all of the available information and leaving the interpretation until we have as much information as possible. The more information we have, the better our interpretations will be. That does not mean that we should not think about what we see. If we analyze what an observation might mean at the time, we may be led to look more closely for further evidence.

4. *We know that our memories are faulty.* While observing, it is important to write down and photograph as much information as possible (Figure 1-5). This will become very important later when we, or our investigating team members, are using our observations to try to piece together a crime. Documentation also is important when acting as an expert witness. A judge will only accept hair evidence that has been documented in writing and with photographs taken at the crime scene. The verbal testimony of a forensic scientist alone may not be entered into evidence without the proper documentation.

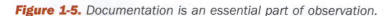

Figure 1-5. *Documentation is an essential part of observation.*

©AP Photo/Mary Altaffer

Digging Deeper
with *Forensic Science e-Collection*

Observation is often as much about finding evidence as it is about spotting patterns of criminal behavior. We know that, on average, most thieves who come in through a window will leave by a door. Search the Gale Forensic Science eCollection on school.cengage.com/forensicscience for the two articles by Carl S. Klump and Kim Rossmo. Write a brief essay comparing the two contrasting views on investigating crimes through observing patterns.

Carl Stanton Klump. "Taking your cue from the clues." (using deductive reasoning in investigations) *Security Management* 41.9 (Sept. 1997): p. 123(3). From *Forensic Science Journals.*

D. Kim Rossmo. "Criminal investigative failures: avoiding the pitfalls." *The FBI Law Enforcement Bulletin* 75.9 (Sept. 2006): p. 1(8). From *Forensic Science Journals.*

OBSERVATIONS IN FORENSICS

Forensics: the word conjures up images of *CSI: Miami*, lab coats, and darkly lit laboratories, but that is not where the word comes from. Forensic derives from the Latin word, forensis, which means "of the forum." The forum was an open area where scholars would gather to debate and discuss issues. The forum is the historical equivalent of modern-day courts. Two thousand years ago, crimes were solved by debate. Sides for the suspect and victim would give speeches, and the public would decide who gave the best argument. Today, debating is often still called forensics.

However, debating and arguing is not forensic science. Forensic science is strictly concerned with uncovering evidence that stands as fact. It is using science to help in legal matters, such as crimes. A forensic investigator is not interested in making the suspect look guilty; he or she is only interested in collecting and examining physical evidence, reporting this to investigators, and possibly later to the courts. The lawyers then partake in a more Roman-style forensics and try to convince the jury by constructing a plausible story around these facts.

Figure 1-6. *A forensic scientist acting as an expert witness in court.*

WHAT FORENSIC SCIENTISTS DO

So what do forensic scientists do? Their first task is to find, examine, and evaluate evidence from a crime scene. One of the key skills in doing this job well is observation. Forensic science and observation go hand in hand. Forensic scientists also act as expert witnesses for the prosecution lawyers (Figure 1-6). Generally, specialists deal with certain types of evidence. Ballistics experts work with bullets and firearms; pathologists work with bodies to determine the cause of death through the examination of injuries. Textile experts, blood-spatter experts, vehicle experts, and animal experts all rely on observation skills to do their jobs.

Police officers and examiners are trained to have good observation skills. This does not always come naturally, even to police officers. Part of their training is learning to take in the entire scene before making a final assessment based on their observations. They are told to avoid tunnel vision when they observe a crime scene, and they learn the same things that you are learning in this chapter. Police are trained to not only observe but also to carefully analyze what they see. The ability to solve a crime depends on observing all of the evidence left at a crime scene. **Analytical skills** of this type require patience and practice.

The character Sherlock Holmes had excellent observation skills that made him a phenomenal detective. He could look at a situation and find clues in the ordinary details that others missed. Then, he worked backward from the evidence to piece together what happened leading up to the crime. Holmes used **deductive reasoning** to verify the actual facts of the case. The abilities to observe a situation, organize it into its component parts, evaluate it, and draw appropriate conclusions are all valuable analytical skills used by forensic examiners. Forensic scientists are all, in their own way, modern-day Sherlock Holmeses.

Did You Know?

High-ranking police officers in New York City are trained in observation skills at a local art gallery, the Frick Museum. The police learn to identify details in the paintings and draw conclusions about the paintings' subjects. They apply their new skills out in the real world.

SUMMARY

- Our ability to observe is affected by our environment and the natural filters of sensory information in our brains.

- The observations of witnesses to crimes can be partial and faulty, but in some cases also precise.

- The Innocence Project has found that 87 percent of their wrongful conviction cases resulted from flawed eyewitness testimony.

- Police officers and crime-scene investigators are trained in good observation practices.

- Forensic scientists find, examine, and evaluate evidence from a crime scene and provide expert testimony to courts.

CASE STUDIES

Carlo Ferrier (1831)

In 1831, three men aged 33, 30, and 26 were tried in a court in London, England, for the murder of a 15-year-old Italian immigrant, Carlo Ferrier. John Bishop, James May, and Thomas Williams brought the body in a sack to a local university, King's College, seeking money in exchange for the corpse. It was common practice at the time for universities and hospitals to buy bodies that had died from natural causes to use for anatomy lessons and

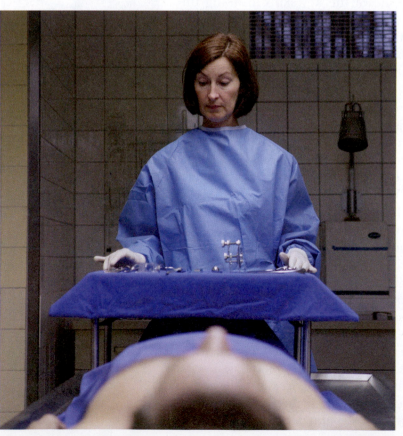

©Zefa Royalty Free/Jupiter Images

research. However, the university staff member noticed that this body looked particularly fresh, and he turned the three men over to the police because of his suspicion. The conviction of the suspects rested on a variety of evidence that was collected because of excellent observation skills. A surgeon carefully examined the body and noticed that all of the organs were healthy; the cause of death did not appear natural. Blood pooled around the spinal cord at the back of the neck was the only sign of violence, and was in keeping with what would be expected from a blow to the back of the neck. Other evidence included bloodstained clothes belonging to the dead boy, which were found buried in the back garden of the accused. These articles were recovered when a policeman inspecting the residence noticed a patch of soft earth in the garden. Bishop, May, and Williams were sentenced to death.

Three Wrongful Convictions

In August 2003, charges were dropped against two men who were wrongly identified and imprisoned for 27 years based on a faulty eyewitness account. In 1976, Michael Evans and Paul Terry were tried and sent to prison for the rape and murder of nine-year-old Lisa Cabassa. They were convicted on the testimony of one purported witness. Recent DNA tests proved that the men were not guilty of the charges.

In a Florida case, death row inmate Frank Lee Smith died of cancer in January 2000 while in prison. He was convicted in 1986 of the rape and murder of an eight-year-old child, even though no physical evidence was found. He was found guilty largely on the word of an eyewitness. Four years after the crime, the eyewitness recanted her testimony, saying she had been pressured by police to testify against Smith. Despite this information, prosecutors vigorously defended the conviction and refused to allow Smith a postconviction DNA test he requested. After his death, the DNA test exonerated him.

©AP Photo/Jamie Rector

Think Critically Review the Case Studies and the information on observation in the chapter. Then state in your own words how eyewitness evidence impacts a case.

Bibliography

Books and Journals

Brennand, Heather. *Child Development*, 2nd edition. London: Hodder Arnold, 2006.

Levine, Martin. *Law and Psychology* (International Library of Essays in Law and Legal Theory, Schools, 16). New York: NYU Press, 1995.

Loftus, E. F. *Eyewitness Testimony*. Cambridge, MA: Harvard University Press, 1996.

Munsterbe, Hugo. *On the Witness Stand: Essays on Psychology and Crime*. South Hackensack, NJ: Fred Rothman and Company, 1981.

Nordby, J. J. "Can we believe what we see, if we see what we believe? Expert disagreement," *Journal of Forensic Science* 37(4): 1115–1124, 1992.

Norman, Donald A. *Learning and Memory*. New York: WH Freeman and Company, 1983.

United States. *Federal Rules of Evidence*. New York: Gould Publications, a division of LexisNexis, 1991.

Web site

Gale Forensic Sciences eCollection, school.cengage.com/forensicscience.

CAREERS IN FORENSICS

Paul Ekman

Very few people can lie to Paul Ekman and get away with it. He can read faces like an open book, spotting the most subtle changes in expression that reveal if a person is lying. A psychologist who has spent the last 50 years studying faces, Ekman is a leading expert on facial analysis and deception. This skill puts him in high demand by law-enforcement groups around the world, such as the Federal Bureau of Investigation (FBI), Central Intelligence Agency (CIA), Scotland Yard, and Israeli Intelligence.

When looking for deception, Ekman watches for inconsistencies, such as facial expressions that do not match what is being said. He can also detect what are called *microexpressions*—rapid changes in expression that last only a fraction of a second but reveal a person's true feelings. It is a rare talent to be able to spot these microexpressions. Only 1 percent of people are able to do so without training.

Ekman was the first to determine that a human face has 10,000 possible configurations and which muscles are used in each. He then created the Facial Action Coding System. This atlas of the human face is used by a variety of people looking to decode human expression, including investigators, psychologists, and even cartoon animators.

Ekman has turned his expert gaze onto many famous faces. He thinks the mysterious Mona Lisa is flirting, and he can identify the exact facial

Courtesy, Paul Ekman

Paul Ekman

muscles Bill Clinton used when he lied about Monica Lewinsky. He has studied tapes of Osama Bin Laden to see how his emotions changed leading up to the 9/11 terrorist attacks.

Ekman first became interested in facial expressions at the age of 14, after his mentally ill mother committed suicide. He hoped to help others like her by understanding emotional disorders. From his experience as a photographer, he realized that facial expressions would serve as a perfect tool for reading a person's emotions.

Ekman's early research led to a major discovery that changed how scientists view human expression. Experts used to believe that facial expressions were learned, but Ekman thought otherwise. He traveled around the world and found that facial expressions were universally understood, even in remote jungles where natives had never before seen a Westerner. It could mean only one thing: our expressions are biologically programmed. This opened the door for Ekman to study human expression in a completely new way.

Fifty years of groundbreaking research followed Ekman's discovery. He served first as Chief Psychologist for the U.S. Army, and then as a professor at the University of California. Now in his seventies, Ekman continues to train others to detect deception and improve safety and security. Liars will never have it easy again!

Learn More About It
To learn more about Paul Ekman and the work of forensic psychologists, go to school.cengage.com/forensicscience.

True or False

1. The word *forensic* refers to the application of scientific knowledge to legal questions.

2. Good observation skills come naturally to investigators; they do not need to be trained.

3. If we remember seeing something happen, we can trust that it happened just as we think it did.

4. Most wrongful convictions seem to be the result of faulty eyewitness testimony.

5. The Innocence Project is an organization that seeks to get convicted killers out of prison.

Multiple Choice

6. A forensic scientist is called to a court of law to provide
   a) fact
   b) opinion
   c) judgement
   d) reflection

7. Our state of mind affects how we observe our surroundings. What mental state is the best for observing?
   a) happy
   b) relaxed
   c) nervous
   d) excited

8. The Innocence Project found that most faulty convictions were based on
   a) out-of-date investigating equipment
   b) poor DNA sampling
   c) inaccurate eyewitness accounts
   d) officers not thoroughly observing a crime scene

9. All of the following are ways to improve our observational skills **except**
   a) be sure to look at the entire area, not just the body, weapons, or signs of break-in
   b) observe everything no matter how big or small
   c) when collecting evidence, record only those things that you are sure are important
   d) write down and photograph everything you find

10. The forensic scientist has many duties. Which of these is **not** a job for a forensic scientist?
   a) give evidence in court
   b) question a suspect
   c) sign a Cause of Death document
   d) search for evidence

Short Answer

11. Why are observation skills important to forensic science?

_____

_____

_____

_____

12. Name three ways you can improve your observation skills.

_____

_____

_____

_____

13. Describe two ways that your brain may alter sensory information.

_____

_____

_____

_____

14. Describe a situation where two different people might perceive a crime scene in different ways.

_____

_____

_____

_____

15. Briefly describe what can be detected by observing facial expressions.

_____

_____

_____

_____

ACTIVITY 1-1
LEARNING TO SEE

Objectives:

By the end of this activity, you will be able to:
1. Describe some of the problems in making good observations.
2. Improve your observational skills.

Time Required to Complete Activity: 25 minutes

Materials:

lab sheets for Activity 1-1
pencil

Safety Precautions:

None

Procedure:

1. Your teacher will provide you with Photograph 1 and a question sheet.
2. Study Photograph 1 for 15 seconds.
3. When directed by your teacher, turn over your question paper and answer as many of the questions as you can in three minutes.
4. Repeat the process for Photographs 2 and 3.
5. Discuss the answers to the questions below with your classmates.

Questions (for class discussion):

1. Did everyone answer all of the questions correctly?
2. If everyone viewed the same photograph, list some possible reasons why their answers differed.

ACTIVITY 1-2
YOU'RE AN EYEWITNESS!

Objectives:

By the end of this activity, you will be able to:
1. Assess the validity of eyewitness accounts of a crime.
2. Test your own powers of observation.

Time Required to Complete Activity: 45 minutes

Materials:

(per student)
A copy of the scene of Jane's Restaurant
A copy of the questionnaire concerning Jane's Restaurant

Safety Precautions:

None

Procedure:

1. Obtain the image of a crime scene from your teacher.
2. Study the image for three minutes.
3. When given the signal, turn over the image, and answer the questions about the crime scene.

Questions:

1. How well did you do in remembering the details in this picture?
2. What do the results of this activity say, if anything, to you about the usefulness of eyewitness accounts in a court?
3. What factors influenced your observations?
4. How could you improve your observation skills?

ACTIVITY 1-3
WHAT INFLUENCES OUR OBSERVATIONS?

Objectives:

By the end of this activity, you will be able to:
1. Test your ability to make observations during events.
2. Design an experiment involving a television or print commercial that demonstrates how different factors influence one's ability to observe.

Introduction:

Familiar TV commercials can be the basis for testing your observational skills.

Time Required to Complete Activity: 45 minutes

Materials:

videotape of a commercial provided by your teacher
question sheets provided by your teacher
pen or pencil

Safety Precautions:

None

Procedure:

1. Watch the commercial taped by your teacher.
2. Answer the questions on the sheet provided.

Questions:

1. How many people are in the video?
2. Describe the main character(s) in the commercial in terms of
   a. Size
   b. Age
   c. Skin color
   d. Height
   e. Weight
   f. Hair: style, color, length
   g. Clothing
   h. Hat
   i. Glasses
   j. Distinguishing features
   k. Jewelry
   l. Beard or no beard
   m. Any physical limitations

3. Describe the other people in the commercial.
4. Describe the area where the video was located.
5. What furniture, if any, was in the commercial?
6. Was the time noted?
7. Was it possible to determine the season?
8. What were the people doing in the commercial?
9. Were there any cars in the commercial? If so, describe the:
    a. Model
    b. Year
    c. Color
    d. License plate number
10. How long was the video?

Student-Designed Commercial Activity:

Design an activity involving commercials that would demonstrate how different factors influence our ability to observe. You should include the following:
1. Question
2. Hypothesis
3. Experimental design
    a. Control
    b. Variable
4. Observations
    a. What you will measure and how you will measure it
    b. Include data tables
5. Conclusion based on your data

Suggested Factors to Be Tested:

1. Will the number of people in the room affect someone's observational skills?

2. Will someone's observational skills be affected if he or she is listening to music while making the observation?

3. Are men less observant of the surrounding environment if the commercial features an attractive woman?

4. Are women less observant of the surrounding environment if the commercial features a handsome man?

5. Are young people less observant of an older person in a commercial as opposed to a younger person?

6. Are older people less observant of younger people in a commercial as opposed to an older person?

7. Will famous people (e.g., actors, actresses, singers, athletes) in a commercial encourage someone to watch the commercial and therefore be more observant of the product information?

8. Does racial background affect someone's ability to recognize someone of a different race?

9. Does the color of someone's clothing make the person more noticeable?

10. Are bald men more difficult to recognize than men who have hair?

11. If the person wears a hat, does that make him or her more difficult to recognize or more likely to be recognized?

12. Does a person's style of clothing make him or her more noticeable or less? (For example, are there differences with responses with a man in a suit as opposed to a man in jeans?)

13. Does the presence of a beard make someone less noticeable or more noticeable?

14. Is an overweight person less likely to be observed than someone of normal weight?

Successful advertisement agencies realize that their commercials need to appeal to that segment of the population that is most likely to purchase the product. The better they target their commercial to the prospective buyers, the greater the chance that particular audience will listen and observe the information given in that commercial.

As a result of these commercial observational activities, students will be able to note how many factors influence our ability to observe. Police collect eyewitness accounts of a crime understanding that this is not the most reliable source of information used in solving a crime.

CHAPTER 2

Crime-Scene Investigation and Evidence Collection

LESSONS FROM THE JONBENET RAMSEY CASE

The 1996 homicide investigation of six-year-old JonBenet Ramsey provides valuable lessons in proper crime-scene investigation procedures. From this case, we learn how important it is to secure a crime scene. Key forensic evidence can be lost forever without a secure crime scene.

In the Ramsey case, the police in Boulder, Colorado, allowed extensive contamination of the crime scene. Police first thought JonBenet had been kidnapped because of a ransom note found by her mother. For this reason, the police did not search the house until seven hours after the family called 911. The first-responding police officer was investigating the report of the kidnapping. The officer did not think to open the basement door, and so did not discover the murdered body of the girl.

Believing the crime was a kidnapping, the police blocked off JonBenet's bedroom with yellow and black crime-scene tape to preserve evidence her kidnapper may have left behind. But they did not seal off the rest of the house,

©AP Photo/Paul Sakuma

The Ramsey Home in Boulder, Colorado.

which was also part of the crime scene. Then the victim's father, John Ramsey, discovered his daughter's body in the basement of the home. He covered her body with a blanket and carried her to the living room. In doing so, he contaminated the crime scene and may have disturbed evidence. That evidence might have identified the killer.

Once the body was found, family, friends, and police officers remained close by. The Ramseys and visitors were allowed to move freely around the house. One friend even helped clean the kitchen, wiping down the counters with a spray cleaner—possibly wiping away evidence. Many hours passed before police blocked off the basement room. A pathologist did not examine the body until more than 18 hours after the crime took place.

Officers at this crime scene obviously made serious mistakes that may have resulted in the contamination or destruction of evidence. To this day, the crime remains unsolved. Go to the Gale Forensic Sciences eCollection for more information on this case.

OBJECTIVES

By the end of this chapter you will be able to:

✔ Summarize Locard's exchange principle.

✔ Identify four examples of trace evidence.

✔ Distinguish between direct and circumstantial evidence.

✔ Identify the type of professionals who are present at a crime scene.

✔ Summarize the seven steps of a crime-scene investigation.

✔ Explain the importance of securing the crime scene.

✔ Identify the methods by which a crime scene is documented.

✔ Demonstrate proper technique in collecting and packaging trace evidence.

✔ Describe how evidence from a crime scene is analyzed.

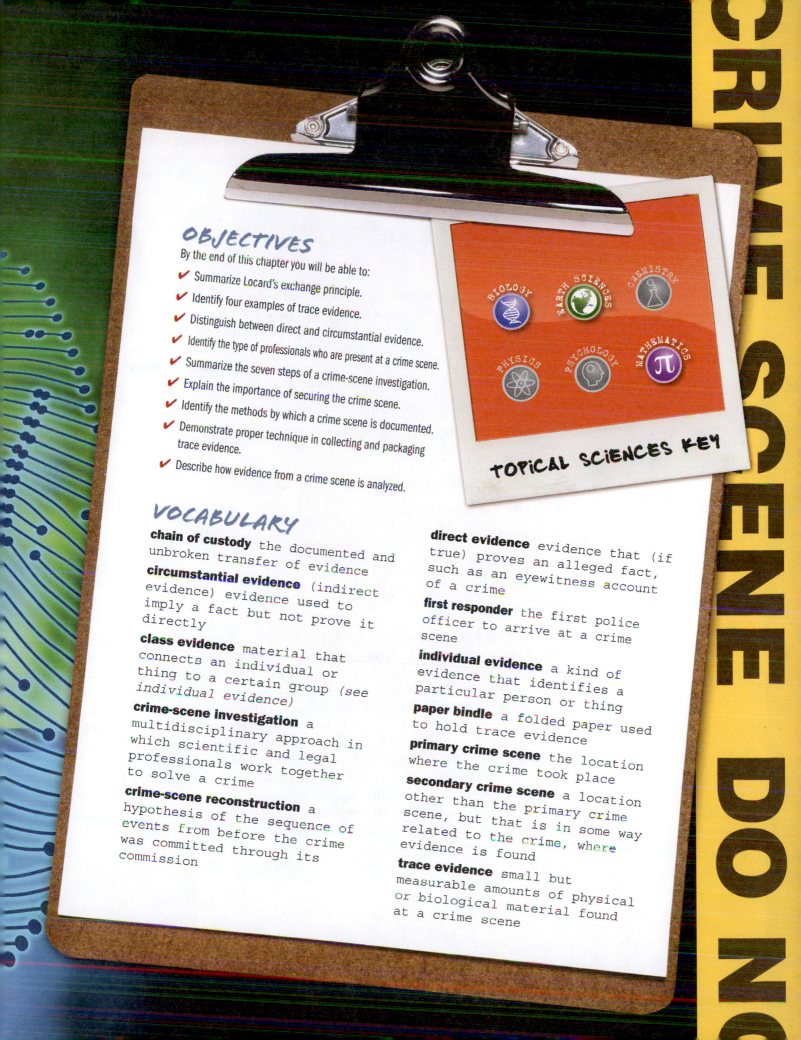

BIOLOGY · EARTH SCIENCES · CHEMISTRY
PHYSICS · PSYCHOLOGY · MATHEMATICS

TOPICAL SCIENCES KEY

VOCABULARY

chain of custody the documented and unbroken transfer of evidence

circumstantial evidence (indirect evidence) evidence used to imply a fact but not prove it directly

class evidence material that connects an individual or thing to a certain group (see individual evidence)

crime-scene investigation a multidisciplinary approach in which scientific and legal professionals work together to solve a crime

crime-scene reconstruction a hypothesis of the sequence of events from before the crime was committed through its commission

direct evidence evidence that (if true) proves an alleged fact, such as an eyewitness account of a crime

first responder the first police officer to arrive at a crime scene

individual evidence a kind of evidence that identifies a particular person or thing

paper bindle a folded paper used to hold trace evidence

primary crime scene the location where the crime took place

secondary crime scene a location other than the primary crime scene, but that is in some way related to the crime, where evidence is found

trace evidence small but measurable amounts of physical or biological material found at a crime scene

INTRODUCTION

How is it possible to identify the person who committed a crime? A single hair or clothing fiber can allow a crime to be reconstructed and lead police to the responsible person. The goal of a **crime-scene investigation** is to recognize, document, and collect evidence at the scene of a crime. Solving the crime will then depend on piecing together the evidence to form a picture of what happened at the crime scene.

PRINCIPLE OF EXCHANGE

Whenever two people come into contact with each other, a physical transfer occurs. Hair, skin cells, clothing fibers, pollen, glass fragments, debris from a person's clothing, makeup, or any number of different types of material can be transferred from one person to another. To a forensic examiner, these transferred materials constitute what is called **trace evidence**. Some common examples of trace evidence include:

- Pet hair on your clothes or rugs
- Hair on your brush
- Fingerprints on a glass
- Soil tracked into your house on your shoes
- A drop of blood on a T-shirt
- A used facial tissue
- Paint chips
- Broken glass
- A fiber from clothing

The first person to note this condition was Dr. Edmond Locard, director of the world's first forensic laboratory in Lyon, France. He established several important ideas that are still a part of forensic studies today. *Locard's exchange principle* states that when a person comes into contact with an object or another person, a cross-transfer of physical evidence can occur. The exchanged materials indicate that the two objects were in contact. Trace evidence can be found on both persons (and/or objects) because of this cross-transfer. This evidence that is exchanged bears a silent witness to the criminal act. Locard used transfer (trace) evidence from under a female victim's fingernails to help identify her attacker.

The second part of Locard's principle states that the intensity, duration, and nature of the materials in contact determine the extent of the transfer. More transfer would be noted if two individuals engaged in a fistfight than if a person simply brushed past another person.

TYPES OF EVIDENCE

Evidence can be classified into two types: direct evidence and circumstantial evidence (Figure 2-1). **Direct evidence** includes firsthand observations such as eyewitness accounts or police dashboard video cameras. For example, a witness states that she saw a defendant pointing a gun at a victim during a robbery. In court, direct evidence involves testimony by a witness about what that witness personally saw, heard, or did. Confessions are also considered direct evidence.

Circumstantial evidence is indirect evidence that can be used to imply a fact but that does not directly prove it. No one, other than the suspect and victim, actually sees when circumstantial evidence is left at the crime scene. But circumstantial evidence found at a crime scene may provide a link between a crime scene and a suspect. For example, finding a suspect's gun at the site of a shooting is circumstantial evidence of the suspect's presence there.

Circumstantial evidence can be either physical or biological in nature. Physical evidence includes impressions such as fingerprints, footprints, shoe prints, tire impressions, and tool marks. Physical evidence also includes fibers, weapons, bullets, and shell casings. Biological evidence includes body fluids, hair, plant parts, and natural fibers. Most physical evidence, with the exception of fingerprints, reduces the number of suspects to a specific, smaller group of individuals. Biological evidence may make the group of suspects very small, or reduce it to a likely individual, which is more persuasive in court.

Trace evidence is a type of circumstantial evidence, examples of which include hair found on a brush, fingerprints on a glass, blood drops on a shirt, soil tracked into a house from shoes, and others (Figure 2-2).

Evidence can also be divided into class evidence and individual evidence. **Class evidence** narrows an identity to a group of persons or things. Knowing the ABO blood type of a sample of blood from a crime scene tells us that one of many persons with that blood type may have been there. It also allows us to exclude anyone with a different blood type. **Individual evidence** narrows an identity to a single person or thing. Individual evidence typically has such a unique combination of characteristics that it could only belong to one person or thing, such as a fingerprint.

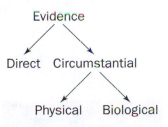

Figure 2-1. *Classification of types of evidence.*

Evidence
→ Direct   Circumstantial
Circumstantial → Physical   Biological

Did You Know?

It is relatively easy to recover DNA from cigarette ends found at the scene of a crime.

Figure 2-2. *Common examples of trace evidence.*

Animal or human hair
Fingerprints
Soil or plant material (pollen)
Body fluids such as mucus, semen, saliva, or blood
Fiber or debris from clothing
Paint chips, broken glass, or chemicals such as drugs or explosives

THE CRIME-SCENE INVESTIGATION TEAM

Who is involved in a crime-scene investigation? The team is made up of legal and scientific professionals who work together to solve a crime. Professionals at the scene of a crime may include police officers, detectives, crime-scene investigators, district attorneys, medical examiners, and scientific specialists. Who is at the scene?

- *Police officers* are usually the first to arrive at a crime scene. A district attorney may be present to determine whether a search warrant is necessary for the crime-scene investigators.

- *Crime-scene investigators* document the crime scene in detail and collect physical evidence. Crime-scene investigators include recorders to record the data, sketch artists to sketch the scene, photographers to take photos of the crime scene, and evidence collectors.

- *Medical examiners* (also called coroners) may be necessary to determine the cause of a death when a homicide has occurred.

- *Detectives* look for leads by interviewing witnesses and talking to the crime-scene investigators about the evidence.

- *Specialists* such as entomologists (insect biologists), forensic scientists, and forensic psychologists may be consulted if the evidence requires their expertise.

Did You Know?

Crime-scene investigation teams do not clean up the scene. This dirty job often falls to the victim's family. Professional crime-scene cleaners can be hired in many places to do this job.

THE SEVEN S's OF CRIME-SCENE INVESTIGATION

SECURING THE SCENE

Securing the scene is the responsibility of the first-responding police officer (**first responder**). The safety of all individuals in the area is the first priority. Preservation of evidence is the second priority. This means the officer protects the area within which the crime has occurred, restricting all unauthorized persons from entering. Transfer, loss, or contamination of evidence can occur if the area is left unsecured (Locard's exchange principle). The first officer on the scene will begin keeping a security log of all those who visit the crime scene. The officer will collect pertinent information and request any additional needs required for the investigation. He or she may ask for more officers to secure the area. Depending on the nature of the crime, the first-responding officer may request various teams of experts to be sent to the crime scene.

SEPARATING THE WITNESSES

Separating the witnesses is the next priority. Witnesses must not be allowed to talk to each other. Their accounts of the events will be compared. This separation is done to avoid witnesses working together to create a story (collusion). The following questions need to be asked of each witness:

- When did the crime occur?
- Who called in the crime?
- Who is the victim?
- Can the perpetrator be identified?
- What did you see happen?
- Where were you when you observed the crime scene?

SCANNING THE SCENE

The forensic examiners need to scan the scene to determine where photos should be taken. A determination may be made of a **primary crime scene** and **secondary crime scene** and priorities assigned regarding examination. A robbery in front of a store might be the primary scene, and the home of a suspect might be the secondary scene. A murder may have taken place at one location (primary scene) and the corpse found at another (secondary scene).

SEEING THE SCENE

The crime scene examiner needs to see the scene. Photos of the overall area and close-up photos with and without a measuring ruler should be taken. Triangulation of stationary objects should be included in the photos as reference points. A view of the crime scene should be taken from several different angles and distances. Several close-up photos of any evidence and bodies should be taken.

SKETCHING THE SCENE

An accurate rough sketch of the crime scene is made, noting the position of the body (if any) and any other evidence. All objects should be measured from two immovable landmarks. On the sketch, north should be labeled and a scale of distance should be provided. Any other objects in the vicinity of the crime scene should be included in the sketch. This includes doors, windows, and furniture. If the crime scene is outdoors, the position of trees, vehicles, hedges, and other structures or objects should be included in the sketch. Later, a more accurate, final copy of the crime scene should be made for possible presentation in court. Computer programs are available to later create a neater and more accurate sketch suitable for use in a court proceeding. The sketch should include the information indicated in Figure 2-3.

Figure 2-3. *A blank crime-scene sketch form showing the information that must be provided with the sketch.*

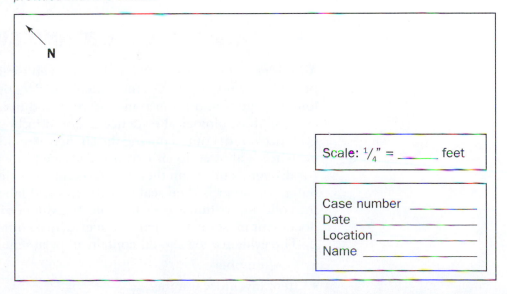

N

Scale: $\frac{1}{4}$" = _____ feet

Case number _____
Date _____
Location _____
Name _____

Digging Deeper
with Forensic Science e-Collection

What happened to Natalee Holloway in Aruba in 2005? This is an unsolved case in which questions have been raised about why crime-scene investigators have not been able to find her body. In fact, investigators searched the island with an array of cutting-edge tools, from a remote-controlled submersible equipped with a video camera and sonar used for probing the water under bridges and in lagoons, to telescoping rods tipped with infrared sensors and cameras used for looking beneath manhole covers and into shadowy caverns. Go to the Gale Forensic Sciences eCollection on school.cengage.com/forensicscience and research the case. Make your own investigation by reading the primary sources available on the Web site. Write a brief explanation that summarizes the forensic tools used to find Holloway's body and any evidence that was discovered during the search.

SEARCHING FOR EVIDENCE

Depending on the number of investigators, a spiral, grid, linear, or quadrant pattern should be walked and location of evidence marked, photographed, and sketched. Single investigators might use a grid, linear, or spiral pattern. A group of investigators might use a linear, zone, or quadrant pattern. These patterns are systematic, ensuring that no area is left unsearched (Figure 2-4).

Additional light sources might be needed to find hair and fibers. A vacuum cleaner with a clean bag is sometimes used to collect evidence but is not the method of choice. The use of a flashlight for examination and forceps for collecting are preferable, because this method avoids picking up extraneous materials.

Figure 2-4. Four crime-scene search patterns.

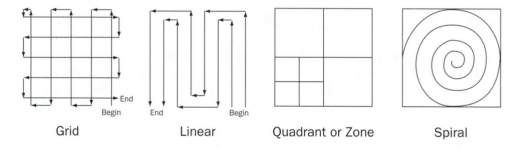

Grid   Linear   Quadrant or Zone   Spiral

SECURING AND COLLECTING EVIDENCE

All evidence needs to be properly packaged, sealed, and labeled. Specific procedures and techniques for evidence collection and storage must be followed. Liquids and arson remains are stored in airtight, unbreakable containers. Moist biological evidence is stored in breathable containers so the evidence can dry out, reducing the chance of mold contamination. After the evidence is allowed to air dry, it is packaged in a **paper bindle**. The bindle (or druggist's fold) can then be placed in a plastic or paper container. This outer container is then sealed with tape and labeled with the signature of the collector written across the tape. An evidence log and a **chain of custody** document must be attached to the evidence container.

The evidence log should contain all pertinent information, including:

• Case number

• Item inventory number

- Description of the evidence
- Name of suspect
- Name of victim
- Date and time of recovery
- Signature of person recovering the evidence
- Signature of any witnesses present during collection

Packaging Evidence

The size of the bindle depends on the size of the evidence. If the evidence is small, the bindle can be constructed from a sheet of paper. If the evidence is large, the bindle might be constructed from a large sheet of wrapping paper. The packaging techniques are demonstrated in Figure 2-5. The steps are as follows:

1. Choose the appropriate-size sheet of clean paper for the bindle.
2. Crease the paper as shown in the figure.
3. Place evidence in the X location.
4. Fold left and right sides in.
5. Fold in top and bottom.
6. Insert the top flap into the bottom flap then tape closed. (Continued on page 28.)

Figure 2-5. *Demonstration of packaging of dry evidence.*

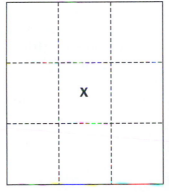
a. Placement of evidence.

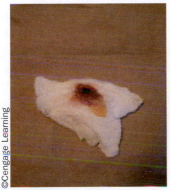
b. Allow evidence to dry.

c. Place dried evidence on bindle paper.

d. Fold bindle. Tuck the top flap into the bottom.

e. Secure bindle in labeled evidence bag using stick-on label.

f. Place evidence in a plastic bag with an inserted evidence label. (Note that this is a different evidence source than the bloody cloth above.)

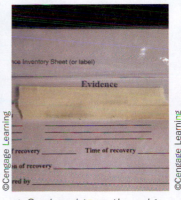
g. Seal and tape the edge of the baggie.

h. Write the collector's signature across the baggie's taped edge.

7. Place bindle inside a plastic or paper evidence bag. Fold the bag closed.

8. Place a seal over the folded edge of the evidence bag.

9. Have the collector write his or her name over the folded edge.

If a wet object to be packaged is large, it should be placed in a paper container and sealed to allow it to air dry. Wet evidence should never be packaged in a plastic container while wet. Any DNA present will degenerate and evidence may become moldy and useless.

There are standards for collecting different types of evidence that describe how to collect and store the evidence. The Federal Bureau of Investigation and state police agencies publish descriptions of the proper procedures.

Control samples must also be obtained from the victim for the purpose of exclusion. For example, blood samples found on a victim or at a crime scene are compared with the victim's blood. If they match, the samples are excluded from further study. If the blood samples do not match, then they may have come from the perpetrator and will be further examined.

CHAIN OF CUSTODY

In securing the evidence, maintaining the chain of custody is essential. The individual who finds evidence marks it for identification and bags the evidence in a plastic or paper container. The final container for the evidence is a collection bag, which is labeled with the pertinent information. The container is then sealed, and the collector's signature is written across the sealed edge.

The container is given to the next person responsible for its care. That person takes it to the lab and signs it over to a technician, who opens the package for examination at a location other than the sealed edge. On completion of the examination, the technician repackages the evidence with its original packaging, reseals the evidence in a new packaging, and signs the chain-of-custody log attached to the packaging. This process ensures that the evidence has been responsibly handled as it was passed from the crime scene to a courtroom (Figure 2-6).

Figure 2-6. *Chain-of-custody procedures.*

©Cengage Learning

a. Original evidence bag

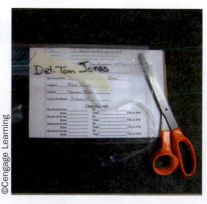

©Cengage Learning

b. Opened evidence bag maintaining signature on first seal

©Cengage Learning

c. Original evidence bag with uncut seal and signature, updated chain-of-custody log in a new sealed and signed evidence bag

ANALYZE THE EVIDENCE

Following a crime-scene investigation, the forensic laboratory work begins (Figure 2-7). The FBI crime lab is one of the largest forensic labs in the world. A forensic lab processes all of the evidence the crime-scene investigation collected to determine the facts of the case. Unlike what television CSI programs portray, forensic lab technicians are specialized and process one type of evidence.

The laboratory results are sent to the lead detective. Test results eventually lead to crime-scene reconstruction; that is, forming a hypothesis of the sequence of events from before the crime was committed through its commission. The detective looks at the evidence and attempts to determine how it fits into the overall crime scenario. The evidence is examined and compared with the witnesses' statements to determine the reliability of their accounts. Evidence analysis can link a suspect with a scene or a victim, establish the identity of a victim or suspect, confirm verbal witness

Figure 2-7. *A modern forensics laboratory.*

©Dean Golja/Getty Images

testimony, or even acquit the innocent. The evidence does not lie, but investigators must consider all possible interpretations of the evidence. Direct evidence is more compelling than circumstantial evidence.

CRIME-SCENE RECONSTRUCTION

Crime-scene reconstruction involves forming a hypothesis of the sequence of events from before the crime was committed through its commission. The evidence is examined and compared with the witnesses' statements to determine the reliability of their accounts. The investigator looks at the evidence and attempts to determine how it fits into the overall crime scenario. The evidence does not lie, but it could be staged. It is important that investigators maintain an open mind as they examine all possibilities.

STAGED CRIME SCENES

Staged crime scenes pose a unique problem. The evidence does not match the testimony of witnesses. Here is a list of some common situations in which a crime scene is staged:

- *Arson.* The perpetrator stages a fire to cover some other crime such as murder or burglary.

- *Suicide/murder.* A victim is murdered, and the perpetrator stages the scene to look like a suicide. The death may be caused by alcohol or drug overdose. The motive could be insurance money, release from an unhappy marriage, or simply theft.

- *Burglary.* A burglary is staged to collect insurance money. In the determination of whether a crime scene is staged, the following points should be considered:

- Initially treat all death investigations as homicides.

- Do the type(s) of wounds found on the victim match the weapon employed?

- Could the wounds be easily self-inflicted?

- Establish a profile of the victim through interviews with friends and family.

- Evaluate the behavior (mood and actions) of the victim before the event.

- Evaluate the behavior (mood and actions) of any suspects before the event.

- Corroborate statements with evidential facts.

- Reconstruct the event.

- Conduct all forensic examinations to determine the facts of the case.

SUMMARY

- Locard's exchange principle states that contacts between people and objects during a crime can involve a transfer of material that is evidence of the crime.

- Evidence may be direct, as in eyewitness accounts, or circumstantial, which does not directly prove a fact.

- Evidence may be physical or biological. Trace evidence is a small amount of physical or biological evidence.

- A crime-scene investigation team consists of police officers, detectives, crime-scene investigators, medical examiners, and specialists.

- A crime-scene investigation consists of recognizing, documenting, and collecting evidence from the crime scene.

- First-responding officers must identify the extent of a crime scene, including primary and secondary scenes, secure the scene(s), and segregate witnesses.

- After walking through the crime scene and identifying evidence, the crime-scene investigators document the scene by taking photographs and preparing sketches of the scene.

- Evidence must be properly handled, collected, and labeled so that the chain of custody is maintained.

- Evidence is analyzed in a forensic laboratory, and the results are provided to detectives, who fit the results into the crime scenario.

CASE STUDIES

Lillian Oetting (1960)

Three Chicago socialites were murdered in Starved Rock State Park, Illinois. All three women had fractured skulls. Their bodies, bound with twine, were found in a cave. Near the bodies of the women, a bloodied tree limb was found and considered to be the murder weapon. Because all three women had been staying at a nearby lodge, the staff of the lodge was questioned. Chester Weger, a 21-year-old dishwasher at the lodge, was asked about a blood stain on his coat. He said it was animal blood. He agreed to take a lie detector test and passed it. He was requestioned and took a second lie detector test and passed it as well. The blood was examined by the state crime lab and found to be animal blood as Weger had indicated at questioning. The case reached a dead end.

Investigators decided to revisit the evidence. The rope used to bind the women was examined more carefully. It was found to be 20-stranded twine sold only at Starved Rock State Park. Identical twine was found in an area accessible to Weger. He again became a prime suspect. The blood on his coat was reexamined by the FBI Crime Lab and found to be human and compatible with the blood of one of the victims. Weger submitted to another lie detector test and failed it. Weger was found guilty for the murder of one of the women, Lillian Oetting, and has spent more than 45 years in prison. He recently petitioned the Governor of Illinois for clemency, saying he was beaten and tortured into making the confession. He still maintains his innocence.

The Atlanta Child Murders (1979–1981)

Wayne Williams is thought to be one of the worst serial killers of adolescents in U.S. history. His victims were killed and thrown into the Chattahoochee River in Georgia. Williams was questioned, because he was seen near where a body had washed ashore. Two kinds of fiber were found on the victims. The first kind was an unusual yellow-green nylon fiber used in floor carpeting. Through the efforts of the FBI and DuPont Chemical Company, the carpet manufacturer was identified. The carpet had been sold in only 10 states, one of them being Alabama, where Williams lived. Thus, the fibers found on the victims were linked to carpet fibers found in Williams' home.

Another victim's body yielded the second type of fiber. This fiber was determined to be from carpeting found in pre-1973 Chevrolets. It was determined that only 680 vehicles registered in Alabama had a matching carpet. Williams owned a 1970 Chevrolet station wagon with matching carpet. The probability of both types of fibers being owned by the same person was calculated. The odds against another person owning both carpet types were about 29 million to one. Williams was convicted and sentenced to two life terms.

 Think Critically Review the Case Studies and the information on investigating crime scenes in the chapter. Then explain how evidence obtained at a crime scene is crucial to a successful case.

CAREERS IN FORENSICS

Crime-Scene Investigator

The crime-scene investigator has a challenging job. His or her specialty is in securing and processing a crime scene. To be well versed in the field, extensive study, training, and experience in crime-scene investigations are needed. He or she must be knowledgeable in the areas of recognition, documentation, and preservation of evidence at a crime scene to ensure that those recovered items will arrive safely at the lab. Investigators generally turn in the evidence to forensic specialists for analysis. However, they may have to testify in court about the evidence collected, the methods used to recover it, and the number of people who came into contact with the evidence.

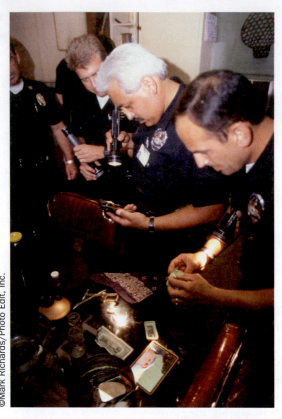

©Mark Richards/Photo Edit, Inc.

Crime-scene investigators at work in the field.

Is the job of a crime-scene investigator the way it is portrayed on television? Let's ask a real-life CSI. Carl Williams of Jupiter, a retired Pennsylvania state police detective, has 25 years of crime-scene investigation experience. Carl says, "The television shows are for entertainment, not reality. The crime scene doesn't wrap up in an hour, never mind an entire investigation. That can take months. Also, television doesn't show the real horror of what one human being can do to another. Not a lot of people can stomach it. But if you take it, the job can be fascinating work. Every day was different. It was interesting. I helped stop the people who committed horrendous acts before they could

do it again. I'm proud of the work I've done."

What is a typical day like? Here is one scenario: At the beginning of a shift, you might be given a list of calls that have come in from police officers overnight. You will need to prioritize them and plan to investigate them in a logical order. Once you arrive at the crime scene, you will work with the first-responding police officer and decide what the best methods are for you to obtain evidence. You will then record the scene using photography and video, and gather evidence such as shoe prints, clothing fibers, blood, and hair. You may discover fingerprint evidence by brushing surfaces with special powders and take impressions of fingerprints from anyone who has accessed the crime scene. Finally, you will secure all of your samples in protective packaging and send them to forensic laboratories for analysis.

What does it take to become a crime-scene investigator? It is usually necessary to obtain a degree in crime-scene investigation through college degree programs or certification programs. The crime-scene investigator should have an associate's or bachelor's degree either in an area of science, with emphasis in law enforcement and crime-scene processing, or a criminal justice degree with an emphasis in science.

Learn More About It

To learn more about crime-scene investigation, go to
school.cengage.com/forensicscience.

Multiple Choice

1. Locard's exchange principle implies all of the following **except**
   a) Fibers can be transferred from one person to another.
   b) Blood spatter can be used to identify blood type.
   c) Cat hair can be transferred to your pants.
   d) Soil samples can be carried from the yard into your home.

2. Transfer evidence can include all of the following **except**
   a) the victim's own blood gushing from a wound
   b) hair that was transferred to a hairbrush
   c) the blood of the victim found on a suspect
   d) a footprint

3. The reason it is important to separate the witnesses at the crime scene is to
   a) prevent contamination of the evidence
   b) prevent fighting among the witnesses
   c) prevent the witnesses from talking to each other
   d) protect them from the perpetrator

4. Correct collection of evidence requires which of the following?
   a) documenting the location where the evidence was found
   b) correct packaging of evidence
   c) maintaining proper chain of custody
   d) all of the above

5. A crime-scene sketch should include all of the following **except**
   a) a scale of distance
   b) date and location of the crime scene
   c) a north heading on the diagram
   d) the type of search pattern used to collect the evidence

Short Answer

6. Distinguish between circumstantial evidence and direct evidence, and provide an example of each type.

   _____

   _____

   _____

7. Blood type is considered to be class evidence. Although it may not specifically identify the suspect, explain how it still could be useful in helping to investigate a crime.

_____

_____

_____

8. The recorder at the crime scene needs to work with all of the police personnel at the crime scene. What type of information would the recorder need to obtain from each of the following persons?

a. first-responding officer

_____

_____

b. photographer

_____

_____

c. sketch artist

_____

_____

d. evidence collection team

_____

_____

9. When the crime-scene investigators arrive at a crime scene, one of their duties is to try to collect all evidence from the victim's body. However, due to the location of the crime scene, some evidence will need to be collected off the body at a later time in the crime lab. For each type of situation below, describe the type(s) of evidence that could be obtained by:

a. transporting the body in a closed body bag

_____

_____

b. taking nail clippings from the deceased

_____

_____

c. placing a plastic bag over the hands of the deceased before transporting the person to the morgue

_____

_____

d. brushing the clothing of the victim with a clothes brush

_____

_____

10. Identify the error in each of the following scenarios:

Case 1

A dead body and a gun were found in a small room. The room was empty
except for a small desk and a chair. The room had two windows, a closet,
and a door leading into a hallway. The crime-scene sketch artist measured the perimeter of the room and drew the walls to scale. He sketched
the approximate position of the dead body and the gun. He sketched the
approximate location of the chair and the desk. What did he forget to do?

Case 2

At the scene of the crime, the evidence collector found a damp, bloody
shirt. The evidence collector quickly wrapped the shirt in a paper bindle. He
inserted the paper bindle with the shirt into an evidence bag. The bag was
sealed with tape, and the collector wrote his name across the tape. The
evidence collection log was completed and taped to the evidence bag. What
did he do incorrectly?

Case 3

A single hair was found on the back of a couch. The evidence collector
placed it in a paper bindle. He then inserted the paper bindle into a plastic
evidence bag. Using tape, the evidence collector sealed the bag. After completing the evidence log and the chain-of-custody form, he brought the evidence bag to the crime lab. What did he do incorrectly?

Case 4

Often, several different labs need to share a very small amount of evidence.
It is important that the chain of custody be maintained. If the chain of custody
is broken, then the evidence may not be allowed in a court proceeding.
Identify the error in the following case. After obtaining the evidence, the first
lab technician removed the tape that contained the signature of the crime
scene evidence collector. On completion of her examination of the evidence,
the lab technician put the evidence back into a paper bindle, and inserted the

bindle into an evidence bag. The technician resealed the bag in the same place as the original crime-scene investigator. After carefully sealing the bag, the lab technician signed her name across the tape. She completed the chain-of-custody form on the outside of the evidence bag and brought the evidence to the next lab technician at the crime lab.

Bibliography

Books and Journals

Bennett, Wayne W. and Karen M. Hess. *Criminal Investigations, 8th ed.* Belmont, CA: Wadsworth Publishing, 2006.

"Crime Scene Response Guidelines," in the California Commission on Peace Officer Standards and Training's Workbook for the Forensic Technology for Law Enforcement Telecourse, 1993.

Kirk, P. L. *Crime Investigation*. New York: Interscience, John Wiley & Sons, 1953.

L.A. Department of Public Safety and Corrections, Office of State Police, Crime Laboratory, "Evidence Handling Guide." Los Angeles, CA.

Lee, Henry. *Physical Evidence in Forensic Science*. Tucson, AZ: Lawyers & Judges Publishing, 2000. Company.

Locard, E. *L'Enquete Criminelle et les Methodes Scientifique*. Paris: Ernest Flammarion, 1920.

Web sites

Byrd, Mike. "Proper Tagging and Labeling of Evidence for Later Identification," www.crime-scene-investigator.net.

Gale Forensic Sciences eCollection, school.cengage.com/forensicscience.

Hencken, Jeannette. "Evidence Collection: Just the Basics," www.theforensicteacher.com.

Ruslander, H. W., S.C.S.A. "Searching and Examining a Major Case Crime Scene," www.crime-scene-investigator.net.

Schiro, George. "Collection and Preservation of Evidence," www.crime-scene-investigator.net.

Also check state police Web sites for evidence handling guides.

ACTIVITY 2-1
LOCARD'S PRINCIPLE

Introduction:

Locard's exchange principle states that trace evidence can be exchanged between a crime scene, victim, and suspect, leaving trace evidence on all three.

Objectives:

By the end of this activity, you will be able to:
1. Demonstrate how transfer of evidence occurs.
2. Identify a possible crime-scene location based on trace evidence examination.

Materials:

(per group of four students)
Activity Sheet 2-1
3 fabric squares each about 2½ inches in a separate evidence resealable plastic bag
1 white sock in an evidence plastic bag
4 pairs of tweezers (forceps)
1 permanent marker
2 hand lenses or microscopes
1 roll of clear ¾-inch-wide adhesive or masking tape
2 pencils
4 sheets white paper (8½ × 11")
4 sheets of paper for bindling (8½ × 11")
4 pairs of plastic or latex gloves
4 resealable plastic bags
4 sheets of paper for bindles
1 pair of scissors
4 copies of the Evidence Collection label

Safety Precautions:

Wash your hands before starting work.
Refrain from touching hair, skin, or clothing when collecting evidence.
Wear gloves while collecting evidence.

Scenario:

A dead body has been found. The crime-scene investigators determined that the body has been moved after the killing. Trace evidence was found on the victim's sock. It was determined that the crime could have occurred in three possible locations. Can you match the trace evidence found on the victim's sock with trace evidence collected from three different locations and determine which location was the crime scene?

Procedure:

Part A: Evidence Collection

1. After washing your hands and putting on your gloves, visit the school library.
2. Open one of the resealable plastic bags, and rub the floor with a fabric square three times. Place the fabric square in a paper bindle, then into a plastic bag, and seal the plastic bag. Label your plastic bag with the location from which your sample was taken.
3. Complete the evidence label, and either attach the label to the plastic bag or place it inside the plastic bag. Seal the plastic bag.
4. Place a piece of adhesive or masking tape over the sealed edge of the plastic bag and write your name across the tape so that your signature begins on one side of the tape and ends on the other side.
5. Repeat steps 1 through 4 at collection site 2 (determined by your instructor).
6. Repeat steps 1 through 4 at collection site 3 (determined by your instructor).
7. Return to your classroom with the three labeled samples.
8. Be sure to maintain the chain of custody with all samples collected. When an evidence bag is opened for examination, the person handling the evidence must open the bag at a location other than the sealed edge (see Figure 2-6).
9. On completion of the examination, the cut plastic bag and all former contents must be resealed into another plastic bag, and the chain-of-custody log attached to the new evidence container must be updated and attached (see Figure 2-6).

Part B: Evidence Examination and Data Collection
Examination of evidence samples

1. Students should wear gloves while examining all evidence.
2. Open a sample bag and bindle from location 1 as previously described by cutting along an edge other than the signed, sealed one.
3. Using forceps and a hand lens or microscope, examine and identify items found on the sample.
4. Record your findings on the data table provided. Be sure to include:
   a. Who collected the sample
   b. When it was collected
   c. Why it was collected
   d. Date
   e. Exact site of collection
5. Press a piece of adhesive tape onto the surface of the fabric to remove any additional evidence that the tweezers cannot pick up. Tape the evidence on white paper and examine it. Add items found to your list of evidence.
6. Return the fabric square for location 1 and all evidence examined to the correct bindle and plastic bag. Seal the plastic bag, relabel it with the chain-of-custody list, and sign off on the plastic bag as described previously.
7. Repeat steps 1 through 7 for location 2 evidence plastic bag.
8. Repeat steps 1 through 7 for location 3 evidence plastic bag.

Return to collect more evidence

1. Choose one member from your group to return to one of the three previous areas examined (i.e., location 1, 2, or 3).

2. The chosen group member should then decide which of the three previous sites should be considered the crime scene. He or she should then return to that location and put on gloves. This group member will not divulge the crime-scene location to his or her fellow examiners.

3. The group member puts on the sock from the plastic bag over his or her own sock. The group member walks around in the selected location. This sock will serve as the victim's sock, which is now covered with trace evidence from the crime scene.

4. While at the crime scene, the chosen team member carefully removes the sock and places it in a bindle and then a plastic bag. It should then be sealed and labeled with "crime scene," date, time, and collector's name, etc. as before.

5. The group member returns to the meeting room to have his or her partners examine the sock evidence.

6. Crime-scene trace evidence should now be treated as described in steps 1 through 7, "Examination of evidence samples."

7. Your team must try to determine which of the three original locations matches the crime-scene location.

8. Complete the Crime Scene report, listing all evidence collected from the sock with your partner investigators.

Questions:

1. Based on your examinations of the trace evidence, which of the three sites was probably the crime scene? Justify your answer.

2. Did your team correctly identify the crime scene?

3. How might the adhesive tape interfere with your evidence collection?

4. Why were gloves necessary in the collection and handling of trace evidence?

5. What other instruments could be used to improve on your ability to identify evidence?

6. A suspect's shoes and clothing are confiscated and examined for trace evidence. What kind of trace evidence might be found on the clothes or shoes? List at least five examples of trace evidence from the shoes or clothing that might be useful in linking a suspect to a crime scene.

7. A home burglary has occurred. It appears the perpetrator entered after breaking a window. A metal safe had been opened by drilling through its tumblers. A suspect was seen running through the garden. Three suspects were interrogated and their clothing examined. List at least three examples of trace evidence that might be found on the suspect.

8. Some examples of trace evidence are listed. For each item, suggest a possible location where the trace might have originated. For example, broken glass fragments—headlight from a hit-and-run accident.

Example: glass fragment	car accident
sand	
sawdust	
pollen	
makeup	
hair	
fibers	
powders or residues	
metal filings	
oil or grease	
gravel	
insects	

Evidence Inventory Label

Case # _____   Inventory # _____

Item #   Item description
_____   _____
_____   _____
_____   _____

Date of recovery _____   Time of recovery _____

Location of recovery _____

Recovered by _____

Suspect _____

Victim _____

Type of offense _____

Chain of custody

Received from _____   By _____
            Date _____   Time _____ AM or PM
Received from _____   By _____
            Date _____   Time _____ AM or PM
Received from _____   By _____
            Date _____   Time _____ AM or PM
Received from _____   By _____
            Date _____   Time _____ AM or PM

ACTIVITY 2-2
CRIME-SCENE INVESTIGATION

Objectives:

By the end of this activity, you will be able to:
1. Explain the correct procedure for securing and examining a crime scene.
2. Demonstrate the correct techniques for collecting and handling evidence.

Introduction:

The crime scene presents a wonderful hands-on way to review many of the skills described in this chapter. A crime has occurred, and you and your investigative team must secure the area and properly collect the evidence.

Time Required to Complete Activity: 60 to 90 minutes (six students per team)

Scenarios:

Two crime scenes prepared in advance by your instructor

Materials:

(Per group, with six students in each group)
Checklists 1–5
evidence Label
10 evidence inventory labels of sheets
10 resealable plastic bags, 6-gallon size
10 resealable plastic bags, 6-quart size
4 paper collection bags
2 marking pens
4 pairs plastic gloves
1 roll crime-scene tape
4 compasses
1 videocamera (optional)
"bunny suit" (optional)
6 forceps (one pair per person)
4 flashlights or penlights (one per person)
2 floodlights
1 digital camera
10 bindle paper sheets, both large and small
6 hand lenses
sketch paper
2 photographic rulers
1 25-foot tape measure
1 roll ¾-inch masking tape

Procedure:

Your crime-scene team is composed of six students. Each team of students has a first officer, a recorder, a photographer, a sketch artist, and two designated evidence collectors.

By the completion of this part of the activity, each team of students must submit the following:

- A log maintained by the first responder

- Checklists 1 through 5 completed, dated, and signed

- Two sketches—a rough sketch and a quality sketch, both with accurate measurements

- A series of 8 × 10 photographs that adequately encompass the crime-scene location; close-up shots of any evidence, evidence numbered and photographed next to a ruler

- Evidence bags properly packaged, labeled, and sealed

Part A: Securing and Preserving the Crime Scene

1. The crime scene is secured by the first officer to arrive. His or her job is to limit access to the crime scene and preserve the scene with minimal contamination. He or she has primary responsibility for:

    - Securing the safety of individuals at the scene; approach the scene cautiously (look, listen, smell) and determine if the site poses any danger

    - Obtaining medical attention for anyone injured at the scene; call for medical personnel for the injured

    - Calling in backup help, including medical personnel to help the injured and/or lab personnel

    - Separating the witnesses so they may be interrogated separately to see if their stories match.

    - Performing an initial walk-through of the area (scan the scene) to provide an overview of the crime scene

    - Searching the scene briefly (scan the scene) to notify lab personnel what equipment is needed

    - Collecting information, including the crime-scene address/location, time, date, type of call, and the names and addresses of all parties involved and present

    - Securing the integrity of the scene by establishing the boundaries of the crime scene by setting up a physical barrier (tape) to keep unauthorized personnel (and animals, if present) out of the area

    - Protecting the crime scene by remaining alert and attentive

    - Documenting the entry and exit of all authorized personnel

    - Providing a brief update to the next-of-command officer to arrive on the scene

The first-responding officer can use checklist 1 to complete all necessary procedures.

Note: Later-arriving police or CSI will set up barricades to prevent unauthorized persons from entering the crime-scene area.

Part B: Search and Evidence Collection

Once your designated crime-scene specialists arrive, evidence collectors will actually collect the evidence for processing back in the lab.

2. The recorder has the responsibility of working with the primary officer to maintain updated records. The recorder will complete checklist 2. The recorder will:

    - Document by date, time, location, and name of collector all evidence that is found.

    - Work with the sketch artist to measure and document the crime scene.

    - Help search for evidence, if necessary.

3. The sketch artist has the responsibility of drawing accurate and detailed sketches of the area designated as the crime scene. At the crime scene, a rough sketch is made, complete with accurate measurements. At a later time, a neater (or computer-generated) sketch is completed. Checklist 3 outlines those responsibilities. The sketch artist working with the recorder will complete that checklist.

4. The photographer has the responsibility to:

    - Work with the sketch artist and recorder to document the crime scene.

    - Photograph any victims and possible suspects.

    - Take photos of the crime scene, noting the four points of the compass, the entrance and exit points in the area, any disturbances (damage) at the scene, etc.

    - Note and photograph any evidence encountered both with and without a ruler.

    - Complete photographer's checklist.

5. The evidence collectors have the responsibility to:

    - Mark off the area around the victim and keep all unnecessary spectators out.

    - Work within the crime scene, wearing gloves to collect evidence.

    - Walk an appropriate search pattern in the crime-scene area. The pattern will be chosen by your instructor. It may be a spiral, grid, or linear pattern, or the area may be divided into zones for examination.

    - Properly handle, bindle, and package any materials considered to be evidence. Remember that the size of the bindle can vary from very small to large enough to package evidence as large as an overcoat.

    - Complete evidence collector's checklist.

6. The proper handling of evidence includes being aware that:

    - Wet or damp evidence should be placed in a paper bag and sealed.

    - Dry evidence should be placed in a paper bindle and then packaged in plastic bags or envelopes and sealed.

    - Liquid evidence should be stored in sealed, unbreakable containers.

    - Care must be taken to prevent any contamination or damage to the evidence collected.

- Flashlights and penlights can be used to search for hair, fibers, and other small or fine trace evidence.

- All evidence containers should be identified with an evidence label or Evidence Inventory Sheet taped to the container or placed inside the container. Such labels or inventory sheets will be provided by your instructor. The name or initials of the collector should be written over the tape sealing the container. The last page in this activity has a copy of an evidence label.

- If for any reason an evidence container is opened, it should be opened at a location other than the sealed edge. It must be repackaged and resealed with the names of all those who have handled the evidence, along with the original packaging. The name of the new packager should be written over the new seal. This chain-of-custody information is also located on the Evidence Inventory Sheet.

Examining the Evidence

Thorough examination of the crime scene will hopefully lead to a comprehensive collection of evidence. After careful examination of all the evidence and after interviewing the suspects, each team of investigators will collect information helpful in solving the crime.

Checklist 1: First Responder's Responsibilities

Place a check mark by each of the following responsibilities as completed:

❑ I approached the scene cautiously (look, listen, smell) and determined if the site poses any danger.

❑ I checked to see if medical attention was needed by anyone injured at the scene.

❑ I called in backup to help the injured.

❑ I secured and separated any witnesses present.

❑ I completed an initial walk-through of the area (scan the scene) to provide an overview of the crime scene.

❑ I notified superiors of the need for additional police officers and CSI technicians at the crime scene.

❑ I secured the integrity of the scene by establishing the boundaries of the crime scene by setting up a physical barrier (tape) to keep unauthorized personnel (and animals, if present) out of the area.

❑ I collected and recorded information, including my name and badge number, case number, address/location of crime scene, time, date, type of call, names of all involved and present parties, as well as the names of everyone present.

❑ I protected the crime scene by remaining alert and attentive.

❑ I documented the entry and exit of all authorized personnel.

❑ I provided the next-in-command officer with a brief update of the situation.

Date_____Signed_____

Checklist 2: Recorder's Checklist

Place a check mark by each of the following responsibilities as completed:

❑ I documented by date, time, location, and name of collector all evidence that was found by completing an Evidence Summary Sheet for each piece of evidence recovered.

❑ I documented weather conditions, available light, unusual odors, and other environmental conditions.

❑ I worked with the sketch artist to measure and document the crime scene.

❑ I helped search for evidence.

❑ I helped document the location and direction of what was photographed.

❑ I helped document the location and direction of what was sketched.

Date_____Signed_____

Checklist 3: Sketch Artist's Checklist

Place a check mark by each of the following responsibilities as completed:

I will prepare two sketches of the crime scene—a rough sketch and a carefully detailed sketch—each of which includes:

❑ All directions of the compass correctly labeled

❑ All objects and landmarks within the crime scene labeled in correct position and to scale (each sketch should contain two immovable objects at a measured distance)

❑ A series of carefully measured distances to add to the accuracy of my sketches

❑ Working with the photographer to document the exact location and direction from which photographs were taken

Date_____ _____Signed_____

Checklist 4: Photographer's Checklist

Place a check mark by each of the following responsibilities as completed:

❑ I worked with the sketch artist, recorder, and evidence collectors to document the crime scene.

❑ I took photos of the crime scene, noting the four points of the compass, the entrance and departure points into the area, any disturbances (damage) at the scene, etc.

❑ I took photographs of any injured persons at the crime scene.

❑ I took close-up photographs of the victim and/or immediate location of the crime.

❑ I took a series of distance photos to give perspective to the crime scene.

❑ I noted and photographed any evidence encountered, both with and without a ruler, and had the recorder and sketch artist also record the location of the evidence.

❑ I took a series of at least eight to ten photographs pertinent to the crime scene. These are of sufficient quality that they could be used in a courtroom reconstruction.

Date_____Signed_____

Checklist 5: Evidence Collector's Checklist

Place a check mark by each of the following responsibilities as completed:

❑ I marked off the area around the victim and kept all unnecessary spectators out.

❑ I worked within the crime scene, wearing gloves to collect evidence.

❑ I walked an appropriate search pattern in the crime-scene area. The pattern walked was _____.

❑ I properly handled and packaged all materials considered evidence into a bindle.

❑ I properly bindled and packaged all materials considered evidence into a bag or plastic bag and completed the Evidence Inventory Sheet for each evidence bag.

❑ I properly sealed and labeled all evidence containers.

❑ I wrote my signature across the seals on all evidence I collected.

❑ I completed the chain-of-custody information for each evidence bag.

Date_____Signed_____

Evidence

Case # _____   Inventory # _____

Item #   Item description
_____   _____
_____   _____
_____   _____

Date of recovery _____   Time of recovery _____

Location of recovery _____

Recovered by _____

Suspect _____

Victim _____

Type of offense _____

Chain of custody

Received from _____   By _____
        Date _____   Time _____ AM or PM
Received from _____   By _____
        Date _____   Time _____ AM or PM
Received from _____   By _____
        Date _____   Time _____ AM or PM
Received from _____   By _____
        Date _____   Time _____ AM or PM

CHAPTER **3**

The Study of Hair

NEUTRON ACTIVATION ANALYSIS OF HAIR

In 1958, the body of 16-year-old Gaetane Bouchard was discovered in a gravel pit near her home in Edmundston, New Brunswick, across the Canadian–U.S. border from Maine. Numerous stab wounds were found on her body. Witnesses reported seeing Bouchard with her boyfriend John Vollman prior to her disappearance. Circumstantial evidence also linked Vollman with Bouchard. Paint flakes from the place where the couple had been seen together were found in Vollman's car. Lipstick that matched the color of Bouchard's lipstick was found on candy in Vollman's glove compartment.

At Bouchard's autopsy, several strands of hair were found in her hand. This hair was tested using a process known as neutron activation analysis (NAA). NAA tests for the presence and concentration of various elements in a sample. In this case, NAA showed that the hair in Bouchard's hand contained a ratio of sulfur to phosphorus that was much closer to Vollman's hair than her own. At the trial, Vollman confessed to the murder in light of the hair analysis results. This was the first time NAA hair analysis was used to convict a criminal.

©Stephen J. Krasemann/Photo Researchers, Inc.

Investigators search for clues in a gravel pit similar to the one in which Gaetane Bouchard was buried.

OBJECTIVES

By the end of this chapter you will be able to

✔ Identify the various parts of a hair.

✔ Describe variations in the structure of the medulla, cortex, and cuticle.

✔ Distinguish between human and nonhuman animal hair.

✔ Determine if two examples of hair are likely to be from the same person.

✔ Explain how hair can be used in a forensic investigation.

✔ Calculate the medullary index for a hair.

✔ Distinguish hairs from individuals belonging to the broad racial categories.

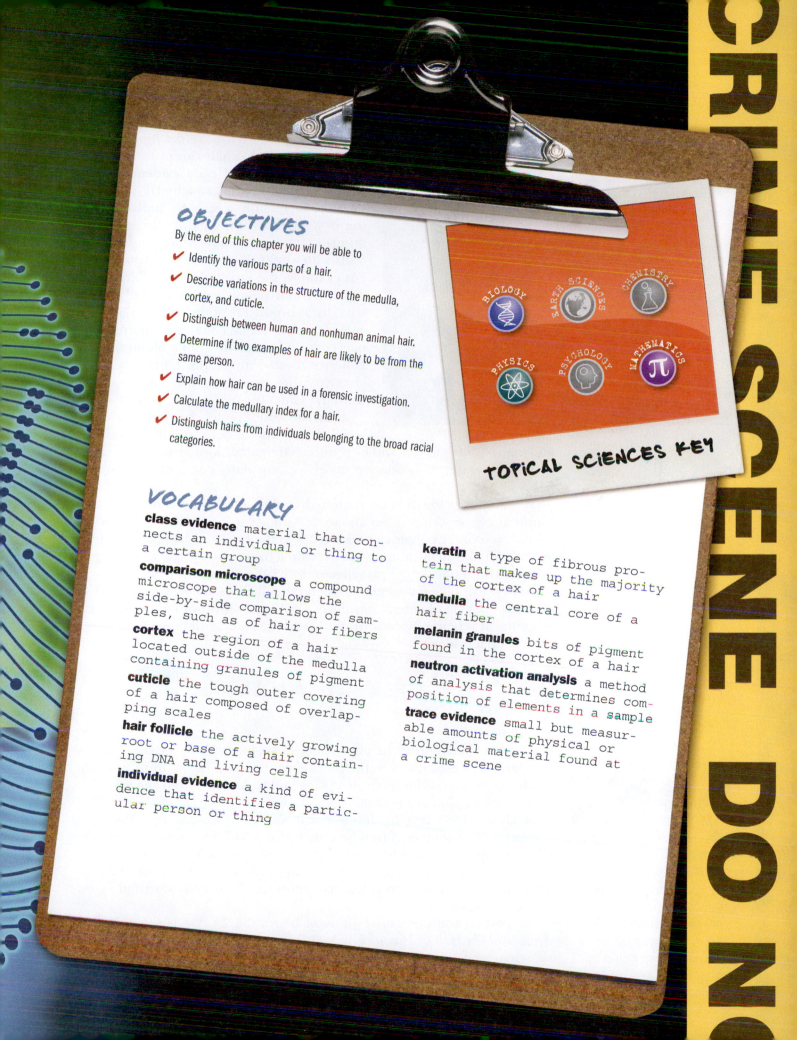

BIOLOGY EARTH SCIENCES CHEMISTRY

PHYSICS PSYCHOLOGY MATHEMATICS

TOPICAL SCIENCES KEY

VOCABULARY

class evidence material that connects an individual or thing to a certain group

comparison microscope a compound microscope that allows the side-by-side comparison of samples, such as of hair or fibers

cortex the region of a hair located outside of the medulla containing granules of pigment

cuticle the tough outer covering of a hair composed of overlapping scales

hair follicle the actively growing root or base of a hair containing DNA and living cells

individual evidence a kind of evidence that identifies a particular person or thing

keratin a type of fibrous protein that makes up the majority of the cortex of a hair

medulla the central core of a hair fiber

melanin granules bits of pigment found in the cortex of a hair

neutron activation analysis a method of analysis that determines composition of elements in a sample

trace evidence small but measurable amounts of physical or biological material found at a crime scene

INTRODUCTION

An investigator finds a blond hair at a crime scene. She thinks that it might help solve her case. What information could be gained from analysis of that hair (Figure 3-1)? What are the limitations of the information that hair can provide?

Figure 3-1. A forensic scientist prepares a hair for analysis.

©AP Photo/Ric Feld

Hair is considered **class evidence**. Alone (without follicle cells attached), it cannot be used to identify a specific individual. In the best case, an investigator can identify a group or class of people who share similar traits who might share a certain type of hair. For example, the investigator can fairly confidently exclude people with Asian and African ancestry as producers of the blond hair found at a crime scene. She could also compare the hair collected with hair from a blond suspect. However, even though the hairs may share characteristics, they may not necessarily be from the same source.

Hair can easily be left behind at a crime scene. It can also adhere to clothes, carpets, and many other surfaces and be transferred to other locations. This is called *secondary* transfer. Secondary transfer is particularly common with animal hair.

Because of its tough outer coating, hair does not easily decompose. Hair found at crime scenes or secondary locations can be analyzed. The physical characteristics of hair can offer clues to the broad racial background of an individual. Chemical tests can provide a history of the use of drugs and other toxins, indicate the presence of heavy metals, and provide an assessment of nutritional deficiencies. When the follicle of a hair is present, DNA evidence may be obtained. Results of DNA analysis is not considered class evidence. It is better, because it can lead to individual identification, thus it is **individual evidence**.

HISTORY OF HAIR ANALYSIS

Investigators recognized the importance of analysis of hair as **trace evidence** in criminal investigations in the late 1800s. The case of the murder of the Duchesse de Praeslin in Paris in 1847 is said to have involved the investigation of hairs found at the scene.

A classic 1883 text on forensic science, *The Principles and Practice of Medical Jurisprudence* by Alfred Swaine Taylor and Thomas Stevenson, contains a chapter on using hair in forensic investigations. It includes drawings of human hairs under magnification. The various parts of human hair are identified. The book also references cases in which hair was used as evidence in England.

In 1910, a comprehensive study of hair titled *Le Poil de l'Homme et des Animaux (The Hair of Man and Animals)* was published by the French forensic scientists Victor Balthazard and Marcelle Lambert. This text includes numerous microscopic studies of hairs from most animals.

Did You Know?

The history of prescription drug-use by Henri Paul was determined by analysis of his hair. Paul was the driver of the car in which he, Dodi Fayed, and Princess Diana died on August 31, 1997.

The use of the **comparison microscope** to perform side-by-side analysis of hairs collected from a crime scene and hairs from a suspect or victim first occurred in 1934 by Dr. Sydney Smith. This method of comparison helped solve the murder of an eight-year-old girl.

Further advances in hair analysis continued throughout the 20th century as technological advances allowed for comparison of hairs through chemical methods. Today, hair analysis includes neutron activation analysis and DNA fingerprinting and is considered a standard tool in trace evidence analysis.

THE FUNCTION OF HAIR

All mammals have hair. Its main purpose is to regulate body temperature—to keep the body warm by insulating it. It is also used to decrease friction, to protect against sunlight, and to act as a sense organ. In many mammals, hair can be very dense, and it is then referred to as fur.

Hair works as a temperature regulator in association with muscles in the skin. If the outside temperature is cold, these muscles pull the hair strands upright, creating pockets that trap air. This trapped air provides a warm, insulating layer next to the skin. If the temperature outside is warm, the muscles relax and the hair becomes flattened against the body, releasing the trapped air.

In humans, body hair is mostly reduced; it does not play as large a role in temperature regulation as it does in other animals. When humans are born, they have about 5 million hair follicles, only 2 percent of which are on the head. This is the largest number of hair follicles a human will ever have. As a human ages, the density of hair decreases.

Did You Know?

All of the hair follicles in humans are formed when a fetus is five months old.

THE STRUCTURE OF HAIR

A hair consists of two parts: a follicle and a shaft (Figure 3-2). The **follicle** is a club-shaped structure in the skin. At the end of the follicle is a network of blood vessels that supply nutrients to feed the hair and help it grow. This is called the *papilla*. Surrounding the papilla is a bulb. A sebaceous gland, which secretes oil that helps keep the hair conditioned, is associated with the bulb. The erector muscle that causes the hair to stand upright attaches to the bulb. Nerve cells wind around the follicle and stimulate the erector muscle in response to changing environmental conditions.

The hair shaft is composed of the protein **keratin**, which is produced in the skin. Keratin makes hair both strong and flexible. Like all proteins, keratin is made up of a chain of amino acids that forms a helical, or spiral, shape. These helices are connected by strong bonds between amino acids. These bonds make hair strong.

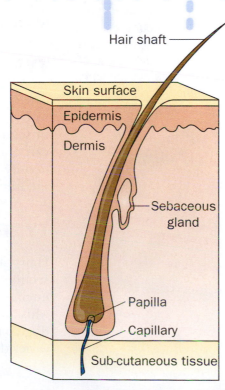

Hair shaft
Skin surface
Epidermis
Dermis
Sebaceous gland
Papilla
Capillary
Sub-cutaneous tissue

Figure 3-2. *This cross-section shows a hair shaft in a hair follicle in the skin. If the root of the hair is present, DNA may be extracted, amplified, and compared to known samples for identification. If no root is present, hair can be matched by other characteristics that can be viewed under a compound microscope.*

Figure 3-3. *The structure of a hair shaft is similar to that of a yellow pencil.*

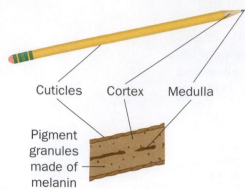

Cuticles   Cortex   Medulla

Pigment granules made of melanin

Figure 3-4. *This scanning electron photomicrograph shows the cuticle of a human hair with the overlapping scales.*

©Dee Breger/Photo Researchers, Inc.

Figure 3-5. *The pigment granules in the red hair on the left are evenly distributed throughout the cortex. Commonly, the pigment is denser near the cuticle of human hairs, as shown on the right.*

Courtesy, FBI; photos by Sandra Koch & Douglas W. Deedric

The hair shaft is made up of three layers: an inner **medulla**, a cortex, and an outer cuticle. A good analogy for the structure of a hair shaft is the structure of a pencil (Figure 3-3). The painted yellow exterior of the pencil is similar to the cuticle. The graphite in the middle of the pencil is similar to the medulla. The wood of the pencil is analogous to the cortex of a hair. Human hair has cuticle scales that are flattened and narrow, also called *imbricate.* Animal hair had different types of cuticles that is described and pictured later in the chapter under animal hair.

THE CUTICLE

The **cuticle** is a transparent outer layer of the hair shaft. It is made of scales that overlap one another and protect the inner layers of the hair (Figure 3-4). The scales point from the proximal end of the hair, which is closest to the scalp, to the distal end, which is farthest from the scalp. When examining a section of hair under a microscope, noticing the direction the scales point shows the younger and older ends of the hair. This information can be used when an investigator needs to analyze hair for the presence of different toxins, drugs, or metals at specific points in time. Human hair has cuticle scales that are flattened and narrow, also called imbricate. Animal hair has different types of cuticles that are described and pictured later in the chapter under animal hair.

TYPES OF CORTEX

In humans, the **cortex** is the largest part of the hair shaft. The cortex is the part of the hair that contains most of the pigment granules (**melanin**) that give the hair its color (Figure 3-5). The pigment distribution varies from person to person. Some people have larger pigment granules within the cortex, giving the cortex an uneven color distribution when viewed under the compound microscope.

TYPES OF MEDULLA

The center of the hair is called the medulla. It can be a hollow tube, or filled with cells. In some people the medulla is absent, in others it is fragmented, or segmented, and in others it is continuous or even doubled. The medulla can contain pigment granules or be unpigmented. Forensic investigators classify hair into five different groups depending on the appearance of the medulla, as illustrated in Figure 3-6.

TYPES OF HAIR

Hair can vary in shape, length, diameter, texture, and color. The cross section of the hair may be circular, triangular, irregular, or flattened, influencing the curl of the hair. The texture of hair can be coarse as it is in whiskers or fine as it is in younger children. Some furs are a mixture as in dog coats, which often have two layers: one fine and one coarse. Hair color varies depending on the distribution of pigment granules and on hair dyes

Figure 3-6. *Five different patterns of medulla are identified in forensic hair analysis.*

Medulla Pattern	Description	Diagram
Continuous	One unbroken line of color	
Interrupted (Intermittent)	Pigmented line broken at regular intervals	
Fragmented or Segmented	Pigmented line unevenly spaced	
Solid	Pigmented area filling both the medulla and the cortex	
None	No separate pigmentation in the medulla	

that might have been used (Figure 3-7). These attributes can all be used for identification or exclusion in forensic investigations.

In humans, hair varies from person to person. In addition, different hairs from one location on a person can vary. Not all hairs on someone's head are exactly the same. For example, a suspect may have a few gray hairs among brown hairs in a sample taken from his head. Because inconsistencies occur within each body region, 50 hairs are usually collected from a suspect's head. Typically, 25 hairs are collected from the pubic region.

HAIR FROM DIFFERENT PARTS OF THE BODY

Hair varies from region to region on the body of the same person (Figure 3-8). Forensic scientists distinguish six types of hair on the human body: (1) head hair, (2) eyebrows and eyelashes, (3) beard and mustache hair, (4) underarm hair, (5) auxiliary or body hair, and (6) pubic hair. Each hair type has its own shape and characteristics.

One of the ways in which hairs from the different parts of the body are distinguished is their cross-sectional shape. Head hair is generally circular or elliptical in cross section. Eyebrows and eyelashes are also circular but often have tapering ends. Beard hairs tend to be thick and triangular. Body hair can be oval or triangular, depending on whether the body region has been regularly shaved. Pubic hair tends to be oval or triangular.

Hairs from different parts of the body have other characteristic physical features. Hair from the arms and legs usually has a blunt tip, but may be frayed at the

Figure 3-7. *Hairs coming from a single area on one person can vary in characteristics.*

©Blend Images/Jupiter Images

Figure 3-8. *The physical characteristics of hairs provide information about which part of the body they came from.*

Courtesy, FBI; photos by Sandra Koch & Douglas W. Deedric.

Pubic hair showing buckling

Beard hair with double medulla

Arm or leg hair with blunt, frayed end

Figure 3-9. *Examples of dyed human hair.*

Courtesy, FBI; photos by Sandra Koch & Douglas W. Deedric

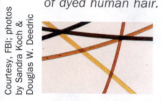

ends from abrasion. Beard hair is usually coarse and may have a double medulla. The diameter of pubic hair may vary greatly, and buckling may be present.

THE LIFE CYCLE OF HAIR

Hair proceeds through three stages as it develops. The first stage is called the *anagen stage* and lasts approximately 1,000 days. Eighty to ninety percent of all human hair is in the anagen stage. This is the period of active growth when the cells around the follicle are rapidly dividing and depositing materials within the hair. The *catagen stage* follows as the hair grows and changes (perhaps turning gray). The catagen stage accounts for about 2 percent of all hair growth and development. The final stage is the *telogen stage*. During this stage the hair follicle is dormant or resting and hairs are easily lost. About 10 to 18 percent of all hairs are in the telogen stage. There is no pattern as to which hairs on the head are in a particular stage at any time.

TREATED HAIR

Hair can be treated in many different ways (Figure 3-9). Bleaching hair removes pigment granules and gives hair a yellowish color. It also makes hair brittle and can disturb the scales on the cuticle. Artificial bleaching shows a sharp demarcation along the hair, while bleaching from the sun leaves a more gradual mark. Dyeing hair changes the color of the hair shaft. An experienced forensic examiner can immediately recognize the color as unnatural. In addition, the cuticle and cortex both take on the color of the dye.

If an entire hair is recovered in an investigation, it is possible to estimate when the hair was last color-treated. The region near the root of the hair will be colored naturally. Human hair grows at a rate of about 1.3 cm per month (approximately 0.44 mm per day). Measuring the length of hair that is naturally colored and dividing by 1.3 cm provides an estimate of the number of months since the hair was colored. For example, if the unbleached root region measured 2.5 cm, then 2.5 cm divided by 1.3 cm per month equals approximately 1.9 months or about 7 weeks. This information can be used to identify hairs from different locations as belonging to an individual.

RACIAL DIFFERENCES

Hair examiners have identified some key physical characteristics that are associated with hair of different broad racial groups. These characteristics are only generalities and may not apply to individuals of certain races. In addition, a certain hair may be impossible to assign to a particular race because its characteristics are poorly defined or difficult to measure. The broad characteristics of hairs from different races are compared in Figure 3-10.

ANIMAL HAIR AND HUMAN HAIR

Animal hair and human hair have several differences, including the pattern of pigmentation, the medullary index, and the cuticle type. The pattern of the pigmentation can vary widely in different animals. While the pigmentation in human hair tends to be denser toward the cuticle, in animals it is denser toward the medulla. Animal pigments are often found in solid masses called *ovoid bodies*, especially in dogs and cattle. Human hairs are

Figure 3-10. *A comparison of some general physical characteristics of hair from different races.*

Race	Appearance	Pigment Granules	Cross Section	Other
European	Generally straight or wavy 	Small and evenly distributed	Oval or round of moderate diameter with minimal variation 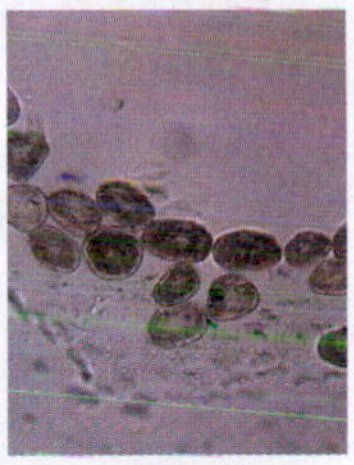	Color may be blond, red, brown, or black
Asian	Straight 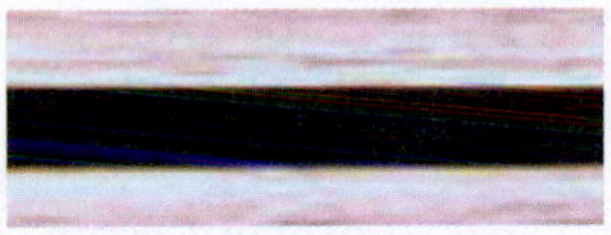	Densely distributed	Round with large diameter 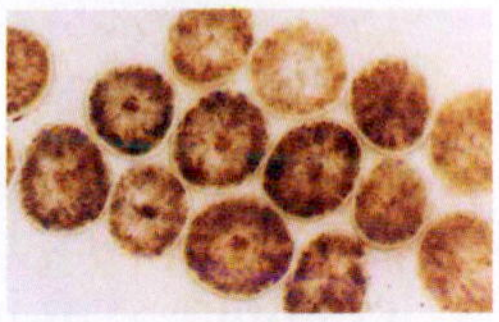	Shaft tends to be coarse and straight Thick cuticle Continuous medulla
African	Kinky, curly, or coiled 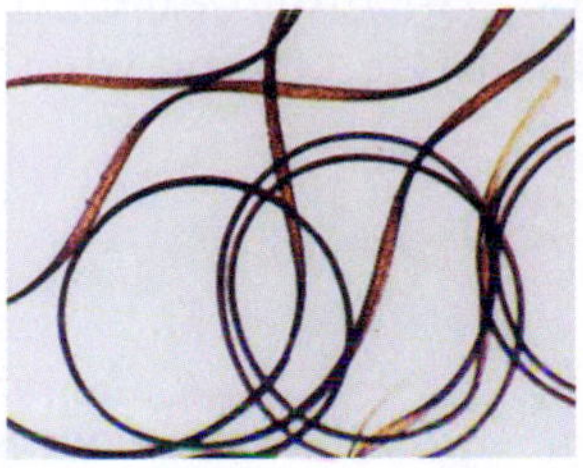	Densely distributed, clumped, may differ in size and shape	Flattened with moderate to small diameter and considerable variation 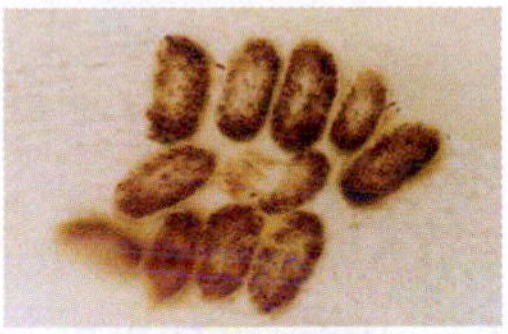	

Courtesy, FBI; photos by Sandra Koch & Douglas W. Deedric

usually one color along the length. Animal hairs can change color abruptly in a banded pattern.

In animals, the medulla is much larger than it is in humans (Figure 3-11). The ratio of the diameter of the medulla to the diameter of the entire hair is known as the *medullary index*. If the medullary index is 0.5 or greater, the hair came from an animal. If the medullary index is 0.33 or less, the hair is from a human.

Figure 3-11. *The medulla of animal hair is much larger than in human hair, and it is always continuous.*

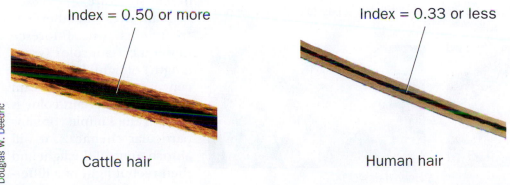

Index = 0.50 or more

Index = 0.33 or less

Cattle hair

Human hair

Courtesy, FBI; photos by Sandra Koch & Douglas W. Deedric

The cuticle of the hair shaft can also help distinguish human hair from animal hair. There are different types cuticles in different mammal hair cuticles. Rodents and bats have a *coronal* cuticle with scales that give the appearance of a stack of crowns. Cats, seals, and mink have scales that are called *spinous* and resemble petals. Human hair has cuticle scales that are flattened and narrow, also called *imbricate*.

USING HAIR IN AN INVESTIGATION

Whenever two objects are in contact, some transfer of material will occur. This is known as Locard's exchange principle. It is the fundamental reasoning behind the use of trace evidence in forensic investigations. If a person is at a crime scene, he or she will leave some trace of his or her presence behind, or pick up some trace evidence from the crime scene. One of the major examples of trace evidence is hair.

When investigators enter a crime scene, they collect trace evidence, including hair. Hair can be collected from evidence by plucking, shaking, and scraping surfaces. It can also be collected by placing tape over a surface so that the hair adheres to it. When surfaces are large, they can be vacuumed. The material that is filtered into the canister can be examined for hair and other trace particles. Investigators are always careful to prevent cross-contamination of evidence by inadvertently transferring hair from one object to another.

If a large number of hairs are collected from a victim or a crime scene, an investigator will compare the sample with hair taken from the six major body regions of the victim or suspect(s). An initial analysis is performed using a low-power compound microscope to determine whether the hair is human or animal.

MICROSCOPY

Hair viewed for forensic investigations is studied both macroscopically and microscopically. Length, color, and curliness are macroscopic characteristics. Microscopic characteristics include the pattern of the medulla, pigmentation of the cortex, and types of scales on the cuticle (Figure 3-12). Medullary index can be measured. Typical magnification for viewing hair is between 40 times and 400 times. A particularly useful microscope for hair analysis is called a comparison microscope. It allows for simultaneous viewing of two different samples.

Several specialized microscopic techniques are also used in hair analysis. Phase contrast microscopy involves using a special objective lens and special condenser with a compound microscope. This configuration focuses light that passes through objects of different refractive indexes. The resulting image shows more contrast, especially when viewing translucent particles. Phase contrast microscopy in hair analysis is useful for observing fine detail in hair structure.

Many dyes and other hair treatments will fluoresce under a certain color (wavelength) of light. In a fluorescence microscope, a beam of light of a certain color is used. If the sample contains particular chemicals, it will absorb some of the light and then reemit light of a differ-

Figure 3-12. *Using microscopy, investigators might link dog hair to a dog owner or deer hair to a hunter.*

Dog hair 400×

Deer hair 400×

Courtesy, FBI; photos by Sandra Koch & Douglas W. Deedric

Digging Deeper
with Forensic Science e-Collection

Edmund Locard was a French criminalist who played a pivotal role in expanding the field of forensic science. Originally trained as a physician, Locard became fascinated with police work and using laboratory techniques to help solve crimes. Search the Gale Forensic Science eCollection on school.cengage.com/forensicscience and research Locard's life. Write a short biography of this great researcher explaining how his contributions affected forensic science. What ideas did he have that advanced the field? Who were his most important influences? How has forensics changed as a result of Locard's ideas?

ent color. This is called *fluorescence.* A fluorescence microscope is equipped with filters to detect the fluoresced light, indicating the presence of a dye or other treatment.

Instead of using light to view a sample, electron microscopes direct a beam of electrons at a sample. Electron microscopes provide incredible detail of the surface or interior of the sample (Figure 3-13), magnifying the object 50,000 times or more.

TESTING FOR SUBSTANCES IN THE HAIR SHAFT

Because hair grows out of the skin, chemicals that the skin absorbs can become incorporated into hair. Ingested or absorbed toxins such as arsenic, lead, and drugs can be detected by chemical analyses of hair. During testing, the hair is dissolved in an organic solvent that breaks down the keratin and releases any substances that have been incorporated into the hair. A forensic chemist can perform chemical tests for the presence of various substances. In forensic investigations, this type of analysis can provide evidence of poisoning or drug use.

Because hair does not readily decompose, by testing different parts of the hair, it may be possible to establish a timeline for when exposure to poisons or other toxins might have occurred. The procedure for developing the timeline would be similar to the one used with hair color analysis discussed earlier in the chapter. Human hair grows at the rate of about 1.3 cm per month (approximately 0.44 mm per day). The hair can be analyzed in sections for the specific toxin. If the root is present to identify the base of the hair, these sections can be dated based on their distance from the root. If the toxin occurs 9 cm from the root, dividing this value by 1.3 cm per month provides an estimate of the number of months since the toxin was ingested. In this case, 9 cm divided by 1.3 cm per month equals approximately 7 months.

Neutron activation analysis (NAA) is a particularly useful technique that can identify up to 14 different elements in a single two-centimeter-long strand of human hair. The hair is placed in a nuclear reactor and bombarded with high-energy neutrons. Different elements will give off gamma radiation with different signals. These signals can be recorded and interpreted to determine concentrations of elements in the sample. Elements such as antimony, argon, bromine, copper, gold, manganese, silver, sodium, and zinc can be identified and quantified using NAA. The probability of the hairs of two individuals having the same concentration of nine different elements is about one in a million.

Figure 3-13. *A transmission electron microscope produced this extremely detailed image of a long section of human hair. Notice the overlapping cuticle scales on the left side and the pigment granules in the cortex.*

Courtesy, FBI; photos by Sandra Koch & Douglas W. Deedric

Digging Deeper
with Forensic Science e-Collection

How is DNA collected from hair follicles used in forensic investigations? What sorts of DNA testing are most often performed on hair follicles? What sorts of controversies surround these tests? Are errors ever made? What are the major problems with DNA testing of material associated with hair? Go to the Gale Forensic Science eCollection on school.cengage.com/forensicscience and enter the search terms "DNA hair analysis." Click on Academic Journals and read about some of the problems and successes associated with DNA analysis and hair. Write a brief essay describing your findings, and be sure to cite your resources.

TESTING THE HAIR FOLLICLE

If hair is forcibly removed from a victim, the entire hair follicle may be present. This is called a *follicular tag*. If this occurs, blood and tissue attached to the follicle may be analyzed. For example, blood proteins can be isolated to identify the blood type of a suspect. DNA analyses can also be performed on hair-follicle cells (Figure 3-14). DNA analysis of the hair follicle provides an identification with a high degree of confidence, whereas analysis of the hair shaft usually provides class evidence only. In many cases, a microscopic assessment of the hair is performed initially because it is more cost effective and rapid than blood protein and DNA testing. If a microscopic match between a suspect and a sample is found, then the samples will be forwarded for blood and DNA testing.

Figure 3-14. *DNA can be extracted from cells in the hair follicle for DNA analysis.*

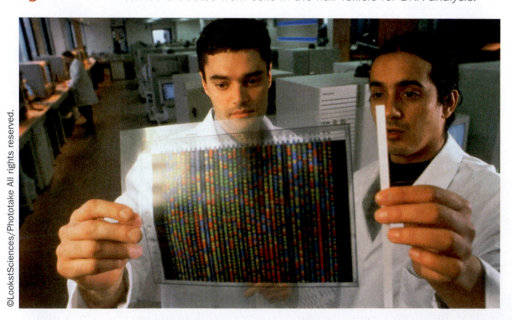

SUMMARY

- Hair is a form of class evidence that has been used in forensic analysis since the late 19th century.

- Hair is a character shared by all mammals and functions in temperature regulation, reducing friction, protection from light, and as a sense organ.

- Hair consists of a follicle embedded in the skin that produces the shaft.

- The shaft is composed of the protein keratin and consists of the outer cuticle, a cortex, and an inner medulla, each of which varies among individuals or species.

- Hair varies in length and cross-sectional shape, depending on where on the body it originates.

- Hair development is broken into three developmental stages, called the anagen (growth), catagen (growth and change), and telogen (dormant) stages.

- Various hair treatments produce characteristic effects that are useful to forensic experts, and some hair characteristics allow them to be grouped into general racial categories.

- Forensic experts examine hair using light (phase contrast, fluorescence, comparing) and electron microscopy, and analyze hair chemically for drugs and toxins.

- Neutron activation analysis allows unique signatures of elements contained in hair to be identified, and the hair follicle can provide DNA for sequencing.

CASE STUDIES

Alma Tirtsche (1921)

Alma Tirtsche's beaten body was found wrapped in a blanket in what is known as Gun Alley in Melbourne, Australia. Because the body was relatively free of blood, the police deduced that she had been murdered elsewhere and brought to the alley. Her body had been washed before being wrapped in the blanket. A local bar owner, Colin Ross, was questioned. Ross admitted seeing Tirtsche in his bar earlier in the day.

Investigators collected blankets from Ross's home and found several strands of long, reddish blond hair on them. The length of the hair implied it had come from a female, and the concentration of pigment in the hair implied a younger woman. Some of the ends of the hair were irregular, implying the hair had been forcibly broken off. The physical similarity of the hair found on the blanket with that of Alma Tirtsche convinced the jury that Ross was the murderer. This was the first time that hair was used to secure a conviction in Australia. Unfortunately, analysis of the hairs 75 years later showed that two of the strands found on the blankets came from different individuals, which throws doubt on Ross's guilt.

©Digital Vision/Getty Images

Eva Shoen (1990)

In Telluride, Colorado, Eva Shoen was found dead from a single gunshot to her head. The police recovered the bullet and expected to solve the case using ballistics information. Unfortunately, they did not have any useful leads. Three years later, the police received a phone call from a man who believed that his brother, Frank Marquis, was responsible for Shoen's death. A gun was found on Marquis, but he had already tampered with its barrel, preventing a ballistic match.

From questioning a companion of Marquis's, police learned that Marquis had been in Telluride when Shoen was murdered. They also discovered that Marquis had thrown two bundles out of his car during his drive home to Arizona. Detectives searched the road until they found a bundle of clothing. One of the shirts in the bundle contained a single strand of hair. The color and structure of the hair matched that of Eva Shoen's hair. When confronted with the evidence, Marquis confessed to the murder and was imprisoned for 24 years.

©The Gallery Collection/Corbis

Napoleon's Hair

Napoleon Bonaparte proclaimed himself emperor of France in 1804 after rising swiftly through the ranks of the French army. Following his defeat at Waterloo, he was exiled on the British island of St. Helena in the Atlantic Ocean. History books proclaim that he died in exile of stomach cancer.

In 2001, a Canadian Napoleon enthusiast, Ben Weider, challenged this theory. He had five strands of Napoleon's hair collected in 1805, 1814, and 1821 tested using neutron activation analysis. The results of the analysis showed that Napoleon's hair contained between 7 and 38 times more arsenic than normal, a fatal dose. In 2002, further analysis of Napoleon's hair showed extremely elevated levels of arsenic, leading researchers to joke that Napoleon should have died twice before his actual death, and suggesting that the hair must have been contaminated during storage.

Eventually, the esteemed chemist, Walter McCrone, tested a sample of Napoleon's hair. His work contradicted the previous reports, stating that the levels of arsenic that had been incorporated into Napoleon's hair were much too low to have killed him. The story continues to cause controversy. Most chemists believe that McCrone's work is the final story, but Napoleon enthusiasts believe that the emperor's death is surrounded by too many questions to disregard the possibility of murder.

Think Critically Do you consider hair evidence important in proving a crime? Explain your answer.

CAREERS IN FORENSICS

William J. Walsh, Chemical Researcher

With a doctorate in chemical engineering and a research record that includes such illustrious laboratories as Atomic Research in Ames, Iowa, Los Alamos National Laboratory in New Mexico, and Argonne National Laboratory in Illinois, William Walsh has spent more than 30 years studying chemical processes involved in nuclear fuel production, liquid metal distillation, and electrochemistry. Dr. Walsh has authored more than 200 scientific articles and reports and made numerous presentations on his research. Dr. Walsh is the Chief Scientist of the Health Research Institute and Pfeiffer Treatment Center, both in Illinois.

Scientist performing hair analysis.

©AP Photo/Ric Field

Dr. Walsh's work in chemistry led to an interest in developing tools and chemical methods for extracting information from hair. Dr. Walsh and his colleagues collected known chemistry information from more than 100,000 people and synthesized it into the world's first standard of known hair composition. Walsh has served as an expert chemist in numerous forensic studies of hair samples in collaboration with medical examiners, coroners, and police groups. Some of the more famous, or infamous, people whose hair chemistry Walsh has studied include Charles Manson (Manson Family murders), Henry Lee Lucas (20th-century serial killer), James Hubeity (McDonald's massacre), William Sherrill (Oklahoma post office slayings), and other notorious criminals. In addition, while volunteering at the Stateville Penitentiary in Joliet, Illinois, Walsh became interested in the way that chemicals can affect behavior. These combined interests—hair forensics and the influence of biochemicals on behavior—made Walsh the perfect candidate to head up one of the most famous hair investigations: that of composer Ludwig van Beethoven.

Walsh was the chief scientist on the Beethoven Research Project in 2000. The goal of the project was to understand whether chemical toxins may have played a role in Beethoven's death. Beethoven developed an illness in his twenties that involved abdominal distress, irritation, and eventually depression. By the age of 31, he began to lose his hearing, and by 42, he was completely deaf. He died of liver and kidney failure. Using highly sensitive techniques—scanning electron microscope energy dispersion spectrometry (SEM/EDS) and scanning ion microscope mass spectrometry (SIMS)—Walsh verified that Beethoven's hair contained extremely high concentrations of lead, which almost certainly contributed to his death.

Learn More About It

To learn more about forensics hair analysis, go to school.cengage.com/forensicscience.

CHAPTER 3 REVIEW

True or False

1. The shaft of the hair is considered class evidence in a trial.

2. Hair is composed of a protein called cellulose.

3. All hairs on the head of a person are identical.

4. The cortex may contain pigment granules.

Multiple Choice

5. The hair shaft is composed of the cuticle, cortex, and
   a) medulla
   b) root
   c) crown
   d) granules

6. Which factors are used to calculate the medullary index of the hair?
   a) scale diameter of cuticle and the length of the hair
   b) width of cortex and the width of the medulla
   c) length of entire hair and the pattern of pigmentation
   d) width of medulla and the width of the hair

7. Which of the following characteristics is found in typical Asian hair?
   a) dark medulla
   b) sparsely distributed pigment granules
   c) flattened cross section
   d) hair is curly

8. Human hair has which type(s) of cuticle?
   a) imbricate
   b) spinous
   c) coronal
   d) pigmented

9. Neutron activation analysis can check hair for the presence of
   a) silver
   b) DNA
   c) water content
   d) hair dye

10. Which part(s) of a hair can be analyzed for DNA?
    a) root
    b) cuticle
    c) medulla
    d) cortex

11. The cuticle scales of the hair always point toward the
    a) root
    b) medulla
    c) tip of the hair
    d) follicle

12. The period of active hair growth is called the _____ stage.
    a) catagen
    b) telogen
    c) anagen
    d) imagen

13. Although variations can occur, which of the following best describes northern European hair?
    a) kinky with dense, unevenly distributed pigment
    b) straight with evenly distributed granules
    c) round cross section with a large diameter
    d) coarse with a thick cuticle and a continuous medulla

14. Which of the following is most likely a result of hair bleaching?
    a) increased number of disulfide bonds
    b) a yellowish tint to the hair
    c) a more triangular cross section
    d) thickened scales on the cuticle

Short Answer

15. Why is hair considered class evidence?

    _____

    _____

    _____

    _____

16. Describe the structure of hair. Include in your answer the terms *follicle*, *medulla*, *cortex*, and *cuticle*.

    _____

    _____

    _____

    _____

17. Crime-scene investigators collected hair from a dead person's body. One of the first things that needs to be established is if this hair is human or animal. Describe two ways that animal hair differs from human hair.

_____

_____

_____

_____

18. The body of a woman was found in the woods. Some hair fibers found on the body were sent to the crime lab for analysis. The ends of the hair attached to the body were gray, but the tips of the hair showed that it had been dyed. The distance from the root of the hair to the beginning of the dyed area measured 8 mm. Investigators determined that the victim's hair had last been dyed on August 1, 2004. Assuming the hair grows at the rate of 0.44 mm per day, on approximately what date did the woman die? Explain your answer.

_____

_____

_____

_____

19. Calculate the medullary index of a hair whose diameter is 110 microns wide and whose medulla measures 58 microns. Is this a human or animal hair?

_____

_____

_____

_____

20. A woman with long hair is a suspect in a burglary case. At the crime scene, several long hairs were found attached to a broken lock of the safe. The police obtain a warrant and request a sample of 25 to 50 hairs from this woman. They tell the woman it is important that they pull the hairs from her head rather than to merely cut the hairs. The police suspect that the woman was stealing to help support a drug habit.

   a. Why is it important that the police pull the hairs from her head rather than cut her hair?

   b. Why is it necessary to obtain 25 to 50 hairs from this woman?

   c. The woman denies that she is currently taking drugs and states that she stopped using drugs a year ago. Explain how the police can determine if the woman has been off drugs for over one year.

   d. Suppose the hairs of the woman match the hairs found at the crime scene. Why does this not necessarily prove that she was the guilty party?

Bibliography

Books and Journals

Baden, Michael, and Marion Roach. *Dead Reckoning.* London: Arrow Books, 2002.

Lee, Henry. *Physical Evidence in Forensic Science.* Tucson, AZ: Lawyers & Judges Publishing, 2000.

Saferstein, Richard. *Criminalistics: Introduction to Forensic Science*, 7th ed. Englewood Cliffs, NJ: Prentice Hall, 2000

Taylor, Alfred Swaine. *The Principles and Practice of Medical Jurisprudence.* London: Churchill Livingstone, 2007

Watkins, Ronald J. *Birthright: Murder, Greed, and Power in the U-Haul Family Dynasty.* New York: William Morrow & Company, 1993

Web sites

"Alma Tirtsche." http://www.history.com/this-day-in-history.do?printable=true&action=tdihArticlePrint &id=982

Deedrick, Douglas W. "Hair, Fibers, Crime and Evidence." Quantico, VA: FBI. http://www.fbi.gov/hq/lab/fsc/backissu/july2000/deedric1.htm.

Deedrick, Douglas W., and Sandra L. Koch. "Microscopy of Hair. Part 1: A Practical Guide and Manual for Human Hairs." Quantico, VA: FBI.

Deedrick, Douglas W., and Sandra L. Koch. "Microscopy of Hair. Part 2: A Practical Guide and Manual for Human Hairs." Quantico, VA: FBI. http://www.fbi.gov/hq/lab/fsc/backissu/jan2004/research/2004_01_research01b.htm

Gale Forensic Sciences eCollection, school.cengage.com/forensicscience.

"Thread of Evidence." The Discovery Channel, Netherlands, Forensic Institute. http://www.forensis-chinstituut.nl/NFI/en/Typen+onderzoek/Items/Forensic+examination+of+hair.htm.

"Thin Blue Line." http://www.policensw.com/info/forensic/forensic7a.html.

ACTIVITY 3-1
TRACE EVIDENCE: HAIR

Objectives:

By the end of this activity, you will be able to:
1. Describe the external structure of hair.
2. Distinguish between different hair samples based on color, medulla types, cuticle types, thickness, and length.
3. Compare a suspect's hair with the hair found at a crime scene.
4. Form a hypothesis as to which suspect could have been present at a crime scene.
5. Justify whether or not a suspect's hair sample matches the hair sample left at a crime scene.

Time Required to Complete Activity: 60 minutes

Introduction:

In this laboratory exercise, you will work with hair evidence that was collected at a crime scene. Your task is to try to match the hair evidence that was collected at the crime scene with hair collected from four suspects.

Materials:

Activity 3-1 Lab Sheet
plastic microscope slides
clear plastic tape
compound microscope
prepared slides of hair samples
2 glass slides
glass cover slips
scissors
clear nail polish

Safety Precautions:

Always carry a microscope using both hands.
Do not get nail polish on the lens.

Scenario:

A murder was committed. To dispose of the body, the suspect(s) tossed the body from the car into a ditch. When crime-scene investigators arrived, they photographed the crime scene and drew sketches of the body. Hair evidence was found on the victim. Hair samples were collected from the four suspects, as well as a sample of hair taken from the victims head. At the crime lab, a comparison microscope was used to examine each of the hair samples. Your task is to examine all hair samples under the compound microscope and record your observations. After reviewing all samples, determine if any of the suspects' hair matches the hair found at the crime scene. You will need to justify your decision.

Procedure:

Part 1: Cuticle Impression

1. Obtain a clean glass slide.
2. Place the slide along the edge of the desk.
3. Wipe a thin layer of nail polish on the slide the length and width of a cover slip.
4. Either pull out or cut a hair from your head.
5. While holding onto the hair between two fingers in front of the slide, slowly lower the hair onto the slide being careful not to wiggle the hair back and forth. Pull the hair down into the nail polish and let go of the hair.
6. Wait 10 minutes to remove the hair.
7. After 10 minutes, grasp the lose end of the hair and pull straight up to completely remove the hair from the nail polish.
8. Observe the slide under 100×. Sketch your cuticle.

Part 2: Observation of Your Own Hair

1. Obtain a plastic slide. Write your initials on the end of the slide.
2. Remove a hair from your head, preferably a hair that contains a root. You may pull it out or use scissors to cut it.
3. Place the hair on your desk.
4. Fold the tape with the sticky side facing the hair on the table. Hold the tape near the hair, but do not touch the hair. The hair should be attracted to the sticky surface of the tape.
5. Place the tape with the attached hair to the plastic slide. Use your finger to press down on the tape to squeeze out any air pockets. Cut off the excess tape. You now have a permanent slide.

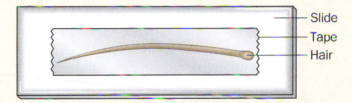

Slide — Tape — Hair

The finished slide.

6. Label the slide with your name using a permanent marking pen.
7. Focus the hair using 100× magnification.
   a. Draw your hair in the space provided on Data Table 1.
   b. Identify the type of medulla, cuticle, color, and any other distinguishing features.

Data Table 1

Source of Hair	Sketch	Color	Medulla	Cuticle	Straight or Curly	Other Characteristics
Your name						

Part 3: Comparative Analysis of Suspect and Crime Scene Hair

1. Obtain a slide of the victim's hair from the envelope prepared by your instructor. Draw a sketch of the victim's hair, and record all of the information in Data Table 2. Return the slide to the envelope as soon as you are finished so that someone else can use the slide.

2. Look at each of the four suspects' hairs. Draw sketches and record all required information in Data Table 2. Please take only one slide at a time!

3. You will need to rule out that the hair found on the victim did not come from the victim's own head. You will need to examine the sample entitled "Victim's Own Hair."

4. Compare your results with another classmate. If you find you have different answers, it might be necessary to examine more than one hair sample from any individual. Recall that not all hairs are exactly alike.

5. Is it possible to match any of the suspects' hair with the evidence hair that was found on the victim? Be prepared to justify your answer using forensic evidence.

6. Record your results in Final Analysis.

Final Analysis:

1. Does your crime scene hair match any of the suspects' hairs? If yes, which particular suspect?

2. Cite three different characteristics of hair that can be used to support your answer to question number 1. Use complete sentences and correct terminology.

Data Table 2

Source of Hair	Sketch	Color	Medulla	Cuticle	Straight or Curly	Other Characteristics
Crime Scene Hair						
Suspect 1						
Suspect 2						
Suspect 3						
Suspect 4						
Victim's Own Hair						

ACTIVITY 3-2
HAIR MEASUREMENT AND MATCH

Objectives:

By the end of this activity, you will be able to:
1. Describe how to measure the diameter of a hair that is viewed under a compound microscope.
2. Measure hair samples and determine if the diameter of the hair samples from different sources are the same.

Time Required to Complete Activity: 60 minutes

Introduction:

Hair is an example of trace evidence that can be left at a crime scene or removed from a crime scene (Locard's exchange principle). Although hair is not unique to a specific person, it can be used to identify a class of individuals. (The exception to this occurs if the root of the hair is present and DNA can be extracted and a match made with a crime-scene sample.)

Materials:

(per group of 2 students)
Activity 3-4 Lab Sheet
compound microscope
clear plastic mm ruler
2 glass slides
dropper of fresh water
pencil
pre-made slide of crime scene hair
pre-made slide of the victim's hair
pre-made slide of suspect #1 hair
pre-made slide of suspect #2 hair
pre-made slide of suspect #3 hair

Safety Precautions:

Always carry the microscope with both hands.
No special safety concerns

Scenario

You might say that some people's hair is very fine. Others may have hair that is very coarse. The diameter of the hair provides us with another way to compare a suspect's hair to the crime scene hair. In this lab activity, you will compare the crime scene hair with three suspect's hairs by comparing their medulla, cortex, and cuticle types, as well as compare the diameter of the hair samples.

Procedure:

1. Measure the size of the diameter of the microscope under 100×.
   a. If an ocular micrometer is available, measure the diameter of the field of view. (Most microscopes have a field of view of approximately 1.2 mm.)
   b. If an ocular micrometer is not available:
      - Place a small, clear plastic ruler under the microscope under 100×.
      - Focus on the metric side of the ruler.
      - Measure the diameter of the field of view to the nearest tenth of a millimeter.
      - Record your answer in Data Table 1.

2. Pull out one of your hairs and place it in a drop of water on a microscope slide.

3. Place a cover slip over the hair and view under LOW power (100×).

Plastic ruler viewed under 100×.

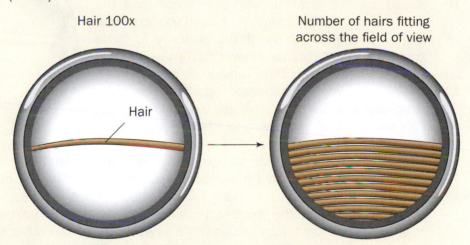

©Cengage Learning

4. Note the following characteristics of your hair and record the information in Data Table 2:
   - Color of cortex
   - Type of medulla (e.g., continuous, interrupted, fragmented, solid, none)
   - Type of cuticle (e.g., spinous, coronal, or imbricate)

5. Measure or estimate the width of the hair using the diameter of your field of view as a reference. Record your answer in Data Table 2. For example: Center your hair so that it is in the middle of the field of view. Estimate how many hairs would fit across the field of view (100×).

Hair 100x

Number of hairs fitting across the field of view

Hair

It appears that about 10.5 hairs fit across half of the diameter of the field of view (100×). Therefore, it would take about twice as many hairs (or 2 × 10.5 = 21 hairs) to fit across the field of view.

The diameter of the single hair is ¹⁄₂₁ of the diameter of the field of view.

If the diameter is 1.2 mm, or 1200 microns, then the size of a single hair is:

Diameter	= ¹⁄₂₁ of 1.2 mm	Diameter	= ¹⁄₂₁ of 1200 microns
	= 0.05 × 1.2 mm		= 0.05 × 1200 microns
	= 0.06 mm		= 60 microns

6. Focus your hair under 400×. Draw a sketch of your hair. Record your answer in Data Table 2.

7. The diameter of the high-power (400×) field of view is ¼ of the diameter of the field of view under 100×, or approximately 300 microns. Calculate the diameter of your field of view under 400× in microns. Record your answer in Data Table 3.

8. Obtain a pre-made slide of a hair sample from the crime scene from your teacher. Measure (or estimate) the diameter of the hair in microns. Record your observations and sketch the hair sample in Data Table 4.

   You will need to record the following information:
   - Sample number
   - Width of the hair in microns
   - Color of cortex
   - Type of medulla
   - Type of cuticle
   - Straight, curly, or kinky

9. Obtain a premade slide of a suspect's hair sample from your instructor. Measure (or estimate) the diameter of the hair in microns. Record your observations and sketch the hair sample in Data Table 4.

   You will need to record the following information:
   - Sample number
   - Width of the hair in microns
   - Color of cortex
   - Type of medulla
   - Type of cuticle
   - Straight, curly, or kinky

Diameter of the field of view at high power = (¼) × diameter at low power.

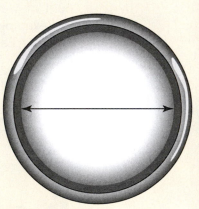

10. Based on the forensic analysis of hair and the size of the hair's diameter, would you consider the suspect's hair to match the evidence or crime scene hair? Justify your answer using the information recorded in your Data Table 4.

11. Check with your classmates regarding the other suspects' hair sample analysis. Did anyone find a hair sample that did seem to match the hair evidence left at the crime scene? Does more than one hair sample match the hair sample left at the crime scene?

12. Record the data obtained from your classmates regarding the other suspects' hair samples to Data Table 5. You do not need to view these slides under the microscope since your team of classmates is sharing their data with you. Indicate whether these two other suspects' hair matches the crime-scene hair and justify your answer.

Bonus:

Describe how you can determine that the hair sample left at the crime scene is definitely a human hair and not an animal's hair. Include calculations in your answer. Record your answer on the last page of the data sheet.

Data Table 1: Size of Field of View Under 100×

Diameter of Field of View under 100×	
(millimeters)	(microns)

Data Table 2: Your Own Hair

Your Name	Color Cuticle	Type of Medulla	Type of Cuticle	Straight, Curly, or Kinky	Width in Microns	Sketch

Data Table 3: Size of Microscope Diameter under 400×

Diameter of Field of View 100× in Microns	Calculations ¼ Diameter of Field of View under 100×	Diameter of Field of View under 400× Microns

Data Table 4: Whose Hair Matches the Crime-Scene Hair?

Hair Sample	Color Cuticle	Type of Medulla	Type of Cuticle	Straight, Curly, or Kinky	Width in Microns	Sketch
Crime Scene						

Hair Sample #	Match? or Not a Match	Justification

Data Table 5: Check with Two Other Classmates

Hair Sample #	Match? or Not a Match?	Justification
Sample # ——		
Sample # ——		

Thought Questions:

Explain each of your answers.
1. Is it possible that none of the hair samples matches the hair found at the crime scene?
2. Is it possible that more than one person's hair matches the crime scene?
3. If someone's hair does match the crime-scene evidence, does that mean that he or she committed the crime?
4. If someone's hair did match the crime scene, what type of evidence could be obtained to indicate that the DNA at the crime scene is a match to their DNA and not to anyone else's DNA?

Return all materials, complete this sheet, and hand it in during your lab. Explore further information at links on the Forensic Fundamentals and Investigations web site at school.cengage.com/forensicscience.

Bonus:

Is the last sample animal hair or human hair? Explain your answer.

ACTIVITY 3-3
HAIR TESTIMONY ESSAY

Objectives:

By the end of this activity, you will be able to:
1. Write a clear and organized essay.
2. Describe the basics of forensic hair analysis.
3. Explain why hair is considered class evidence.
4. Write a convincing argument stating your case that the suspect's hair either matches the hair found at the crime scene or that the hair does not match hair found at a crime scene

Time Required to Complete Activity: 1.5 to 2 hours

Background:

Your task is to write an essay. You are an expert witness called on to testify in a court case. You are asked to prepare a presentation to the jury that will demonstrate that a particular suspect can be linked to the crime scene. You should assume that the jury knows nothing about hair. Your paper should be typed (double-spaced), with paragraphs separating major ideas. Use spellcheck to correct any spelling errors.

Procedure:

You should prepare:
1. An introductory paragraph addressing the following questions:
   a. Who are you?
   b. Why are you here?
   c. Remember: do not cite specific information about hair within your opening statement to the jury.
2. A body paragraph in which you educate the jury about hair.
   a. Include a graphic or visual aid. Cite the source of your picture.
   b. Define all terms.
   c. Describe what characteristics or traits to look for when analyzing hair.
      • Macroscopically
      • Microscopically
3. Another body paragraph in which you convince the jury why you believe a particular suspect is a match to the hair found at the crime scene.
   a. Recall that hair is class evidence, and describe how it pertains to your argument.
   b. Recall that the hair could have been left at the crime scene prior to the murder.
   c. Your job is to convince the jury that the crime-scene hair evidence is a match to a particular suspect.
4. A concluding paragraph in which you:
   a. Summarize your findings.
   b. Remind them you are an expert.
   c. Restate your conclusion about the evidence hair and the crime-scene hair.
   d. Remember: do not introduce any new information in your conclusion.

CHAPTER **4**

A Study of Fibers and Textiles

A THREAD OF EVIDENCE

In the 1980s, a string of murders left the African American youth of Atlanta in a state of fear. For 11 months, someone was kidnapping and disposing of victims in and around Atlanta's poor neighborhoods. The victims were asphyxiated either by rope or smothering, and the bodies were disposed of in dumpsters or wooded areas. Although the police had no suspects, they were gathering a collection of unusually shaped fibers from the victims. When the fiber evidence hit the news, the bodies began to turn up in the river.

One night, two police officers were staking out a bridge over the Chattahoochee River, where many victims had been found, when a white station wagon stopped on the bridge. The car was seen driving off after something had been tossed over the bridge. The officers followed and stopped the car, and the driver, 33-year-old Wayne Williams, was arrested on suspicion of murder.

The problem faced by the police was a lack of a pattern and motive. There seemed to be no reason for the killing spree. Williams was an unsuccessful music producer and a pathological liar, but to many people, he did not seem like a killer. However, the prosecution's fiber evidence seemed to suggest otherwise.

©AP Photo/Charles Kelly

Wayne Williams.

Fibers of an unusual type that matched the carpeting in Williams's house were found on many of the victims. What the court did not hear was that no fiber evidence from the victims was found in Williams's home, other than a single red cotton fiber. Could Williams be guilty and have removed every trace of his crimes, or was he innocent? The jury chose guilty. Of the 29 murders, Williams was convicted of two and sentenced to life imprisonment. The life sentence rested entirely on his choice of carpeting.

Did they get the right man? Some do not think so. After all, there were no witnesses to the crime, no motive, and no confession.

OBJECTIVES

By the end of this chapter you will be able to:

✔ Identify and describe common weave patterns of textile samples.

✔ Compare and contrast various types of fibers through physical and chemical analysis.

✔ Describe principal characteristics of common fibers used in their identification.

✔ Apply forensic science techniques to analyze fibers.

BIOLOGY EARTH SCIENCES CHEMISTRY

PHYSICS PSYCHOLOGY MATHEMATICS π

TOPICAL SCIENCES KEY

VOCABULARY

amorphous without a defined shape; fibers composed of a loose arrangement of polymers that are soft, elastic, and absorbing (for example, cotton)

crystalline regularly shaped; fibers composed of polymers packed side by side, which make it stiff and strong (for example, flax)

direct transfer the passing of evidence, such as a fiber, from victim to suspect or vice versa

fiber the smallest indivisible unit of a textile, it must be at least 100 times longer than wide

mineral fiber a collection of mineral crystals formed into a recognizable pattern

monomer a small molecule that may bond to other monomers to become a polymer

natural fiber a fiber produced naturally and harvested from animal, plant, or mineral sources

polymer a substance composed of long chains of repeating units

synthetic fiber a fiber made from a man-made substance such as plastic

secondary transfer the transfer of evidence such as a fiber from a source (for example, a carpet) to a person (suspect), and then to another person (victim)

textile a flexible, flat material made by interlacing yarns (or "threads")

yarn fibers that have been spun together

INTRODUCTION

Fibers are used in forensic science to create a link between crime and suspect (Figure 4-1). For example, a thief may own a jacket made of a material that happens to match the type of fiber found at the crime scene. It does not mean he was there, but a jacket like his was. If a jacket fiber, sock fiber, and shirt fiber all from items the thief owns are found at the crime scene, then the chances that the suspect was actually there are high or increased.

Figure 4-1. *Fiber evidence is used in criminal cases because it shows links between suspects and victims.*

©AP Photo/Joe McLaughlin

If we wear clothes, we shed fibers. As we walk on carpet, sit on couches, or pull on a sweater, fibers will fall off or be picked up. Check your socks; if you have carpets or pets, you will likely have many fibers from home on you right now. The forensic scientist looks for these small fibers that betray where a suspect has been and with whom he or she has been in contact.

Unlike fingerprints and DNA evidence, fibers are not specific to a single person. Criminals may be aware of police methods and may wear gloves to prevent leaving evidence at the scene of a crime. However, very small fibers shed from most textiles easily go unnoticed, and can therefore provide a very important source of evidence for police.

Fibers are a form of trace evidence. They may originate from carpets, clothing, linens, furniture, insulation, or rope. These fibers may be transferred directly from victim to suspect or suspect to victim. This is called **direct transfer**. If a victim has fibers on his person that he picked up and then transferred to a suspect, this is called **secondary transfer**. Secondary transfer might also occur when fibers are transferred from the original source to a suspect and then to a victim. For example, if a carpet fiber were transferred from the clothing of a victim to his attacker, that would be considered secondary transfer. The carpet fiber went first to the clothing of the victim and then, secondarily, to the clothing of his attacker.

Did You Know?

Police no longer cover dead bodies with cotton sheets because the cotton fiber sheeting may contaminate other fiber evidence on the victim.

Early collection of fibers in an investigation is critical. Within 24 hours, an estimated 95 percent of all fibers may have fallen from a victim or been lost from a crime scene. Thorough examination of the crime scene and the victim's body should be made for fiber evidence. Only fibers you would not expect to find are investigated. If pink fibers were found on the victim's clothes and the victim lived in a house with wall-to-wall pink carpeting, the forensic scientist would not examine these.

HOW FORENSIC SCIENTISTS USE FIBERS

Evidence of any kind must be evaluated, and this is especially important for fibers because they are so plentiful in the environment. The value of fiber evidence in a crime investigation depends on its potential uniqueness. For instance, a white cotton fiber will have less value than an angora fiber, because cotton is so common. A forensic scientist will ask questions about the following:

- *Type of fiber.* What is the composition of the fiber? How common or rare? What suspects or victims or part of the crime scene had this type of fiber on them?

- *Fiber color.* Do the fibers from the suspect's clothes match the color found in the victim's house? Is the type of dye the same?

- *Number of fibers found.* How many fibers were found—one or hundreds? More fibers suggest possible violence or a longer period of contact.

- *Where the fiber was found.* How close can you place the suspect to the scene of the crime—in the house, or close to a victim's body?

- *Textile the fiber originated from.* Are these carpet fibers, or upholstery from a car?

- *Multiple fiber transfers.* Is there only one type of fiber transferred at the crime scene? Or are there fibers from numerous sources from carpets and clothes and bedding? More sources suggest longer contact or possible violence.

- *Type of crime committed.* Was the crime violent, a break-and-enter, a kidnapping? Each type of crime has an expected pattern of contact between suspect, victim, and crime scene that will be reflected in the transfer of fibers.

- *Time between crime and discovery of fiber.* How long ago did the transfer take place—an hour, a day, a week? Unless the fiber location is undisturbed (such as a bagged jacket or locked room), the value of found fiber is greatly reduced with the passage of time because fibers will be expected to fall off, or fibers not related to the crime can be picked up.

SAMPLING AND TESTING

Fiber evidence is gathered with special vacuums, sticky tape, and forceps. It is important to be very accurate in recording where the fibers are found. Inaccurate or incomplete recording may cause evidence to be inadmissible in court.

Often, the forensic scientist will obtain small amounts of fibers from a crime scene, perhaps even just a single fiber. The first task is to identify the type of fiber and its characteristics (such as color and shape) (Figure 4-2). Then the investigator attempts to match it to fibers from a suspect source, such as a car or home. When you have only one fiber as evidence, you cannot do tests that damage or alter the fiber in any way. Two methods that can analyze fibers without damaging them are polarizing light microscopy and infrared spectroscopy.

Polarizing light microscopy uses a microscope that has a special filter in it that allows the scientist to look at the fiber using specific light wavelengths. How the fiber appears can tell the scientist the type of fiber. Natural fibers, such as wool or cotton, require only an ordinary microscope to view characteristic shapes and markings. Infrared spectroscopy emits a beam that bounces off the material and returns to

Figure 4-2. Collecting fiber evidence.

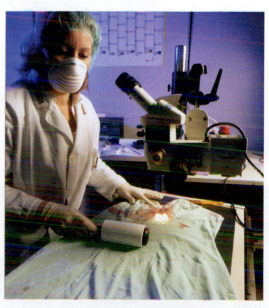

Courtesy, Dr. Vaughn Bryant

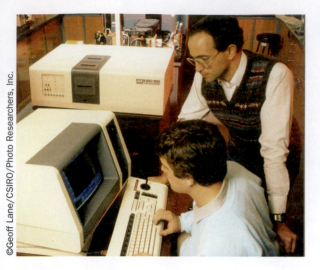

Figure 4-3. *An infrared spectrometer is used to identify unknown fibers.*

©Geoff Lane/CSIRO/Photo Researchers, Inc.

the instrument (Figure 4-3). How the beam of light has changed reveals something of the chemical structure of the fiber, making it easy to tell the difference between fibers that look very much alike.

If large quantities of fibers are found, some of the fibers may be subjected to simple, but destructive, testing—burning them in a flame or dissolving in various liquids. In the lab activities, you will have an opportunity to examine and compare fibers using a microscope. You will also perform burn testing to help identify fibers. Ultimately, you are asked to test your ability to solve a crime by comparing fibers found on different suspects with a fiber found at the crime scene.

Digging Deeper
with *Forensic Science e-Collection*

Forensic scientists can solve crimes because fibers adhere to other surfaces, such as a suspect's car seats or a victim's clothes. They also stick to hair. In fact, the nature of violent crime may mean that fibers found in a victim's hair are the only fibers recovered. Read the following articles from the Gale Forensic Sciences eCollection. You can find the Gale Forensic Science eCollection on school.cengage .com/forensicscience. Use the information in the articles to make a table that compares the fiber evidence found on clothes to the fiber evidence found in hair.

M.T. Salter and R. Cook. "Transfer of fibers to head hair, their persistence and retrieval." *Forensic Science International* 81.2-3 (August 15, 1996): pp. 211–221.

R. Cook, M. T. Webb-Salter, and L. Marshall. "The significance of fibers found in head hair." *Forensic Science International* 87.2 (June 6, 1997): pp. 155–160.

FIBER AND TEXTILE EVIDENCE

The most common form of fiber transfer to be encountered at a crime scene is shedding of a textile. **Textiles** are things like clothing, carpets, and upholstery. Many textiles are constructed by weaving, or intertwining, together **yarns**. Yarns in turn are made up of fibers that have been "spun" together.

FIBER CLASSIFICATION

Fibers are classified as either natural fibers or synthetic fibers. It is important for a forensic scientist to be able to distinguish between different kinds of fibers because this can reveal critical information about the suspect and his or her environment.

Natural Fibers

Natural fibers come from animals, plants, and minerals that are mined from the ground.

Animal fibers Animals provide fibers from three sources: hair, fur, and webbing. All animal fibers are made of proteins. They are used in clothing, carpets, decorative hangings such as curtains, and bedding.

Fur is a good donor of fibers, but it is not a textile. Rather, an animal such as a beaver or fox is trapped, and the skin removed and treated. This results in a flexible skin that retains the fur. Fur is used almost exclusively for coats and gloves.

Hair fibers are the most popular of animal fibers. Animal hair is brushed out of the animal's coat, shed naturally and collected, or clipped. The most common animal hair used in textiles is wool from sheep (Figure 4-4), but there is also cashmere and mohair from goats, angora from rabbits, as well as hair from members of the camel family—alpacas, llamas, and camels. Hair fibers are used for articles of clothing, bedding, heavy coats, carpets, bags, and furniture upholstery. When animal hair fibers are made into textiles, they are often loosely spun to feel more comfortable, making textiles that shed fibers easily.

Silk, another natural fiber, is collected from the cocoons of the caterpillar *Bombyx mori*. The caterpillars are reared in captivity, and each cocoon must be carefully unwound by hand. The shimmering appearance of silk is caused by the triangular structure of the fiber, which scatters light as it passes through, just like a prism. Fabrics made from silk are commonly used in clothing and some bedding. Because silk fibers are very long, they tend not to shed as easily as hair fibers.

Plant fibers are specialized plant cells. They are grouped by the part of the plant from which they come. Seeds, fruits, stems, and leaves all produce natural plant fibers. Plant fibers vary greatly in their physical characteristics; some are very thick and stiff, whereas others are very smooth, fine, and flexible. Some are **amorphous**, a loose arrangement of fibers that are soft, elastic, and absorbent. However, all plant fibers share the common polymer cellulose. Cellulose is a polymer that is made up of simple glucose units, and is not protein. Proteins and cellulose have very different chemical and physical properties that allow a forensic scientist to tell animal and plant fibers apart. For example, cellulose can absorb water but is insoluble (will not dissolve) in water. It is very resistant to damage from harsh chemicals and can only be dissolved by very strong acids, such as sulfuric acid. Cotton is the most common plant fiber used in textiles (Figure 4-5).

Plant fibers are often short, two to five centimeters, and become brittle over time. This means that small pieces of fibers are common as trace evidence at a crime scene.

Seed fibers Cotton is found in the seedpod of the cotton plant. Because of the ease with which cotton can be woven and dyed, it has been used extensively for clothing and household textiles.

Fruit fibers Coir is a coarse fiber obtained from the covering surrounding coconuts. The individual cells of the coir fibers are narrow, with thick walls made of cellulose. When woven together, they

Figure 4-4. *A polarized light micrograph of wool fibers.*

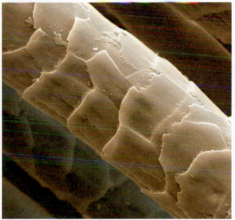

©Eye of Science/Photo Researchers, Inc.

Figure 4-5. *A ripe cotton boll showing the cotton fibers.*

©Susan Van Etten

Figure 4-6. *Coir fibers are often used in things like floor mats because they are so durable.*

Figure 4-7. *The rough fibers of jute are made into rope and twine.*

©Susan Van Etten

©Susan Van Etten

are stronger than flax or cotton. Coir fiber is relatively waterproof, which makes it ideal for such things as doormats and baskets (Figure 4-6).

Stem fibers Hemp, jute, and flax are all produced from the thick region of plant stems (Figure 4-7). They do not grow as single, unconnected fibers like cotton, but in bundles. These bundles may be six feet in length and extend the entire length of a plant. During processing, the bundles are separated from the stem and beaten, rolled, and washed until they separate into single fibers.

Flax is the most common stem fiber and is most commonly found in the textile linen. This material is not as popular as it once was because of its high cost. Linen is a very smooth and often shiny fabric that resists wear and feels cool in hot weather. Pants, jackets, and shirts are the most common garments made from linen. It is also common as tablecloths and bedding. Linen is unique because it is highly **crystalline**, so it is a dense, strong fiber that resists rot and light damage.

Other stem fibers include jute and hemp. Jute fibers produce a textile that is too coarse for garments and is instead used to make rope, mats, and handbags. Hemp is similar to flax and has been used for a long time in Asia for clothing. It has recently become a popular alternative to cotton in North America.

Leaf fibers Manila is a fiber extracted from the leaves of abaca, a relative of the banana tree. The fiber bundles are taken from the surface of the leaves. A fiber bundle, composed of many fiber cells bound together, can reach a length of ten feet. Sisal, a desert plant with succulent leaves, also provides fibers, which

Digging Deeper
with Forensic Science e-Collection

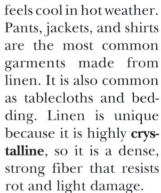

Sometimes a murder victim's body is burned to hide the evidence, but rarely are the remains so destroyed that no fibers remain. Are heated and charred fibers useless to forensics? Explore what happens to heated fibers by reading the following two articles from the Gale Forensic Sciences eCollection. You can find the Gale Forensic Science eCollection on school.cengage.com/forensicscience. Then, write a one- to two-page essay defending whether burning a body successfully hides the evidence. Provide reasons for your position.

Jolanta Was. "Identification of thermally changed fibers." *Forensic Science International* 85.1 (Feb. 7, 1997): pp. 51–63.

Jolanta Was-Gubala and Wolf Krauss. "Damage caused to fibers by the action of two types of heat." *Forensic Science International* 159.2-3 (June 2006): p. 119(8).

are used for making ropes, twines, and netting. It is commonly found as a green garden twine, or on farms as the twine on hay bales. These uses take advantage of the fiber's quick deterioration.

Figure 4-8. *Asbestos fibers.*

©Andrew Syred/Photo Researchers, Inc.

Mineral fibers are neither proteins nor cellulose (Figure 4-8). They may not even be long, repeating polymers. Fiberglass is a fiber form of glass. Its fibers are very short, very weak, and brittle. Rolls of fiberglass batting (layers or sheets of fiberglass) are used to insulate buildings. The fibers are very fine and easily stick to the skin, causing an itchy skin rash.

Asbestos is a mineral naturally occuring in different types of rocks with a crystalline structure composed of long, thin fibers. Asbestos is very durable. Its many uses include pipe coverings, brake linings, ceiling tiles, floor tiles, fire-resistant work clothes, shingles, home siding, and insulation for building materials.

Synthetic (Man-made) Fibers

Until the 19th century, only plant or animal fibers were used to make clothing and textiles. Half of the fabrics produced today are **synthetic fibers** (man-made). They are categorized as regenerated fibers and polymers. In simple terms, the fibers are produced by first joining many **monomers** together to form **polymers**. This is done in large vats. This polymer "soup" is then drained out of the bottom of the vats through tiny holes called *spinnerets* to make fibers that can then be spun into yarns. Man-made fibers include rayon, acetate, nylon, acrylics, and polyesters. By changing the size and shape of the spinneret, the qualities (for example., shine, softness, feel) of the textile can be altered. Check your classmates' clothing labels: what man-made fibers are in your classroom?

Regenerated fibers (or modified natural fibers) are derived from cellulose and are mostly plant in origin. The most common of this type is rayon. It is a fiber that can imitate natural fibers and generally is smooth and silky in appearance. Cellulose chemically combined with acetate produces the fiber Celanese® that is used in carpets. When cellulose is combined with three acetate units, it forms polyamide nylon (such as Capron®)—a breathable, lightweight material, used in high-performance clothing.

Synthetic polymer fibers originate with petroleum products and are non-cellulose-based fibers. The fibers are totally man-made polymers that serve no other purpose except to be woven into textiles, ropes, and the like. These fibers can have very different characteristics. They have no definite shape or size, and many, like polyester, may be easily dyed. Distinguishing among the synthetic fibers is easy in a forensics lab, using either a polarizing microscope or infrared spectroscopy.

Synthetic fibers may be very long, or cut and used short. Their shape is determined by the shape of the spinneret and may be round, flat, cloverleaf, or even more complex. However, under magnification, all synthetic fibers have very regular diameters. They do not have any internal structures, but may be solid or hollow, twisted, and pitted on the surface. Depending on what is put into the mix, they may be clear or translucent.

Polyester A very common synthetic fiber, polyester represents a very large group of fibers with a common chemical makeup. It is found in polar fleece, wrinkle-resistant pants, and is also added to many natural fibers to provide additional strength.

Nylon Nylon has properties similar to polyester, except it is easily broken down by light and concentrated acid. Polyester is resistant to both of these

Did You Know?

Asbestos is known to cause cancer. When broken, the fibers shatter into tiny fragments that are light enough to float in air. If we breathe them in, they make tiny cuts in our lungs with every breath we take, and the resulting scar tissue easily becomes cancerous.

Figure 4-9. A polarized light micrograph of nylon fibers.

© Sidney Moulds/Photo Researchers, Inc.

agents. Nylon was first introduced as an artificial silk, and synthetic pantyhose still go by the name nylons. Nylon fibers are shown in Figure 4-9.

Acrylic Often found as an artificial wool or imitation fur, acrylic has a light, fluffy feel. However, acrylic clothing tends to ball or pill easily. This is an inexpensive fiber.

Olefins Olefins are used in high-performance clothing, such as thermal socks and carpets, because they are very quick drying and resistant to wear.

Comparison of Natural and Synthetic Fibers

The synthetic fibers are stronger than the strongest natural fibers. Unlike natural fibers, man-made fibers are not damaged by microorganisms. A disadvantage of man-made fibers is that they can deteriorate in bright sunlight and melt at a lower temperature than the natural fibers. The table shown in Figure 4-10 shows the different characteristics of various textile fibers.

Figure 4-10. Descriptions of some common textile fibers as seen under magnification.

Fiber	Characteristics
Cotton	• "flattened hose" appearance • up to 2 inches long, tapering to a blunt point • may have a frayed "root" • hollow core not always visible
Flax	• "bamboo stick" appearance • straight with angles but not very curved • "nodes" are visible as an X every inch or so • often occur in bundles of several fibers
Silk	• do not taper, yet exhibit small variations in diameter • may be paired (raw silk) with another fiber • no internal structures
Wool	• surface scales may be visible • hollow or partially hollow core • fibers up to 3 inches long tapering to fine point
Man-made (Synthetic)	• vary widely in cross-sectional shape and diameter • generally straight to gentle curves • very uniform in diameter • may have surface treatment that appears as spots, stains, or pits

YARNS

Fibers too short in their raw state to be used to make textiles in their raw state may be spun together to make **yarns**. Short cotton fibers only two centimeters long can be twisted into very strong yarn of any length. Rope is simply a very big yarn. Depending on their use, yarns may be spun thick or thin, loose or tight. Some may be a blend of fibers, such as wool and polyester, to give desired qualities such as strength or wrinkle resistance. Any given yarn will have a direction of twist. Forensic scientists identify the twist direction as part of their identification (Figure 4-11).

TEXTILES

Weaving originated with basket making. Stone Age man used flax fibers to weave linen cloth. Wool fabrics have been found dating to the Bronze Age. The oldest loom for weaving fabric was found in an Egyptian tomb dating to 4400 B.C. In the early 1700s B.C., the people of China and India developed complicated patterns of weaving fabrics of both silk (China) and cotton (India).

Fibers are woven into textiles or fabrics. Weaving consists of arranging lengthwise threads (the *warp*) side by side and close together (Figure 4-12). Crosswise threads (the *weft*) are then woven back and forth in one of several different patterns. Ancient weavers used a frame to stretch and anchor the warp and either threaded the weft by hand or used a shuttle to alternate the strands of fibers. Machines first performed weaving in the early 1700s.

The pattern in which the weft passes over and under the warp fibers is called the weave pattern. Weave patterns have names like tabby, twill, and satin. Satin is not a type of fiber, it is a type of weave. Look at your shirtsleeve or your pants, and try to identify the yarns that travel in one direction and those that travel at right angles to them.

The simplest weave pattern is the plain, or tabby, weave. It forms a checkerboard, and each weft passes *over* one warp before going *under* the next one. Patterns can be expressed in numbers. A plain weave is a 1/1 weave. The weft yarn goes *over one* warp yarn, then under one warp yarn, then over one warp, and so on.

Twill weaves are used in rugged clothing such as jeans. Twill is a 3/1 weave. The weft travels *over three* warp yarns, then *under one*, with each successive row shifting over one thread. This creates a diagonal texture on the surface. The two sides of this textile look a little different. Look at the cuff of your jeans and compare the inside to the outside.

A satin weave is a 3/1, 4/1, 5/1, 6/1, or more weave, with the weft traveling *over three or more* warps and *under one*. If the warp and weft yarns are different colors, the textile will be different colors on each side. These and other weave patterns are pictured in Figure 4-13.

Weave pattern is one way that fabrics differ, but it is not the only way. The number of threads that are packed together for any given amount of

Figure 4-11. When fibers are spun into yarn, the twist direction may change as the yarn gets larger.

©Untitled Image/Jupiter Images

Figure 4-12. An industrial loom used to weave textiles.

©ImageState/Jupiter Images

Figure 4-13. *Weave patterns.*

Type of Weave	Diagram	Description	Characteristics
Plain	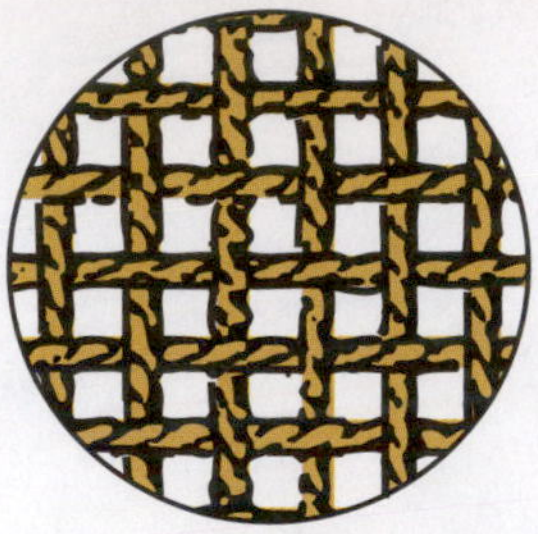	Alternating warp and weft threads	• firm and wears well • snag resistant • low tear strength • tends to wrinkle
Basket		Alternating pattern of two weft threads crossing two warp threads	• an open or porous weave • does not wrinkle • not very durable • tends to distort as yarns shift • shrinks when washed
Satin	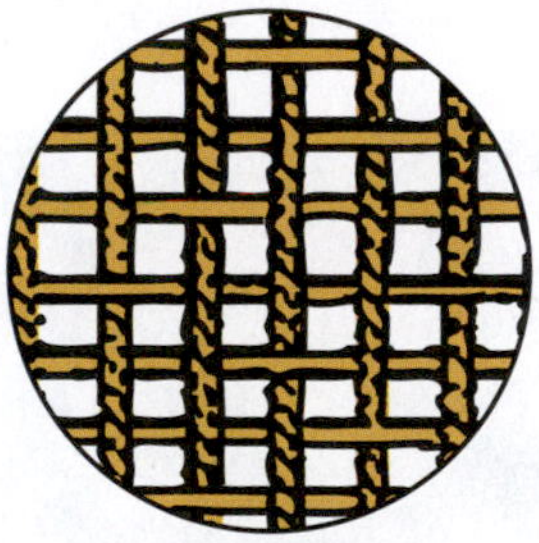	One weft crosses over three or more warp threads.	• not durable • tends to snag and break during wear • shiny surface • high light reflectance • little friction with other garments
Twill		Weft is woven over three or more warps and then under one. Next row, the pattern is shifted over one to the left or right by one warp thread	• very strong • dense and compact • different faces • diagonal design on surface • soft and pliable
Leno		This uses two warp threads and a double weft thread. The two adjacent warp threads cross over each other. The weft travels left to right and is woven between the two warp threads.	• open weave • easily distorted with wear and washing • stretches in one direction only

fabric is another characteristic, which is known as *thread count*. Every package of bed sheets includes information on thread count, as well as the type of fiber used to make them. The price of sheets varies a great deal, and high prices tend to come with all-natural fibers and high thread counts. A high thread count is more costly to manufacture and provides a smoother finish. Thread count is often written as threads per inch. Typical sheets will have a thread count between 180 and 300 threads per inch, but high-quality sheets can have thread counts of 500 threads per inch.

SUMMARY

- Fibers are a form of class evidence used by crime-scene investigators; they are also a form of trace evidence.
- Forensic scientists will try to determine the type of a fiber, its color, how many fibers of each kind were found, where they were found, what textile the fiber came from, and whether there were transfers of multiple types of fibers.
- Fiber evidence may be gathered using special vacuums, sticky tape, or with tweezers.
- Fibers may be analyzed using polarized light microscopy, infrared spectroscopy, burn tests, or tests for solubility in different liquids.
- Fibers may be classified as natural or synthetic.
- Natural fibers include animal hair, plant fibers from seeds, fruit, stems, or leaves, and mineral fibers.
- Synthetic fibers include regenerated or modified natural fibers and synthetic polymer fibers.
- Fibers are spun into yarns that have specific characteristics.
- Yarns are woven, with different patterns, into textiles.

CASE STUDIES

The Murder of George Marsh (1912)

Four bullets were found in millionaire George Marsh's body. Evidence indicated that he had not been robbed. A piece of cloth and a button were found near the corpse. In the rooming house where Marsh lived, an overcoat missing all of the buttons was found in the abandoned room of Willis Dow. The weave of the overcoat matched the weave pattern of the piece of cloth found at the crime scene. Based on this fiber evidence, Dow was convicted of the murder and sentenced to death.

Roger Payne (1968)

Bernard Josephs arrived home to find his wife dead. She had been wearing a purplish-red (cerise) woolen dress. On examination, it was determined that Claire Josephs had been choked into unconsciousness and then had her throat cut with a serrated knife. There was no forcible entry, and Claire appeared to have been in the middle of cooking. This indicated to the police that the murderer was probably someone Claire knew.

©Creatas/Jupiter Images

Suspicion fell to an acquaintance of the Josephses named Roger Payne. On examination of his clothing, more than 60 of the unusual cerise-colored fibers were found. These fibers led to the further examination of Payne's clothing, and fibers from a red scarf similar to Payne's were found under Claire's thumbnail. Additional evidence led to the conviction of Payne and the sentence of life imprisonment.

John Joubert (1983)

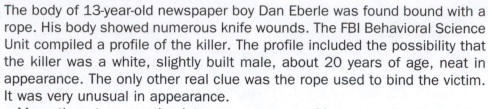

©Brand X Pictures/Jupiter Images

The body of 13-year-old newspaper boy Dan Eberle was found bound with a rope. His body showed numerous knife wounds. The FBI Behavioral Science Unit compiled a profile of the killer. The profile included the possibility that the killer was a white, slightly built male, about 20 years of age, neat in appearance. The only other real clue was the rope used to bind the victim. It was very unusual in appearance.

More than two months later, a woman working at a daycare center noticed a man watching the children from his car. She wrote down his license number, which led police to John Joubert, a slightly built radar technician at Offutt Air Base. Joubert seemed to fit the profile provided by the FBI On examination of his possessions, a hunting knife and a length of rope was found. The rope was unique, having been brought back from Korea. It matched rope found at the crime scene. When confronted by the evidence, Joubert confessed. He was found guilty of Eberle's murder and two others.

Think Critically **Based on these case studies, explain why fiber evidence may be crucial to solving a crime.**

Bibliography

Books and Journals

Burnham, Dorothy. *Warp and Weft, A Textile Terminology*. Toronto: Royal Ontario Museum, 1980.

David, Shanntha K. "Classification of Textile Fibres," in J. Robertson and M. Grieve, eds. *Forensic Examination of Fibres*, 2nd ed. London: Taylor & Francis, 1999.

Gaudette, B. D. "The Forensic Aspects of Textile Fiber Examinations," in R. Saferstein, ed. *Forensic Science Handbook*, Vol. II. Englewood Cliffs, NJ: Prentice Hall, 1988.

Giello, Debbie Ann. *Understanding Fabrics: From Fiber to Finished Cloth*. New York: Fairchild Publications, 1982.

Joseph, Marjory L. *Introductory Textile Science*. New York: Holt, Rinehart and Winston, 1986.

Springer, Faye. "Collection of Fibre Evidence at Crime Scenes," in J. Robertson and M. Grieve, eds. *Forensic Examination of Fibres*, 2nd ed. London: Taylor & Francis, 1999.

Web sites

FBI Trace Evidence and DNA Analysis, http://www.teachingtools.com/HeadJam/index.htm.

Gale Forensic Sciences eCollection, school.cengage.com/forensicscience.

Ramsland, Katherine. "Trace Evidence," http://www.crimelibrary.com/criminal_mind/forensics/trace/1.html.

Irene Good

Irene Good spends her days with fabrics that come from times and places we can only imagine. She is a textile expert who uses her knowledge of fibers and weaving to understand the lives of people who lived long ago. Just how much can be told from a single fiber might be surprising to many archaeologists, although a forensic scientist might understand.

For example, silk threads found in the hair of a 2,700-year-old corpse buried in Germany were once thought to be evidence of trading with China, whose people were manufacturing silk at that time. Good, however, used her fiber analysis skills to test the theory. Using chemical tests, she looked at the protein of the silk threads very closely. So closely, in fact, that she looked at the building blocks of the protein—the amino acids. This told her that the silk found in Germany was not from *Bombyx mori*—the silkworm—and hence was not from China after all. It was from a wild type of silkworm found in the Mediterranean. At once, Good dispelled this key evidence of a trade route between China and Europe hundreds of years ago and revealed new evidence of an ancient European silk industry.

Good has examined ancient textiles of all kinds. On 3,000-year-old mummies from a site in China,

Irene Good.

©Susan Van Etten

she found garments made of cashmere—the oldest known cashmere threads in the world. She was able to identify the hair by its fibers' shape, fineness, and diameter. Not only does the discovery show that the people in China were farming goats to use their hair to make clothes this long ago, but it also reveals that they were highly skilled at spinning.

Good remembers being fascinated by textiles as a child, while growing up on Long Island. She learned to crotchet from her grandmother. Her parents encouraged the fascination by giving her a loom, and Good made her own cloth at home and also spun her own wool. But she pursued a career as an archaeologist with nothing to do with textiles at all. Then one day, by chance, a colleague showed her a fragment of cloth he had found at an excavation site and asked if she could shed any light on the object. She has been using her passion for fibers ever since to solve the mysteries of past cultures.

Good now works at the Peabody Museum at Harvard University. Among other things, part of her work has been to examine a huge collection of ancient garments and fabrics from Peru. Under Good's keen eye, the fabrics are sure to reveal all kinds of secrets about the people of the Andes, the Incas, and how they lived.

Learn More About It
To learn more about Irene Good and forensic fiber analysis, go to school.cengage.com/forensicscience.

CHAPTER 4 REVIEW

Multiple Choice

1. Natural fibers can be harvested from

   a) plants and animals
   b) only from plants
   c) only from animals
   d) plants, animals, and minerals

2. The shiny nature of silk can be related to

   a) its hollow core
   b) its ability to refract light
   c) its smooth, round fibers
   d) mucus secretions from the silkworm

3. What characteristics of cotton make it a great source of fiber for clothing?

   a) It is very easy to grow.
   b) It is resistant to staining.
   c) The fibers are easily woven and dyed.
   d) The fibers are extremely long.

4. Mineral fibers such as asbestos are very durable. These fibers are used in all of the following **except**

   a) rope
   b) shingles
   c) floor tiles
   d) brake liners

5. All of the following are characteristics of a synthetic fiber **except**

   a) They are formed by combining monomer compounds into polymer molecules.
   b) They are man-made.
   c) They are used in the production of carpet fibers.
   d) They do not contain any natural fibers.

6. A characteristic of natural fibers is that they

   a) are stronger than synthetic fibers
   b) will not break down when exposed to bright light
   c) melt at a lower temperature than synthetic fibers
   d) are affected by microscopic organisms

7. Which of the following observations are used to help identify a specific fiber?

   a) smell of the burned fiber
   b) on contact with the flame, does the fiber coil or melt?
   c) color and structure of the residue left after the fiber burns
   d) all of the above

8. Fibers are an excellent source of trace evidence because
   a) They are easily transferred from victim to suspect.
   b) They are often overlooked by a suspect.
   c) They can be easily collected and stored.
   d) all of the above

9. A fiber is collected at a crime scene. When viewed under a compound microscope, what two traits would indicate that the fiber was a human hair and not a piece of fiber obtained from an article of clothing? (Choose 2)
   a) the presence of a cuticle
   b) a medullary index of 0.33 or less
   c) a wide diameter
   d) its ability to dissolve in water

10. Describe the weave patterns of each of the fabrics pictured below. Justify your answer for each.

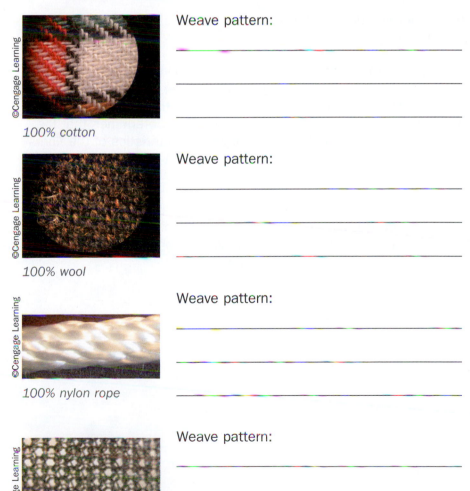

©Cengage Learning

100% cotton

Weave pattern:

_____

_____

_____

©Cengage Learning

100% wool

Weave pattern:

_____

_____

_____

©Cengage Learning

100% nylon rope

Weave pattern:

_____

_____

_____

©Cengage Learning

100% spandex nylon

Weave pattern:

_____

_____

_____

©Cengage Learning

Weave pattern:

_____

_____

_____

100% Cotton blend (hint: more than one pattern)

Questions to Research

11. Explain how the inhalation of asbestos fibers can lead to lung cancer.

12. Explain why roofers removing old asbestos roofs are more at danger of developing lung cancer than a person who installs asbestos flooring.

13. Describe three sources of natural fibers. Provide an example of each type of natural fiber.

14. Silk is a natural fiber produced by the silkworm. How is silk produced by the body of the silkworm?

15. A crime-scene investigator views two small, red fibers. One fiber was obtained from the crime scene off the victim's body, and the other red fiber was removed from the cuff of the suspect's pants. Although the two fibers appear to be from the same fabric, the crime-scene investigator determines that the two fibers are indeed very different. List five other characteristics of the fibers that can be detected under a compound microscope that could be used to distinguish the two red fibers.

16. Fibers collected as trace evidence are often considered to be class evidence and not individual evidence. For example, the presence of a white cotton fiber found on a suspect and found on a victim at a crime scene is not enough evidence to convict the suspect. Justify this statement.

ACTIVITY 4-1
MICROSCOPIC FIBER

Objectives:

By the end of this activity, you will be able to:
1. Identify fibers using microscopic examination.
2. Collect and record data.
3. Apply data collected to solve a forensic problem.
4. Think critically about how well your testing solved the problem and identify possible sources of ambiguity.

Time Required to Complete Activity: 60 minutes

Materials:

Activity 4-1 Lab Sheet
6 microscope slides labeled as follows:
- fibers from Car of Suspect 1
- fibers from Car of Suspect 2
- fibers from Car of Suspect 3
- fibers from Car of Suspect 4
- fibers from Car of the Victim
- fibers found on Victim's Body

colored pencils
microscope
forceps

Safety Precautions:

Always carry a microscope using two hands.

Scenario:

Carpet fibers were found on a murder victim along a roadside. It is speculated that the victim was carried to the location using a car belonging to one of the suspects. Some of the carpet fibers from the floor of the car were transferred to the victim's body. Your task is to match the carpet sample found on the victim's body with one of the carpet samples taken from the cars of four different suspects.

Procedure:

1. View the carpet samples provided by your instructor under the microscope using 100× magnification. Draw sketches of the microscopic view of each carpet fiber sample in the data table.
2. Record the following information:
   Within a sample are the colors the same or multicolored?
   Color(s) of the fibers
   Number of fibers (for example, single, few, numerous)
   Relative thickness of the fibers (for example, thick, thin, or variable)
   Shape of fiber (for example, twisted or straight)

Data Table

Source of Fiber	Sketch of Fiber	Single-color or Multi-colored?	Color(s) of Fiber(s)	Relative Number of Fibers (single, few, numerous)	Relative Thickness of Fibers (thin, thick, variable)	Shape of Fiber (twisted or straight)
Car of Suspect 1						
Car of Suspect 2						
Car of Suspect 3						
Car of Suspect 4						
Car of Victim						
Fibers Found on Victim's Body						

Questions:

1. Did any of the suspects' carpet samples match the carpet sample found on the victim? If so, which one?

2. Using specific characteristics from the data table, explain why you thought a particular suspect's carpet sample is a match with the sample found on the victim.

3. Suppose you found a carpet fiber from a suspect's car that did match the fiber found on the victim. What arguments could the defense attorney cite to demonstrate that the matching of the car fibers alone does not necessarily prove that his or her client murdered the victim?

Further Research:

Research what other tests are performed on fiber samples to help match a fiber found at the crime scene with fiber found on a suspect.

ACTIVITY 4-2
BED SHEET THREAD COUNT

Objectives:

By the end of this activity, you will be able to:
1. Determine the thread count of a fabric.
2. Apply knowledge of thread counts, and use critical thinking skills, to solve a forensic problem scenario.

Scenario:

A robbery occurred within a well-to-do neighborhood. The thief grabbed an expensive satin, cream-colored pillowcase to carry out the jewelry that he stole from the jewelry box in the bedroom. Knowing that the items would be listed as stolen goods, the thief immediately took the jewels to a pawnshop to exchange the jewels for money. He carelessly tossed the pillowcase onto the backseat of his car.

Feeling elated at having gotten so much money for the stolen goods, the thief and some of his friends celebrated at the local bar. Having had too much alcohol, the thief was driving erratically. The police stopped the man to give him a ticket for DWI and noticed the cream-colored pillowcase in the back of his car. The dispatcher at the police headquarters had sent out a message for all patrol officers describing the robbery that had occurred that night. No one saw the robber. The only description given was that the robber used a cream-colored satin pillowcase.

Was this the pillow taken from the home where the robbery occurred? Was it satin? Was the color the same? Because many people may purchase cream-colored pillowcases, what other characteristics would the forensic examiner use to match this pillowcase with the other pillowcases found on the bed where the incident occurred?

In this activity, you will examine textile samples and use characteristics such as weave pattern and number of threads to help match a fabric from a crime scene with a fabric found on a suspect.

Background:

The price of sheets may vary tremendously. One company advertises a sale on sheets. Included in the package are a fitted sheet, a flat sheet, and two pillowcases. Total price is only $40. Another company is selling a sheet that appears to be the same as the sheets on sale, except the two pillowcases alone cost $40! How can this be possible?

The difference between expensive sheets and the bargain brand results from several factors that affect how soft the sheets feel. Factors affecting this could include:
a. Type of textile:
     Muslin
     Cotton
     Flannel
     Satin

b. Thread count per inch
  The greater the number of threads, the more comfortable the sheet, and thus the higher the price.
  Thread count is often listed as 180 threads per inch, 200 threads per inch, or 400 threads per inch.
c. Weave pattern of sheet
  Weave patterns of sheets will vary.
  The greater the number of threads, the stronger the fabric.

Time Required to Complete Activity: 40 minutes

Materials:

3 × 3 square inch sample of bed linens with 180 thread count
3 × 3 square inch sample of bed linens with 200 thread count
3 × 3 square inch sample of bed linens with 400 thread count
3 × 5 card with 1 × 1 square inch cut-out to be used as a thread counter
magnifying glass or stereomicroscope
scissors

Safety Precautions:

No safety precautions are needed for this lab.

Procedure:

Obtain the three bed linen samples.
1. Place the 3 × 5 card with the 1 × 1 square inch cut-out over the middle of each fabric sample.
2. With the aid of a magnifying glass, count the number of warp threads within the square inch.
3. Record your data in the data table provided.
4. With the aid of a magnifying glass or stereomicroscope, count the number of weft threads within a square inch.
5. Record your data in the data table.
6. List other distinguishing characteristics of the bed linens in the data table.

Data Table

Sample Number	Warp Thread Count	Weft Thread Count	Characteristics
1.			
2.			
3.			

Questions:

Compare the fabric samples.

1. Are there any distinguishing characteristics that can help identify one fabric over another?

2. If sample one was obtained from the crime scene evidence in the car and sample two came from another pillowcase from the bed of the house that was robbed, would you consider them to be a match? Explain your answer.

Further Research:

The cost of bed linens can vary depending on the quality. Do an investigation comparing less expensive bed linens to those that are much higher priced. Include in your comparison:
- Thread count
- Type of fabric used
- Softness of the fabric
- How well the linens should wear
- Design of the fabric

ACTIVITY 4-3
WEAVE PATTERN ANALYSIS

Objectives:

By the end of this activity, you will be able to:
1. Compare textiles based on their physical characteristics.
2. Identify the weave patterns of textile samples.
3. Apply comparative data to solve a forensic science problem scenario.

Time Required to Complete Activity: 30 minutes

Materials:

Activity 4-3 Lab Sheet
6 different textile samples
label samples 1-5 as being from five different subjects
label sample 6 as crime scene sample

Safety Precautions:

No safety precautions are needed for this lab.

Scenario:

Weave patterns can help identify a fabric associated with a crime scene. In this lab, you are investigating an assault, in which the victim tore off a piece of his attacker's shirt. Five suspects have been taken in for questioning, and a judge has issued a warrant to allow the forensics investigators to look for shirts in the suspects' homes that might match the torn sleeve obtained during the assault. Your task will be to examine each of the suspects' shirts and determine if any of the fabrics match the torn piece from the crime scene.

Procedure:

1. Obtain different textile samples.
2. Using a magnifying glass, examine and identify the weave pattern of each fabric. You may use your textbook.
3. Gently tug on each fabric. Note any difference in stretchability.
4. Record your answers in the data table.

Data Table

Six Different Textile Samples	Weave Pattern	Stretchability (no stretch, some stretch, easily stretches)	Other Characteristics
Suspect 1			
Suspect 2			
Suspect 3			
Suspect 4			
Suspect 5			
Crime Scene Textile			

Questions:

1. Did you successfully match the shirt of one suspect to the fabric from the crime scene?

2. If so, which physical characteristics were most helpful? If not, why not?

ACTIVITY 4-4
TEXTILE IDENTIFICATION

Objectives:

By the end of this activity, you will be able to:
1. Observe and record physical characteristics of fabric samples.
2. Apply knowledge of fabric characteristics to a forensic science problem.
3. Communicate your findings to a lay audience, as though in a court of law.

Time Required to Complete Activity: 45 minutes

Materials:

Activity 4-4 Lab Sheet
(students work in pairs)
four samples of textiles labeled Suspect 1, 2, 3, and Crime Scene, in plastic bags.
masking tape to reseal evidence bags
marking pen to sign evidence bag
compound microscope
3 × 5 card with 1 × 1 square inch cut-out
magnifying glass or stereomicroscope
set of colored pencils
transparency square (or use a 3 × 5 card): Cut out a square 1 × 1 inch. This will be used as a standard to do thread counts.

Safety Precautions:

Always carry a microscope using two hands.

Procedure:

1. Obtain the three chain-of-custody samples of fabric from your instructor.
2. Correctly open the Sample 1 packaging.
3. Examine the sample under 40 × magnification using a compound microscope. (A hand lens or a stereomicroscope may be substituted.) Note the weave pattern for sample one on your data sheet.
4. Using the colored pencils, sketch the weave pattern in the space provided on your data sheet.
5. If a linen tester is available, measure the number of threads per inch in the sample.
6. If a linen tester is not available, use the 3 × 5 card with the 1-inch cut-out, and a stereomicroscope or a magnifying glass, to count the number of threads per inch in the fabric sample.
7. Record the thread count under your sketch.
8. Correctly reseal the evidence in a new evidence envelope and place your signature across the label.
9. Repeat the process for each of the other two suspect fabric samples.

10. Repeat the process for the crime-scene evidence.
11. Return all samples to your instructor.

Linen tester

View through a tester

1 × 1 square inch cut-out

©Susan Van Etten

©Cengage Learning

Data Observations and Measurements:

Sample #1

Threads per inch_____

Weave pattern_____

Sample #2

Threads per inch_____

Weave pattern_____

Sample #3

Threads per inch_____

Weave pattern_____

Sample Crime Scene

Threads per inch_____

Weave pattern_____

Questions:

1. After examining the textile samples from Suspects 1, 2, and 3 and the Crime Scene, were you able to determine if any of the fabric found at the crime scene matches the fabric found on the suspect?

2. If so, which suspect had a fiber on them that matched the crime scene?

3. You have been called to court to appear as an expert witness in textiles. You are to report your findings to the jury. Assume the jurors know nothing about textiles. Therefore, you need to define any terms that are not part of normal, everyday conversation. In your discussion, be sure to include the following information.
   a. Description of fibers
   b. Thread count
   c. Weave pattern
   d. Color of fabric

Remember that your testimony may be used to either link a suspect to a crime scene or to exclude the witness as a suspect.

ACTIVITY 4-5
BURN ANALYSIS OF FIBERS

Objectives:

By the end of this activity, you will be able to:
1. Analyze fibers using a burn test to identify them.
2. Collect and record data.
3. Apply your data to solve a forensic science problem.

Time Required to Complete Activity: 60 minutes

Materials:

Activity 4-5 Lab Sheet
six different labeled fiber samples:
- fibers from car of Suspect 1
- fibers from car of Suspect 2
- fibers from car of Suspect 3
- fibers from car of Suspect 4
- fibers from Victim's Car—Sample 5
- fibers found on the Victim's Body—Sample 6
alcohol or Bunsen burners
forceps

Safety Precautions:

Handle an open flame source with great care. Fibers will ignite suddenly and burn quickly at a very high temperature. Use something that will not conduct heat (forceps are metal and should not be used to hold onto something that is burning) to hold samples in the flame. Wear protective goggles when flame-testing fabrics. Inhaling fumes from burning should be avoided. Work in a well-ventilated area. Make sure a nonflammable surface is placed below the fibers to catch drips or sparks.

Background:

Some very simple tests can be used to determine the type of fibers found at a crime scene. Unfortunately, they are destructive to some of the fibers and consume the evidence. If only a limited number of fibers are recovered from a crime scene, then these tests may not be possible to perform.

Procedure:

Fiber Burn test Tips:
1. Use a sample large enough to determine if the fabric or fiber will continue to burn or die out quickly.
2. Hold the fiber or fabric with the forceps. Try to hold the fabric parrallel to the table so that the flames will move across the fabric or fiber. If you hold the fabric so that the fabric droops, the flame will engulf the entire fabric at once. This makes it difficult to determine if the fabric continues to burn after being held in the flame or if the flame dies out.
3. Hold the fabric in the falme for two seconds to ensure that the fabric is ignited.

4. To detect odor of the smoke, it is important to check immediately after the fabric is ignited. You will most likely see a small amount of smoke that dissappears quickly. Wave the odor toward your nose with your hand. Describe the odor.
5. Note how the synthetic fibers appear to melt and may give a bubbly appearance as it burns.

Data Table

Source of Fiber	Flame Test (does sample curl as it burns?)	Burn Test (burns or melts? burns slowly or quickly?)	When Removed from Flame (goes out or continues to glow?)	Odor while Burning (tar, burning hair or paper, acrid)	Color and Texture of Residue (beads, ash, crusty, fluffy, round)
Car of Suspect 1					
Car of Suspect 2					
Car of Suspect 3					
Car of Suspect 4					
Car of Victim 5					
Fibers Found on Victim's Body 6					

Questions:

1. Use the information in the data tables along with the Analysis Key which follows to help identify the fibers. Obtain Fiber 1 and proceed through each step of the key beginning with Step 1. Choose the correct alternative path to follow until you have identified the fiber. Repeat for Fibers 2 through 6.

   Hint: There should be one fiber of each type. Repeat testing if necessary. Refer to Chapter 4 for reference.

Fiber Burn Analysis Key

When fiber is removed from flame,

1a. It ceases to burn...Go to 2

1b. Fiber continues to burn..Go to 3

2a. Fibers have the odor of burning hair ....Go to 4

2b. Fibers do not smell like hair..polyester

3a. Fibers produce a small amount of light ash residue.................rayon

3b. Fibers produce a gray fluffy ash...cotton

4a. A hard black bead results from burning .... wool

4b. A brittle, black residue results ....silk

2. Which car fibers were matched with those found on the victim?

3. Why were fibers from the victim's car examined?

4. In your analysis, what characteristics make it difficult to distinguish cotton from rayon?

CHAPTER 5

Pollen and Spore Examination

THE SEASON OF DEATH

Pollen is released by different plants at different times of the year, and pollen can survive for many years. Pollen and spore evidence can be an important clue in determining a crime's location and time of occurrence. It can even unlock murders of the past.

In March 1994 in Magdeburg, Germany, a mass grave of 32 male bodies was uncovered at a construction site. Nazi Secret Police may have executed the victims in the spring of 1945. It was also possible the bodies were Soviet soldiers killed by the Soviet secret police for refusing to break up a revolt in East Germany on June 17, 1953.

The answer was found in the nasal passages of the skeletal remains. The nasal passages contained large quantities of plantain pollen inhaled by the victims shortly before their death. Smaller amounts of pollen from lime trees and rye were found as well. Because all of these plants flower and release pollen in June and July, it was determined that the bodies belonged to the Soviet secret police killed in the summer of 1953 and not people killed by the Gestapo during the spring.

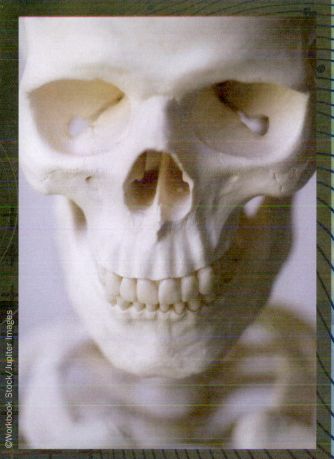

©Workbook Stock/Jupiter Images

Skeletal remains showing nasal passages for forensic analysis.

OBJECTIVES

By the end of this chapter you will be able to

✔ Distinguish between pollen and spores.

✔ Define a pollen "fingerprint."

✔ Classify the different organisms that produce pollen and spores.

✔ Summarize the different methods of pollination in plants and the relevance in solving crimes.

✔ Identify the different ways that spores are dispersed.

✔ State characteristics of pollen and spores that are important for identification in forensic studies.

✔ Summarize how pollen and spore evidence is collected at a crime scene.

✔ Describe how pollen and spore samples are analyzed and evaluated.

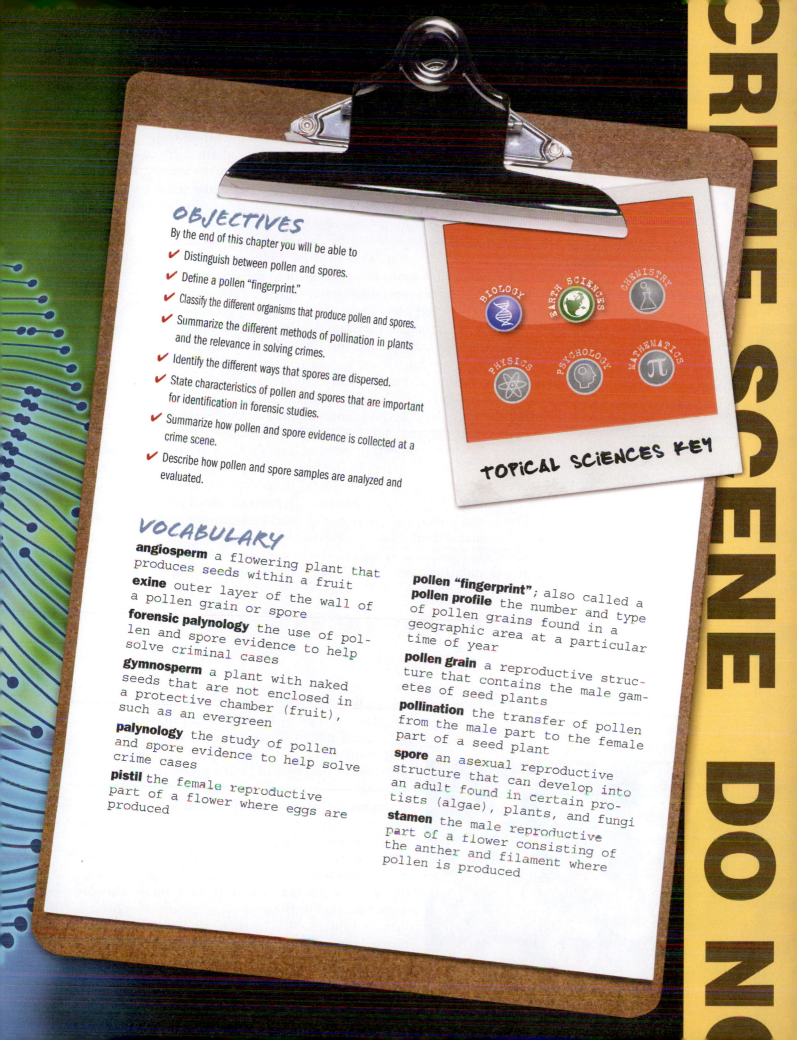

TOPICAL SCIENCES KEY

VOCABULARY

angiosperm a flowering plant that produces seeds within a fruit

exine outer layer of the wall of a pollen grain or spore

forensic palynology the use of pollen and spore evidence to help solve criminal cases

gymnosperm a plant with naked seeds that are not enclosed in a protective chamber (fruit), such as an evergreen

palynology the study of pollen and spore evidence to help solve crime cases

pistil the female reproductive part of a flower where eggs are produced

pollen "fingerprint"; also called a **pollen profile** the number and type of pollen grains found in a geographic area at a particular time of year

pollen grain a reproductive structure that contains the male gametes of seed plants

pollination the transfer of pollen from the male part to the female part of a seed plant

spore an asexual reproductive structure that can develop into an adult found in certain protists (algae), plants, and fungi

stamen the male reproductive part of a flower consisting of the anther and filament where pollen is produced

INTRODUCTION

How do forensic scientists determine where a crime occurred if the body was moved from the scene of the crime? The victim's clothes may actually hold evidence that provides information about the crime scene. Trace evidence, such as plant material, can provide clues about a crime's location, such as whether the crime was committed in the city or country. It can also provide clues that the crime occurred during a particular season of the year. In some instances, the evidence can even help determine when the crime occurred, such as during the day or night.

Several specialized forensic fields are devoted to studying biological evidence at a crime scene. One of these fields is **forensic palynology,** the study of pollen and spore evidence to help solve criminal cases. A **pollen grain** (Figure 5-1) is a reproductive structure that contains the male gametes (or sex cells) of seed plants, such as the oak tree. A **spore** is another kind of reproductive structure that can develop into an adult that is found in certain protists (algae), plants, and fungi. Pollen and spores have different functions, but they have some similar characteristics, such as being microscopic and having a resistant structure. These characteristics make them very useful markers for crime-scene investigations.

The use of both pollen and spores in forensic studies is based on Locard's principle of exchange, which you studied in Chapter 2. Recall that every contact between two objects or persons leaves a trace. In this case, Locard's principle of exchange refers to the transfer of pollen or spores between a victim, a suspect, and/or a crime scene. Suspects are potentially picking up microscopic evidence that helps investigators track them down.

The use of pollen and spore evidence in legal investigations is relatively new. Several unsolved criminal cases might have been solved if investigators had collected pollen and spore evidence. Such evidence is likely to gain greater popularity in crime-scene investigations in the future.

Did You Know?

Pollen grains can be extracted safely from rocks that are millions of years old. This is a valuable characteristic not only for palynologists, but also for oil companies and archaeologists.

POLLEN PRODUCERS

Knowledge of pollen (and spore) production is an important factor in the study of forensic palynology. By understanding pollen production patterns for plants in a given location, one can better predict the type of pollen "fingerprint" to expect in samples that come from that area. A **pollen "fingerprint"** is the number and type of pollen grains found in a geographic area at a particular time of year.

Figure 5-2 shows that the plant kingdom can be classified into two groups based on how they reproduce: nonseed plants or seed plants. The earliest plants were nonseed plants, and they reproduced by dispersing, or spreading out, spores. Some of the more recently evolved seed plants also produce spores, but their primary means of reproduction and dispersal is by seeds. Seeds are an important adaptation for life on land. During their life cycles, seed plants make pollen to disperse male gametes.

Figure 5-1. *A false-color scanning electron microscope (SEM) photograph of pollen grains.*

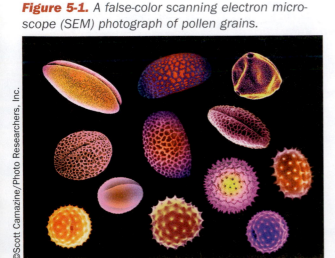

©Scott Camazine/Photo Researchers, Inc.

Figure 5-2. *Examples of seed and nonseed plants.*

Nonseed Plants	Seed Plants
Ferns	Gymnosperms
Mosses	Cycads
Liverworts	Ginkgoes
Horsetails	Conifers
Club mosses	Angiosperms (flowering plants)

Seed plants include two groups: gymnosperms and angiosperms. These two groups are the predominant terrestrial (land) plants. Thus, seed plants are likely to be the type of plants that leave trace evidence at a crime scene.

GYMNOSPERMS

The seeds of **gymnosperms,** the oldest seed plants, are exposed to the outside and are not enclosed in a protective chamber like the angiosperms. Gymnosperms include cycads, ginkgoes, and the conifers. Conifers are the largest group of gymnosperms (Figure 5-3). They include pines, spruces, firs, junipers, and other evergreen plants that usually retain their leaves or needles all year.

Many conifers produce their seeds within a hard, scaly structure, called a *cone.* There are female cones and male cones (Figure 5-4). Female cones, which are typically larger than male cones, contain eggs inside structures called *ovules.* The males cones of conifers occur in clusters and release large amounts of pollen grains to be spread by wind currents. Sperm cells from pollen grains will fertilize the egg inside the ovule of the flower. **Pollination** occurs when pollen is transferred from the male cone to the female cone. The pollen grain germinates, releasing sperm that will fertilize the egg found in the female cone. A seed develops from the ovule. The ability of male cones to release great amounts of pollen is a benefit to forensic palynology. You'll learn more about this later in the chapter.

Figure 5-3. *An evergreen coniferous tree.*

©Brand X Pictures/Jupiter Images

Figure 5-4. *A female cone (left) and a cluster of male cones (right) on a coniferous tree.*

©Paul Whitten/Photo Researchers, Inc.

©Susan Van Etten

Figure 5-5. *An example of an angiosperm, a flowering tree.*

©Susan Van Etten

ANGIOSPERMS

Angiosperms, the most recent plant group to evolve, are known as the flowering plants, and they produce seeds within an enclosed fruit (Figures 5-5 and 5-6). Angiosperms are very diverse and include corn, oaks, maples, and the grasses.

The flowering plants are the most successful and widespread group of plants on Earth today. There are about 300,000 species known, and they are found in almost all habitats. Angiosperm success and diversity are based on several factors, such as the protective nature of the fruits that also help disperse the seeds. Because angiosperm plants are found in so many places, many different crime scene areas are likely to contain samples of angiosperm pollen.

The basic reproductive unit of an angiosperm is the flower. The **pistil** is the female part of a flower that produces eggs. The three main parts of the pistil are the stigma, style, and ovary. The stigma is the area where pollen lands. A pollen tube produced by a germinating pollen grain can grow down the long, thin style until it reaches the ovary. There may be one or hundreds of egg-containing ovules surrounded by the ovary, but the number is species-specific.

The male part of the flower, or **stamen**, is responsible for pollen production and dispersal. The stamen consists of two parts: the anther and the filament. The long, thin filament elevates the anther that produces pollen sacs. Following pollination, one of the released sperm cells successfully fuses with an egg, the ovule becomes a seed, and the ovary develops into a fruit. Figure 5-7 shows apple seeds encased within the fruit.

Figure 5-6. *The basic structure of a flower.*

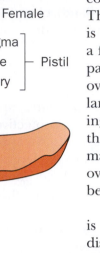

Male — Anther, Filament [Stamen]; Petal

Female — Stigma, Style, Ovary [Pistil]; Ovule

Figure 5-7. *Following pollination, one of the released sperm cells successfully fuses with an egg, the ovule becomes a seed, and the ovary (and other floral parts) develops into a fruit.*

ovary

ovule

©Photos.com Select/2007 www.indexopen.com

TYPES OF POLLINATION

Pollen (and spore) dispersion patterns are important to consider in forensic studies for analyzing palynological samples that come from the crime-scene area. The pollination strategy of the plant is an important factor in determining the presence or absence of pollen in a place or on an object of a crime scene.

Before a sperm can fuse with an egg during fertilization, pollination must occur in all seed plants. Pollination is the transfer of pollen from the male part of a plant to the female part of a seed plant. Pollination can involve one or more flowers (Figure 5-8). In flowering plants, pollination that involves the transfer of pollen from an anther to a stigma within the same flower is known as *self-pollination*, as found in pea plants. If pollination involves two distinct plants, it is known as *cross-pollination*. Note that some plants can both self- and cross-pollinate. Strict self-pollinating plants generally produce less pollen than cross-pollinating ones because of the efficiency of self-pollination. Thus, the pollen of self-pollinating plants is generally of lower value in forensic studies because it exists in such small numbers and is rarely

Figure 5-8. *Note the different path pollen follows for self-pollination versus cross-pollination.*

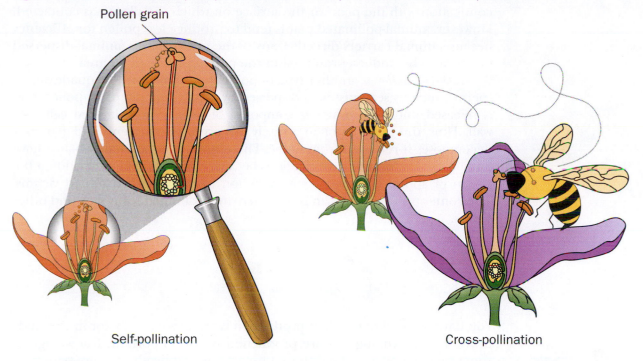

Pollen grain

Self-pollination

Cross-pollination

encountered. Still, one can imagine pollen transfer onto clothing or skin by brushing up against a self-pollinating plant.

Plants have evolved various ways of transporting pollen to the female part of the plant. Pollen can be carried by wind, animals, or water. As you learned earlier in the chapter, many gymnosperms release extensive amounts of pollen that are dispersed long distances by the wind. This is known as *wind-pollination*. Wind-pollinated plants generally have small nonfragrant or colorful flowers. Some species produce as many as 70,000 pollen grains per cone. Producing large amounts of pollen increases the chance of the pollen reaching a female reproductive part. As a result, these plants are often very well represented in the pollen profile of a crime-scene area. However, for the same reason, wind-pollinated plants may actually be overrepresented in collection samples. They may be less effective for determining direct links between individuals and places. For example, traces of pollen from marijuana plants, which are wind-pollinated angiosperms, found on a suspect could originate from a wide array of places due to wind dispersal of this plant's pollen.

Figure 5-9. *Insects are important pollinators of flowering plants.*

METHODS OF POLLINATION

A significant number of flowering plants are also wind-pollinated, but many are pollinated by specific animals, such as insects, birds, bats, and even monkeys. These plants have fragrant or showy flowers to attract the animals (Figure 5-9). Animal-pollinated plants also make adhesive and durable pollen because it must adhere to the animals. Durable pollen grains are more likely to be collected during evidence collection. Moreover, this kind of pollen can provide

©Susan Van Etten

strong evidence of contact because pollen may be transferred only by direct connection with the plant or the surface on which the plant has contacted. However, animal-pollinated plants tend to produce less pollen for efficiency because animal carriers directly transfer the pollen. Thus, animal-dispersed pollen may be underrepresented in the pollen profile of an area.

Water-pollination is another type of pollination. Pollen from aquatic angiosperms, such as sea grasses, is dispersed by water. Aquatic plant pollen that is released under the water is composed of only a single-layered cellulose wall. Thus, this kind of pollen is rarely preserved in sediments and generally decomposes if removed from water. Therefore, water-pollinated plants have limited use for forensic investigations because of the limitation in finding the pollen or spores. However, one exception to this is with drowned victims. The contents of the lung can be emptied and examined for pollen and other debris.

SPORE PRODUCERS

You have learned that pollen production is an important factor in the study of forensic palynology. Spore production is also important. Knowledge of spore production and the different organisms that produce spores can help in making predictions about the expected spore profile (fingerprint) in an area. The spore producers include certain protists (algae), plants, and fungi. Some bacteria also produce a unique type of spore.

Algae, the ancestors of plants, produce reproductive spores that are adapted for dispersion in water or air. Some land plants produce spores from a structure known as a *sporangium, gills,* or *sori* (Figure 5-10). Seedless plants, such as ferns and mosses, release the spores into the air. Thus, seedless plants are the spore-producing plants of interest in forensic studies. Dispersed spores can develop into new individuals under favorable conditions.

Fungi, such as mold, baker's yeast, and mushrooms, produce large numbers of spores. Fungi can produce spores in two ways: asexually and sexually. During asexual reproduction, spores are made by a single fungus, and these spores produce new fungi identical to the original fungus. Fungal structures called sporangia release spores, and these spores can grow into new fungi under good environmental conditions. In contrast, in sexual reproduction, fungi cells from two different fungi fuse to produce unique fungi that are not exactly like either parent. The resulting offspring release genetically diverse spores that can grow into new fungi. Fungal spores, which are dispersed by air currents or water, can be found practically everywhere (Figure 5-11).

Figure 5-10. *Examine the photo of a fern. Note the location of the sori on the underside of the fronds.*

sporangia

frond

©Susan Van Etten

BACTERIA: AN EXCEPTIONAL CASE

When environmental conditions are harsh, some bacteria form thick-walled, resistant spores, called *endospores*. Bacterial spores are different from the other class of spores produced by fungi, algae, and plants because endospores, unlike other spores, are not used in reproduction, and bacteria can produce only one at a time. Endospores have more recently become of interest in forensics because several types of bacteria that make endospores can cause diseases, such as anthrax and botulism. The particular pathogen that causes anthrax has become a deadly agent of bioterrorism. Anthrax will form microscopic spores in the form of a white or cream-colored powder. If you have traveled recently, you have probably had your suitcase or your person checked for the presence of these bacteria.

Figure 5-11. Note the different parts of a mushroom and how spores are released into the wind.

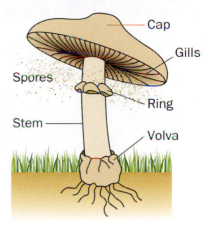

- Cap
- Gills
- Spores
- Ring
- Stem
- Volva

SPORE DISPERSAL

Spores are passively dispersed by wind or water. Some fungi have evolved mechanisms for enhancing spore dispersal, such as spore ejection and animal dispersal. For example, *Pilobus* fungi actively eject masses of spores (Figure 5-12).

Also, the fruiting bodies and spores of the very expensive and edible truffle, as shown in Figure 5-13, are eaten by foraging animals. Spores then pass through their digestive system and are dispersed in excrement deposits.

Overall, spore producers have the same value in forensic studies as the wind-pollinated plants. They have widespread coverage in geographic areas, and they may be over represented in spore fingerprints. This presents a challenge in forensic analyses. Yet, spore analysis has one advantage over pollen in that it is technically possible to grow the organism and identify the species exactly.

Figure 5-12. Pilobus fungi ejecting spores.

Figure 5-13. A very tasty fungus with spores inside, the truffle.

©Kenneth H. Tomas/Photo Researchers, Inc.

©FAN Travelstock/Jupiter Images

POLLEN AND SPORE IDENTIFICATION IN SOLVING CRIMES

Pollen and spore identification can provide important trace evidence in solving crimes, partly because of their unique, species-specific, and diverse structures. The hard outer layer, or shell, of a pollen grain and spore is called an **exine.** It has a unique and complex structure when viewed under the microscope. Outward features include size, shape, wall thickness, and surface textures, such as spines. For example, with regards to size differences, larger pollen grains such as corn pollen cannot travel far and can only drift with the wind about one-half mile. Accordingly, if someone had corn pollen on them, then they were probably close to a cornfield or flower.

Wind-dispersed pollen grains are relatively simple in morphology, have thin walls, and are easily preserved for identification. In contrast, animal-dispersed pollen is usually large, sticky, highly ornamented, thick-walled, and also easily preserved for identification.

Pollen and spores differ in other ways important to forensic scientists. Most importantly, spores are much smaller than pollen grains and are produced in far greater numbers than pollen. Spore evidence can be especially hard to find if there is not a concentration of spores

Identification is generally species-specific, although spores are more difficult to identify than pollen. Moreover, pollen and spore production can be seasonally and geographically specific. Thus, pollen and spore evidence from a crime scene can lead to useful information about the crime, such as a specific season and a specific geographic location. Also, if pollen, spores, or both are found on the victim is not native to the crime scene, it may indicate that the body was moved.

Pollen and spores also play an important role in solving crimes because they are microscopic and difficult to eliminate by a suspect at a crime scene. The resistant nature of most pollen and some spores allows them to avoid dehydration and degradation. In fact, they can be found in sediment from millions of years ago. Also, plant pollen, especially from animal-pollinated angiosperms, usually has sharp edges that enable it to better adhere to the plant and to a suspect. These features can help in determining whether a suspect was present at the crime scene, because suspects can easily pick up evidence on their shoes and clothes that they cannot see and that will not go away.

COLLECTING POLLEN AND SPORE EVIDENCE AT CRIME SCENES

If palynological evidence is suspected to play an important role at a crime scene, pollen and spore collection needs to be done by a forensic palynologist. It is critical that forensic palynologists carefully and methodically collect and store samples, while avoiding contamination. Contamination is a major problem and likely the first choice of a clever defense lawyer when it comes to finding holes in the prosecution's evidence.

Finding Pollen and Spores

Pollen and spores are present everywhere, as shown in the list on the next page. They are in the air we breathe, and they get caught in hair, fur, rope, feathers, clothing, mucous membranes, and sticky surfaces. They are also in soil, plants, and nonliving objects like dust.

The following list identifies some of the common places pollen and spore evidence can be found.

- Living and decaying plant material
- Soil, dirt, mud, and dust
- Hair, fur, and feathers
- Clothing, shoes, blankets, rugs, baskets, carpet, and rope
- Victim's skin, hair, nails, nasal passages, lungs, stomach, intestines, and fecal material
- Paper, money, and packaging material (newspaper, straw, cardboard, plastic)
- Vehicles
- Furniture
- Air filters of cars, homes, airplanes
- Cracks and crevices in floors, walls, roofs, and fences
- Drug resins
- Honey and other food

It is especially important to sample soil, dirt, and dust because they are usually abundant at crime scenes, and they often contain abundant pollen and spores. For example, samples of dirt collected from a victim's garments, shoes, skin, hair, or car can help identify the location of the crime. Clothing and hair from humans and animals are also major traps for pollen and spores. Some unusual materials that can contain pollen and spores include enamel-painted wood, painted works of art, grease on guns, stuffed animals, and dusty foot impressions.

Collecting Pollen and Spores

Every crime scene is unique, and the palynologist needs to consult the entire investigating team to develop an appropriate sampling strategy. Ideally, the forensic palynologist should be called to the crime scene(s) immediately to minimize contamination and destruction of the evidence.

Scientists collect evidence samples as well as control samples at all potential crime scenes during the investigation. Control samples are specimens of surface dirt from the region where a crime was committed. They are taken to document the pollen and spore profile of the entire area so there is a comparison for evidence samples.

All samples are collected by a forensic palynologist who is wearing gloves and using clean tools, such as paintbrushes and cellophane tape (Figure 5-14). The samples are then placed into sterile, sealed containers. Each sampling instrument is cleaned after each sample is taken, or a new one is used each time. Site and vegetation surveys, as well as photographs, should also be taken to help analyze samples, explain how samples were collected, and determine the processing techniques to be used in extracting pollen and spores from the samples.

Photo courtesy of Dr. Vaughn Bryant

All samples should be carefully labeled and sealed in sterile, plastic bags. However, vegetation samples are placed in paper envelopes and dried to prevent degradation. Ideally, samples should be frozen or dried. As you learned in Chapter 2, it is vital to secure the evidence and maintain an accurate chain of custody.

ANALYZING POLLEN AND SPORE SAMPLES

After pollen and spores are processed and chemically extracted from samples in the laboratory, forensic palynologists examine them using microscopes (Figure 5-15). Pollen and spores are best viewed with transmitted light or phase contrast microscopes.

Additional details of surface features may require the use of a scanning electron microscope (SEM). To identify pollen and spores, specialists use pollen and spore reference collections that represent native species and species from other geographic regions. Preliminary identifications may be made by reference to drawings and photographs in atlases, journal articles, Internet web sites, and books. After scientists narrow their search with illustrations, they refer to herbariums to look at actual plant material, pollen, and spores.

Figure 5-15. *A forensic analyst at work examining pollen under the electron microscope.*

©Steve Allen/Photo Researchers, Inc.

Analyzing the evidence is a complex job that requires great skill. Once the pollen evidence has been collected, processed, analyzed, and interpreted, it is presented in court so that a jury can assess the value of the evidence. The probability of a suspect being involved in a crime, as determined from the evidence, is then expressed in court for evaluation. Palynology is best used to confirm certain aspects of the crime. That is, pollen and spore evidence is unlikely to be the premier or only evidence in a crime.

SUMMARY

- Forensic palynology can provide information about the geographic origin of a crime and the time or season when it took place.

- Pollen is a reproductive structure containing male gametes in seed plants. Spores are reproductive cells produced by some protists, plants, and fungi.

- Nonseed plants such as ferns reproduce with spores.

- Seed plants include gymnosperms (plants that do not produce fruits) and angiosperms (plants that do produce fruits). Both gymnosperms and angiosperms produce pollen.

- Plants may disperse pollen on the wind, in the water, or by the movements of animals.

- Pollen from wind-pollinated plants is more common in forensic samples, but pollen from insect-pollinated plants tends to provide more specific information about location.

- Pollen evidence must be collected at a crime scene and include baseline samples from the area for comparison.

- Collection of pollen and spore evidence must be performed carefully to avoid contamination.

- Pollen is examined microscopically and compared with reference collections to identify, if possible, the exact species that produced it.

CASE STUDIES

Dr. Max Frei (1960s)

Dr. Max Frei, a noted Swiss criminalist, often used pollen as a forensic tool to link suspects to events or to crime scenes. In one case, pollen found in the gun oil of a weapon linked the firearm to the murder scene. Frei also used pollen analysis to document forgery. He found fall-pollinating cedar pollen stuck to the ink used to sign a document. The document was dated during the month of June. Because the document contained pollen only visible in the fall, the document could not have been signed in June. Frei examined the Shroud of Turin, a garment believed to have wrapped the body of Jesus after his crucifixion. Frei's pollen analysis did link the shroud to pollen unique to Israel.

Digging Deeper
with Forensic Science e-Collection

Do additional research on the Shroud of Turin, the cloth that is believed to have covered Christ's face. Go to the Gale Forensic Sciences eCollection on school.cengage.com/forensicscience and research the case. The pollen evidence did indicate that the pollen found in the shroud was from Israel. There still remains much controversy over the authenticity of the shroud. Cite some of the arguments used by those who believe that there is not enough evidence to indicate that the Shroud of Turin was the cloth that covered Christ's face.

Dr. Tony Brown (2004)

On September 9, 2004, the BBC News headline read: "Pollen helps war crime forensics." Forensic experts working in Bosnia used pollen to help convict Bosnian war criminals. Professor Tony Brown used pollen analysis to link mass graves in Bosnia to support the claim of mass genocide. Bosnian war criminals tried disguising their acts of genocide by exhuming mass graves and reburying bodies in smaller graves, claiming they were the result of minor battles. Soil samples were taken from skeletal cavities, inside the graves, and from around the suspected primary and secondary burial sites. Pollen from the soil samples was cleaned with powerful chemicals before being analyzed, and the mineralogy of the soil was examined. Once complete, matches could be made between different samples, and this ultimately led to links between primary and secondary burial sites.

Professor Brown noted one primary execution and burial site was in a field of wheat. When bodies were found in secondary burial sites, they were linked to the primary location through the presence of distinctive wheat pollen and soil recovered from the victims.

Think Critically Imagine you are a reporter accompanying either Dr. Frei or Dr. Tony Brown during their investigations. Create a news story—written or script—about their work.

Bibliography

Books and Journals

Berg, Linda. *Introductory Botany: Plants, People, and the Environment.* Belmont, CA: Thomson Higher Learning, 2008.

Murray, Raymond. *Evidence from the Earth.* Missoula, MT: Mountain Press Publishing Company, 2004.

Nickell, Joe and John Fischer. *Crime Science: Methods of Forensic Detection.* Lexington, KY: The University Press of Kentucky, 1999.

Web sites

Bryant, Vaughn M., Jr., and Dallas C. Mildenhall. "Forensic Palynology: A New Way to Catch Crooks," http://www.crimeandclues.com/pollen.htm; http://www.nature.com/news/1998/981008/full/981008-2.html.

Catchpole, Heather. "Pollen Clues Ignored, Says Forensic Expert," ABC News Online, http://abc.net.au/cgi-bin/common/printfriendly.pl?/science/news/enviro/EnviroRepublish_1105992.htm.

Gale Forensic Sciences eCollection, school.cengage.com/forensicscience.

"Meet Dr. Lynne Milne, Palynologist & Forensic Scientist," Science Network. http://www.sciencewa.net.au/science_careers.asp?pg=245.

"Palynology," http://myweb.dal.ca/jvandomm/forensicbotany/palynology.html; http://www.crimeandclues.com/palynologyus.htm.

"Pollen Being Used to Track Criminals," CBS News 42, http://keyetv.com/local/local_story_131221754.html.

Porter, Daniel R. "The Shroud of Turin Story," http://www.shroudstory.com/exhibit/maxfrei01.htm, http://www.shroudstory.com/exhibit/maxfrei01.htm.

Salyers, Abigail A. "Microbes in Court: the Emerging Field of Microbial Forensics," American Institute of Biological Science, http://www.actionbioscience.org/newfrontiers/salyersarticle.html.

Wood, Peter. "Pollen Helps War Crime Forensics," BA Festival of Science. http://news.bbc.co.uk/1/hi/sci/tech/3640788.stm.

Weiss, Danielle M. "Forensic Palynology and Plant DNA: The Evidence That Sticks," American Prosecutors Research Institute, http://www.ndaa.org/publications/newsletters/silent_witness_volume_9_number_4_2005.html.

CAREERS IN FORENSICS

Dr. Lynne Milne, Forensic Palynologist

Dr. Lynne Milne is a palynologist, but she spends most of her time working as a forensic palynologist. She is a lecturer in the Centre for Forensic Science at The University of Western Australia (UWA). There, she supervises Master's and Ph.D. students in the field of forensic palynology. Her work also involves teaching police about forensic palynology, as well as conducting casework for state and federal police.

Milne has spent a great deal of time promoting the field of forensic palynology in the media. She has published a book describing her forensic palynology murder and rape cases. In the book, she also includes other criminal cases and mysteries that palynology has helped solve. Milne states, "like DNA and fingerprinting, palynology can link a person to a crime—but it can also help in investigations for which DNA and fingerprinting are not applicable. For example, palynology can help direct police to an area where a person who committed a crime may live and work, and can determine where illicit drugs and illegally imported goods have come from."

Courtesy, Dr. Lynne Milne, Centre for Forensic Science, University of Western Australia.

Dr. Lynne Milne

Her typical workweek can vary but may include meeting with police; attending a crime scene; working with other forensic scientists, such as soil and fiber experts; collecting samples and doing a vegetation survey at a crime scene; collecting reference pollen samples from the herbarium; processing pollen, soil, and other samples; analyzing prepared pollen samples; writing a police report; and teaching and helping students with their projects in the field or laboratory.

Dr. Lynne Milne enjoys her job because it is never boring and can be extremely rewarding by helping to solve crimes and other mysteries. She says, "Pollen is often very beautiful, and it always has a story to tell. I enjoy the super sleuth aspects—working out past vegetation, patterns in evolution, and helping to solve crimes."

Milne's advice to those wishing to pursue a career in forensic palynology is to obtain an undergraduate degree in botany, geology, geography, or archaeology. Several colleges and universities in the United States and Canada offer courses in Forensic Palynology. A science degree would allow a person to work under the supervision of an experienced palynologist. Much of the real training will be on the job. A Master's or Ph.D. degree in a related area of study is the next step for eventually conducting one's own forensic cases.

Learn More About It

To learn more about the work of a forensic palynologist, go to
school.cengage.com/forensicscience.

True or False

1. Pollen is an example of trace evidence.

2. Through microscopic examination of pollen or spores, it is possible to identify a specific plant that produced that pollen or spore.

3. All pollen and spores are produced continuously throughout the year.

4. With the exception of bacterial endospores, pollen and spores are reproductive structures.

5. Spores are produced only by fungi.

6. Flowering plants or angiosperms are the only plants that produce pollen.

7. Pollen and spore identification can provide important trace evidence in solving crimes due to their structures.

8. Pollen present on the victim that matches pollen from a suspect provides strong evidence that the suspect is guilty.

9. In unsolved cases, it is possible that pollen evidence embedded within a rug could be viewed years after the crime was committed and used to help convict a suspect.

10. A pollen fingerprint is the number and type of pollen grains found in a geographic area at a particular time of year.

Multiple Choice

11. All of the following are true of both angiosperms and gymnosperms **except:**
    a) both form pollen at some time during their life cycle
    b) both form fruits that surround their seeds
    c) both rely on sexual reproduction
    d) both produce sperm and eggs

12. The male part of a flowering plant is called the
    a) seed
    b) exine
    c) pistil
    d) stamen

13. The spore-producing part of fungi is called the
    a) sporangium
    b) endospore
    c) ovule
    d) anther

14. Which type of pollen producer is useful for forensic studies because of the abundance of pollen produced?
    a) aquatic angiosperms
    b) self-pollinators
    c) gymnosperms
    d) all angiosperms

15. Some important criteria for collecting pollen or spore samples include all of the following **except:**

   a) avoiding contamination
   b) using sterile equipment
   c) maintaining a chain of custody
   d) putting all plant samples in a plastic bag

Short Answer

16. Compare and contrast each type of pollination by identifying the strengths and weaknesses of pollination for each group.

   _____

   _____

   _____

17. Describe how pollen can be used to determine if a crime occurred in the city or in the country or during the day or night.

   _____

   _____

   _____

18. Describe the different places at a crime scene where a forensic palynologist might collect pollen or spore evidence to link a suspect to a crime scene.

   _____

   _____

   _____

19. Describe any special devices or instruments that would be needed to analyze pollen and spore evidence.

   _____

   _____

   _____

20. Summarize the characteristics of pollen and spores that make their identification and analysis useful in crime scenes.

   _____

   _____

   _____

ACTIVITY 5-1
POLLEN EXAMINATION: MATCHING A SUSPECT TO A CRIME SCENE

Scenario:

A robbery had taken place at the Huxton's home. Footprints were found throughout the recently watered flower garden leading to the window of a bedroom located at the back of the expensive home. Just as the robber was leaving the house, the owner returned home and caught a glimpse of a young teenage boy dressed in a T-shirt and blue jeans running through the garden.

The police questioned four neighbor boys who live in that vicinity. All four young men denied that they had been anywhere near the Huxton home and stated that they did not rob the home. After obtaining a warrant, the police searched the home of each of the four young men looking for blue jeans that could have been worn during the robbery. All four pairs of jeans were confiscated and taken in evidence bags to the crime lab to be examined for pollen evidence that could link the suspect to the Huxton garden.

Objectives:

By the end of this activity, you will be able to:
1. Prepare wet-mount slides of flower pollen.
2. View pollen under a compound microscope at 100× and 400× magnification.
3. Record information on the individual pollen grains.
4. Determine if pollen evidence from any of the suspects matches pollen from the crime scene.

Time Required to Complete Activity: 40 to 60 minutes

Materials:

Activity 5-1 Lab Sheet
(for each group of two students)
1 microtube labeled "crime scene (CS)" containing pollen from the crime scene or one flower from the crime scene
4 evidence microtubes containing pollen from: Suspect 1, Suspect 2, Suspect 3, Suspect 4
microtube sponge "rack"
1 compound microscope
5 microscope slides
5 flat wooden toothpicks
1 small beaker of tap water
dropper or pipette
5 coverslips
forceps
marker pen
colored pencils (optional)
digital camera (optional)
empty film canister with ends removed (optional)

Safety Precautions:

Any student who has pollen allergies should inform the instructor of the type of allergy and the severity of the allergy. If any students are allergic, those students should prepare the microscope and have their partner handle any flowers and pollen. Handle pollen samples carefully, using a clean toothpick in each microtube. Do not contaminate one pollen sample with another by using the same toothpick in different microtubes.

Procedure:

Part A: Preparation of Pollen Wet-Mount Slides

1. Remove a flower from the "crime scene." Locate the anthers and pollen grains or obtain the microtube containing pollen from the crime scene.
2. Prepare a wet-mount slide of the pollen:
   a. Obtain a clean slide and label it "CS" for crime scene.
   b. Add one or two drops of fresh water to the slide.
   c. Place the flat end of the toothpick into the drop of water to moisten the toothpick.
   d. Using the moist end of the toothpick, touch the anthers of the crime-scene flower to obtain pollen grains, or if using micro tubes of pollen, place the moise end of the toothpick into the crime scene micro tube of pollen.
   e. Swirl the toothpick with the pollen in the drop of water and apply a coverslip.
3. Observe the pollen under 100× of the microscope.
4. Switch the microscope lens to 400× and observe the pollen grain.
5. Complete the data table for the crime-scene pollen sample. Include a sketch of the pollen viewed under 400× and provide a description of the pollen grain.
6. (optional) Using a digital camera and a cut-out film canister as an interface, take a digital photo of the view of the pollen under the microscope.

Part B: Observation of Pollen Collected from the Four Suspects

Evidence samples of pollen have already been collected from the four suspects. The pollen evidence is contained in four different microtubes.

7. Using a marker pen, label four different slides: Suspect 1, Suspect 2, Suspect 3, Suspect 4.
8. Using the same procedure described in Steps 2 and 3, prepare and examine a wet-mount slide of pollen from each of the four samples (suspects).
9. After reviewing the crime-scene pollen and the pollen obtained from the four suspects, can you find a match? Do any of the pollen samples from any of the suspects match the pollen found at the crime scene?
10. Answer the questions.

Questions:

1. Do any of the suspect pollen samples match the pollen collected from the crime scene? If so, which one(s)?
2. Using the information gained from your microscopic examination of the pollen, justify your answer to question 1.
3. Suppose the pollen from one of the suspects did match the pollen found at the crime scene. What arguments could the defense attorney present to try to discredit the evidence?
4. What could you do to improve the reliability of your analysis? Include in your answer any other instruments that you would like to use to compare the pollen samples.

Sample	Color	Shape	Relative Size	Sketch
Crime Scene				
Suspect 1				
Suspect 2				
Suspect 3				
Suspect 4				

Further Study 1

Microscopic Measurement of Pollen

Before you begin the activity, be sure to get a data sheet from your teacher. Several different methods can be utilized to measure the size of an individual pollen grain. If your school has a digital microscope, record the size of the pollen grain. However, if you do not have a digital microscope, it is still possible to estimate the size of each individual pollen grain.

1. Place a clear plastic mm ruler under the field of view at 100×. While looking through the ocular lens, estimate the measurement of the diameter of the field of view at 100×. Estimate this measurement to the nearest $\frac{1}{10}$ mm. Record your answer. (Most microscopes have a field of view of 1.2 mm to 1.4 mm at 100×.)

Diameter of field of view under 100x is 1.6mm

2. To determine the size of the field of view under 400×, take ¼ of the diameter of your field of view under 100×. For example, if the diameter of the field of view under 100× equals 1.6 mm, then the diameter of the field of view under the 400× equals ¼ (1.6 mm) or about 0.4 mm.

3. Record your answer to the nearest $\frac{1}{10}$ mm.

4. To determine the size of one pollen grain, align one pollen grain along the diameter of the field of view under 400×. Estimate how many pollen grains could fit across the diameter of your field of view under 400×.

5. Using the following formula, determine the size of a single pollen grain (e.g., if the size of the 100× field of view is 1.6 mm). To determine

the size of one pollen grain, estimate the number of pollen grains that fit across the diameter of the field of view under 400× while looking through the ocular lens.

Approximately 11 pollen grains fit across the diameter under 400×. If x is equal to the size of one pollen grain, then:

11 pollen grains = diameter of the high-power field of view

$$11x = 0.4 \text{ mm}$$
$$x = 0.4 \text{ mm}/11$$
$$x \approx 0.04 \text{ mm}$$

Further Study 2

Design Your Own Experiment!

Choose one of the following statements. Form a hypothesis and design an experiment to test your hypothesis. Be sure to include a data table, analysis, and conclusion.

1. Is there any similarity between the pollen of two flowers of the same type (e.g., a red tulip and a yellow tulip)?
2. Is there any relationship between the size of the pollen grain and the size of a flower?
3. Is the pollen from gymnosperms larger than the pollen from angiosperms?
4. Does the pollen produced by plants that depend on animals or wind for pollination have a sharper appearance than the pollen produced by flowers that self-pollinate?

Diameter of field of view under 400x

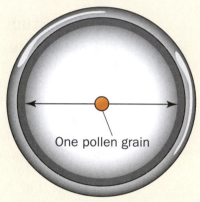

One pollen grain

Field of view under 400x is .4mm

Further Study 3

Before you begin the activity, be sure to get a data sheet from your teacher. Design an experiment to gather and analyze the pollen assemblage (sample of all the pollen from a particular location collected at a specific time) from one of the following:

1. Two different areas collected at the same time
2. From the same area but collected a week apart
3. From the same area but collected before and after a rainfall

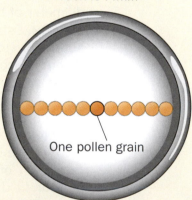

One pollen grain

Determine if there is any difference in the pollen assemblage. Form a hypothesis and design an experiment to test your hypothesis. Be sure to include a data table that describes the different types and amount of pollen collected. Your report should also include an analysis and conclusion for your experiment.

ACTIVITY 5-2
POLLEN EXPERT WITNESS PRESENTATION

Scenario:

Neighbors long suspected that the young college students living in an adjacent apartment were growing their own marijuana. Why else would they need to have so many artificial lights set up in their basement? The parade of people coming and going into their apartment in the late evening hours also made the neighbors suspicious that the students were not only growing marijuana, but selling it too! The police were notified. The police were going to make a surprise visit to the home, arrest the occupants, and confiscate the drugs.

Just before the police arrived, the students got a call from one of their friends telling them what was about to happen. They quickly bagged all of the plants and got rid of the indoor lights an hour before the police arrived. The police had a warrant to search the premises. No plants or grow lights were visible. The occupants denied growing marijuana. With no visible evidence of the plants, the police decided to look for trace evidence of pollen.

Marijuana plants are known to release large quantities of pollen. Although the occupants of the house had removed the plants, they did not eliminate all of the microscopic pollen that was left on the floor and tabletops. The occupants just assumed it was dust from a lack of cleaning. The police sent the trace evidence of pollen to the crime lab. As expected, the dust was identified as marijuana pollen. The amount of pollen in the room indicated that many plants had been present.

Background:

As the expert in pollen analysis, you are called as an expert witness for the prosecutor's office. You are asked to testify at the hearing. Because many jurors lack a scientific background, you will need to begin your presentation with background information. It is important to incorporate photographs, pictures, charts, and other visual aids to help the jury understand your presentation. Your task will be to:

1. Introduce yourself and list your credentials as an expert witness in pollen. (Your research should include what qualifies someone as an expert in pollen identification.)
2. Educate the jurors about pollen:
   a. What is pollen?
   b. Where is it produced?
   c. What does it look like?
   d. How big is pollen?
   e. When is it produced?
   f. What is its function?
   g. How is pollen transferred?
   h. How does pollen differ from plant to plant?

3. Explain about marijuana and pollen:
   a. What does marijuana pollen look like?
   b. Why do marijuana plants produce an abundance of pollen?
4. Explain about pollen and solving a crime:
   a. What is Locard's principle?
   b. How is pollen transferred in marijuana plants?
   c. How can pollen link someone or someplace to a crime scene?
5. Explain about pollen evidence and this case:
   a. Was any marijuana pollen found in this apartment?
   b. How was it collected?
   c. How was the pollen examined in the lab?
   d. How do you know this pollen is definitely from marijuana plants?
6. Conclusion

Based on your expertise, can you conclude that the evidence found in the apartment of these college students is pollen produced from many plants of marijuana?

Objectives:

By the end of this activity, you will be able to:
   1. Research information on pollen.
   2. Prepare expert witness testimony.
   3. Complete either an essay or a PowerPoint presentation.

Materials:

Activity 5-2 Labsheet

Choice 1: Essay Presentation

Essay of at least 250 words
Paragraph format
Introduction, several body paragraphs, and a conclusion
Complete sentences and correct spelling
Typed, double-spaced
Avoid terms such as *it*, *that*, *them*, or *they*.
Properly credit any photos or pictures by writing the source of the picture or photo.

Choice 2: PowerPoint Presentation

You need to address the same information as the essay.
Your presentation will be submitted on a disc (CD or DVD).
You must credit any picture or photo by copying the address of the source or citing where you obtained the picture or photo.
Avoid using sentences. Use short statements or bullets.
Include graphics and photos (either from the Internet or from your own camera).
Please do not include any sounds as you transition from frame to frame.

Scoring Rubric: A maximum of 40 points will be credited for this assignment. Teacher's rubric for grading presentation (circle appropriate value).

	Excellent 5 points	Good 3 points	Needs Improvement 1 point	Total Points
Scientific clarity and accuracy in oral presentation	Material highly accurate and clearly presented; Includes several supporting examples and details	Material mostly accurate with some supporting examples and details	Material lacking accuracy or presented in a confusing way	
Speaks clearly and with sufficient volume	Speaks clearly and with sufficient volume for entire presentation; All terms correctly pronounced	Speaks clearly but is difficult to hear at times; May have some mispronunciations	Is difficult to hear throughout most of the presentation	
Covered all required topics	Thoroughly covered all required material	Covered all material	Incomplete coverage of required topics	
Length of presentation	Thoroughly covered material within the 5 minutes	Covered most of the material within the designated time	Inadequate coverage of material Too brief or too long	
Visuals (PowerPoint, poster, report with projected images)	Extremely well presented; Exceptionally attractive in size, design, layout, and neatness; Supports the text	Well presented; Most visuals attractive and support the text	Confusing or not well presented; Size may be inappropriate; Some visuals did not support the text	
Background information research	Very complete; In-depth background	Completed all required portion of the research	Incomplete research; Needed more information	
Scientific references used in presentation	Used more than three required scientific sources	Used three sources; Most were scientific sources	Most sources are either not scientific or not properly cited; Used less than three sources	
Courtesy	Listens politely to all other presenters, maintains good eye contact, asks pertinent questions	Listens but shows some restlessness; No questions asked of the presenter	Appears not to be listening or may express a lack of interest	

ACTIVITY 5-3
PALYNOLOGY CASE STUDIES PRESENTATION

Objectives:

The purpose of this activity is for you and a partner to research an actual court case that involved either pollen or spore evidence to help solve the crime. After completing the research, you and your partner will prepare an oral presentation.

Time Required to Complete Activity:

Time required to complete this activity will vary depending on if students are expected to complete parts of the research and planning outside of class time. Appropriate time will need to be provided for the students to:
1. Research information
2. Organize and discuss the information
3. Complete the pollen research-planning sheet
4. Prepare the PowerPoint presentation or the poster
5. Present the information to the class

Materials:

reference textbooks or articles
computer with Internet connectivity
CD or DVD for your PowerPoint presentation or poster board for your presentation
projecting device for PowerPoint or flex camera to project images on the poster
Activity 5-3 Lab Sheet

Safety Precautions:

There are no safety concerns with this activity.

Procedure:

1. Students will register for one of the following topics. You are to find a specific court case that used pollen or spores to help solve the crime.
   a. Pollen or spores found in hair
   b. Pollen or spores found in mud or sediment
   c. Pollen or spores found in clothing
   d. Pollen or spores found in animal hair
   e. Pollen or spores found in honey
   f. Pollen or spores found in antique furniture
   g. Pollen or spores found on money
   h. Pollen or spores found in cocaine
   i. Marijuana pollen
   j. Pollen found in the Shroud of Turin
   k. Pollen or spores and bioterrorism
   l. Pollen or spores found on paper (other than money)

2. Research Sources:

   To begin your research, start with the Internet. Perform an Internet search for "pollen" and the name of your topic (e.g., honey, mud) and "forensics" and find numerous references to help you begin your research. Your research should also include books and current articles from magazines, newspapers, or journals (within the past four years).

3. What information do you need to investigate?

   a. Explain the case to the class:

Who?	When?	Why?
What?	Where?	How?

   b. Describe how the case was solved using pollen or spore evidence.

   c. Describe the proper method to collect and preserve pollen or spore evidence.

   d. Describe anything in your case that is not common knowledge. For example, explain what the Shroud of Turin is or what sheep shearing is, how honey is produced, or how cocaine is imported into the United States.

4. Submit preliminary Oral Presentation Planning Sheet. This is due one week before your presentation. See form for the preliminary planning sheet at the end of this activity.

5. Oral Presentation: On completion of your research, each team will be asked to make an oral presentation of their research findings. Your presentation should include items listed above. In addition to your presentation, you will need to provide:

   a. PowerPoint presentation on CD or DVD or a poster

   b. Three-dimensional object that pertains to your topic (e.g., rope, muddy boot, air filter)

Planning the Oral Presentation:

Your presentation should be equally divided among your team members. In the Oral Presentation Planning Chart located at the end of the activity, describe the role each of you will play during your presentation.

Guidelines for Your Presentation:

1. Be prepared: Practice and rehearse and time your presentation in advance.
2. Block out what each person will do for the presentation and when they will do their part of the presentation.
3. Special equipment: If you are showing a PowerPoint presentation, be sure that you are familiar with the equipment before your presentation. If you are using a flex camera, practice setting up your visuals and focusing the camera before your presentation.
4. DO NOT READ YOUR POWERPOINT SLIDES OR YOUR REPORT! You can refer to the PowerPoint or notecards, but you cannot read it word for word.
5. Make eye contact with your audience. Do not turn your back on the audience.
6. Scan the group and speak to the entire group.
7. Speak slowly and with enough volume so the people in the back row can easily hear you! Try to relax and smile!

Extra Credit:

If you provide a handout that pertains to your topic, you will be given extra credit. The handout should contain both information and graphics.

CHAPTER 6

Fingerprints

UNALTERED IDENTITY

"Smiling Gus" Winkler, a gangster, thought he could stay ahead of police by surgically altering his fingerprints.

Augustus "Smiling Gus" Winkler's personal motto was "Take care of Winkler first," and his career as a gangster showed he did just that. Said to be a smooth talker, Gus began his life of crime as a member of Eagan's Rats in St. Louis, Missouri, and by the age of 20, he had earned a reputation as a skilled safe-cracker. Between 1920 and 1926 he served time for assault with a deadly weapon, and on his release he headed north to Chicago. There Gus met up with some of Chicago's most famous gangsters, Fred "Killer" Burke, Al Capone, Bugsy Moran, and Roger Touhy, and was rumored to have participated in the St. Valentine's Day Massacre. He also had connections with police, always keeping his best interest in mind.

In 1933, looking out for himself, Winkler turned in evidence on his buddies, and was key to returning some of the loot from the Lincoln Trust Bank robbery. That act did not sit well with his friends, and in 1933 he was gunned down by unknown assailants. Winkler was laid to rest in a $10,000 silver coffin wearing clothes covered in gems. Winkler was one of the many gangsters who tried to disguise his identity by trying to alter his fingerprints. Below are the fingerprints of Gus Winkler's left middle finger before and after alteration.

Reprinted by special permission of Northwestern Universtiy School of Law, The Journal of Criminal Law and Criminology

Winkler's fingerprints before (left) and after (right) he had a doctor remove a narrow strip down the center.

OBJECTIVES

By the end of this chapter you will be able to

✔ Discuss the history of fingerprinting.

✔ Describe the characteristics of fingerprints.

✔ Identify the basic types of fingerprints.

✔ Describe how criminals attempt to alter their fingerprints.

✔ Determine the reliability of fingerprints as a means of identification.

✔ Explain how fingerprint evidence is collected.

✔ Describe the latest identification technologies.

✔ Determine if a fingerprint matches a fingerprint on record.

✔ Use the process of lifting a latent print.

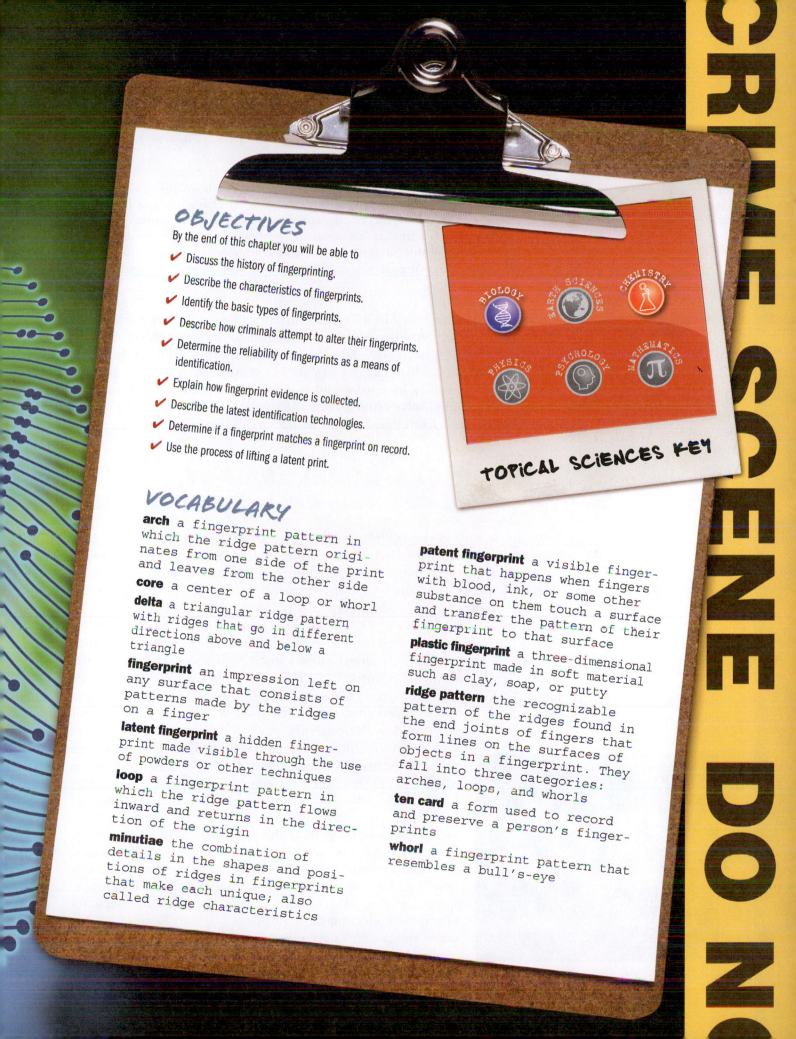

BIOLOGY

EARTH SCIENCES

CHEMISTRY

PHYSICS

PSYCHOLOGY

MATHEMATICS

TOPICAL SCIENCES KEY

VOCABULARY

arch a fingerprint pattern in which the ridge pattern originates from one side of the print and leaves from the other side

core a center of a loop or whorl

delta a triangular ridge pattern with ridges that go in different directions above and below a triangle

fingerprint an impression left on any surface that consists of patterns made by the ridges on a finger

latent fingerprint a hidden fingerprint made visible through the use of powders or other techniques

loop a fingerprint pattern in which the ridge pattern flows inward and returns in the direction of the origin

minutiae the combination of details in the shapes and positions of ridges in fingerprints that make each unique; also called ridge characteristics

patent fingerprint a visible fingerprint that happens when fingers with blood, ink, or some other substance on them touch a surface and transfer the pattern of their fingerprint to that surface

plastic fingerprint a three-dimensional fingerprint made in soft material such as clay, soap, or putty

ridge pattern the recognizable pattern of the ridges found in the end joints of fingers that form lines on the surfaces of objects in a fingerprint. They fall into three categories: arches, loops, and whorls

ten card a form used to record and preserve a person's fingerprints

whorl a fingerprint pattern that resembles a bull's-eye

INTRODUCTION

Pudd'nhead Wilson is a lawyer created by Mark Twain in the novel of the same name, published in November 1894. In his final address to a jury, Lawyer Wilson exhibits his knowledge of the cutting-edge technology of the day:

> Every human being carries with him from his cradle to his grave, certain physical marks which do not change their character, and by which he can always be identified—and that without shade of doubt or question. These marks are his signature, his physiological autograph, so to speak, and this autograph cannot be counterfeited, nor can he disguise it or hide it away, nor can it become illegible by the wear and mutations of time.

No one is sure how Mark Twain learned that fingerprints made good forensic evidence, but he used them in his book to dramatically solve a case in which identical twins were falsely accused of murder. Fingerprints as a means to identify individuals was a major breakthrough in forensic science in real life, as well as in novels, and it gave law enforcement around the world a new tool to solve crimes, clear the innocent, and convict the guilty. Fingerprint cards from Pudd'nhead Wilson are shown in Figure 6-1.

Figure 6-1. *Early, though fictional, fingerprint cards from Twain's* Pudd'nhead Wilson.

THUMB PRINTS
From Pudd'nhead Wilson's Collection
"Every human being carries with him from his cradle to his grave certain physical marks which do not change their character."

Public Domain

HISTORICAL DEVELOPMENT

For thousands of years, humans have been fascinated by the patterns found on the skin of their fingers. But exactly how long ago humans realized that these patterns could identify individuals is not at all clear. Several ancient cultures used fingerprints as markings (Figure 6-2). Archaeologists discovered fingerprints pressed into clay tablet contracts dating back to 1792–1750 B.C. in Babylon. In ancient China, it was common practice to use inked fingerprints on all official documents, such as contracts and loans. The oldest known document showing fingerprints dates from the third century B.C. Chinese historians have found finger and palm prints pressed into clay and wood writing surfaces and surmise that they were used to authenticate official seals and legal documents.

In Western culture, the earliest record of the study of the patterns on human hands comes from 1684. Dr. Nehemiah wrote a paper describing the patterns that he saw on human hands under the microscope, including the presence of ridges. Johann Christoph Andreas Mayer followed this work in 1788 by describing that "the arrangement of skin ridges is never duplicated in two persons." He was probably the first scientist to recognize this fact. In 1823, Jan Evangelist Purkyn described nine distinct fingerprint patterns, including loops, spirals, circles, and double whorls. Sir William Herschel began the collecting of fingerprints in 1856 (Figure 6-3). He noted the patterns were unique to each person and were not altered by age.

Figure 6-2. *This ancient seal shows the fingerprint of a person who lived hundreds of years ago.*

©Photograph by Bruce & Kenneth Zuckerman and Maarily Lundberg, West Semitic Research. Courtesy USC Archaeological Research Collection.

Figure 6-3. *Sir William Herschel*

Figure 6-4. *Sir Francis Galton*

©Bettmann/Corbis

©Bettmann/Corbis

In 1879, Alphonse Bertillon, an assistant clerk in the records office at the Police Station in Paris, created a way to identify criminals. The system, sometimes called Bertillonage, was first used in 1883 to identify a repeating offender.

In 1902, he was credited with solving the first murder using fingerprints. Building on this success, Sir Francis Galton (1822–1911) verified that fingerprints do not change with age. In 1888, Galton, along with Sir E. R. Henry, developed the classification system for fingerprints that is still in use today in the United States and Europe. Galton is shown in Figure 6-4.

Did You Know?

Alphonse Bertillon was the first person to document incoming prisoners with a photograph, the forerunner of the modern mug shot.

Iván (Juan) Vucetich improved fingerprint collection in 1891. He began to note measurements on the identification cards of all arrested persons, as well as adding all 10 fingerprint impressions. He devised his own fingerprint classification system and invented a better way of collecting the impressions. Beginning in 1896, Sir Edmund Richard Henry, with the help of two colleagues, created a system that divided fingerprint records into groups based on whether they have an arch, whorl, or loop pattern. Each fingerprint card in the system was imprinted with all 10 fingerprints of a person and marked with individual characteristics called a **ten card** (Figure 6-5).

Figure 6-5. *An early example of a ten card.*

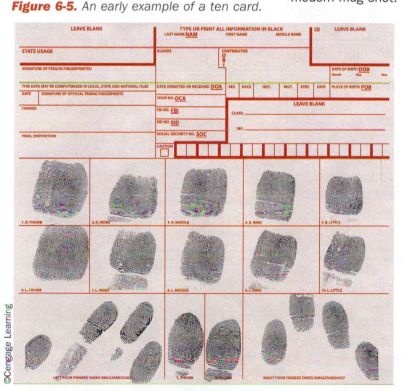

©Cengage Learning

Fingerprints can be taken from dead bodies by chemically treating the fingertips to help them puff out. Another method involves surgically removing the finger skin and placing it like a glove onto the finger of someone else, who can then roll the print.

Digging Deeper
with Forensic Science e-Collection

Who was the first person to discover the individuality of fingerprints and their application to solving crimes? Asking this question in some scientific circles will likely spark a highly charged, spirited argument. The evidence is murky in some areas and clear in others. Search the Gale Forensic Science eCollection on school.cengage.com/forensicscience to find a review of the history of fingerprinting in solving crimes. After you read the review, make your own investigation by reading the primary sources available on that web site. Write a brief explanation of your findings. Make sure to back up your argument with sources, carefully check the dates of the publications, and apply logic to make your conclusions.

WHAT ARE FINGERPRINTS?

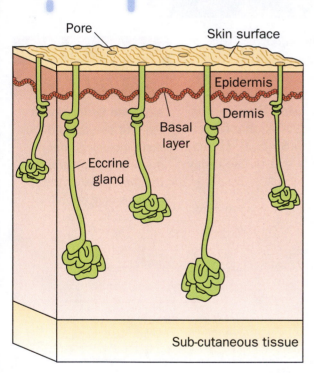

Figure 6-6 *Our fingertips are covered with hundreds of microscopic sweat pores, which make our fingers moist and able to grip better.*

Take a look at the surface of your fingers. Are they smooth and shiny surfaces? No. All fingers, toes, feet, and palms are covered in small ridges. These are raised portions of the skin, arranged in connected units called dermal, or friction, ridges. They help us with our grip on objects that we touch. When these ridges press against things, they leave a mark, an impression called a **fingerprint**.

The imprint of a fingerprint consists of natural secretions of the sweat glands that are present in the friction ridge of the skin (Figure 6-6). These secretions are a combination of mainly water, oils, and salts. Dirt from everyday activities is also mixed into these secretions. Anytime you touch something, you leave behind traces of these substances in the unique pattern of your dermal ridges.

FORMATION OF FINGERPRINTS

The individual nature of fingerprints has been known for about 2,000 years, but scientists only recently understood how fingerprints form in the womb. The

Digging Deeper
with Forensic Science e-Collection

Recent information about the fingerprints of Leonardo da Vinci has suggested that the ancient rumors that his mother was from a Middle Eastern country are most likely accurate, because his fingerprints display those genetic traits. To learn more about the forensic studies of da Vinci's fingerprints, go to the Gale Forensic Science eCollection on school.cengage.com/forensicscience and write a brief evidence report stating your conclusions about the question of da Vinci's ancestry.

latest information suggests that the patterns are probably formed at the beginning of the 10th week of pregnancy, when the fetus is about three inches long. Similar prints are formed in many other areas of the body, such as the palms of the hands, the soles of the feet, and the lips.

The creation of fingerprints happens in the basal layer, a special layer within the epidermis where new skin cells are produced. In a fetus, this layer grows faster than the epidermis on the outside and the dermis on the inside. Because it grows faster, the layer collapses and folds in different directions, creating intricate shapes between the other layers of skin. The pattern cannot be altered or destroyed permanently by skin injuries, because the outer layer protects it.

CHARACTERISTICS OF FINGERPRINTS

Fingerprint characteristics are named for their general visual appearance and patterns. These are called **loops, whorls**, and **arches** (Figure 6-7). About 65 percent of the total population has loops, 30 percent have whorls, and 5 percent have arches. Arches have ridges that enter from one side of the fingerprint and leave from the other side with a rise in the center. Whorls look like a bull's-eye, with two deltas (triangles). Loops enter from either the right or the left and exit from the same side they enter.

Figure 6-7. *There are three basic fingerprint patterns occurring at different frequencies in the population.*

©Hans van den Nieuwendijk

Arches 5%   Whorls 30%   Loops 65%

Two things a forensic examiner looks for on a fingerprint are the presence of a core and deltas. The **core** is the center of a loop or whorl. A triangular region located near a loop is called a **delta**. Some of the ridge patterns near the delta will rise above and some will fall below this triangular region. Sometimes the center of the delta may appear as a small island. A ridge count is another characteristic used to distinguish one fingerprint from another. To take a ridge count, an imaginary line is drawn from the center of the core to the edge of the delta. In Figure 6-8, the red line shows the area used in the ridge count from the delta to core area.

The basic fingerprint patterns can be further divided. Whorl patterns may be plain whorl (24%), central pocket loop whorl (2%), double loop whorl (4%), or accidental whorl (0.01%). The plain whorl (Figure 6-9A, next page) has one or more ridges that make a complete spiral. There are two deltas, and if a line is drawn between them, at least one ridge in the inner pattern is touched or cut by the line. The central pocket loop whorl (Figure 6-9B, next page) has one or more ridges that make a complete circle. There are two deltas, and if a line is drawn between them, no ridges in the inner pattern

Figure 6-8. *The red line is called the core area and exists between a delta and the center of a loop or whorl.*

Core

Ridge count area

Delta

©Hans van den Nieuwendijk

Figure 6-9. *Types of whorl pattern.*

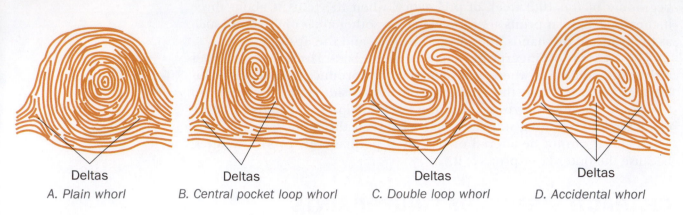

Deltas
A. Plain whorl

Deltas
B. Central pocket loop whorl

Deltas
C. Double loop whorl

Deltas
D. Accidental whorl

Figure 6-10. *Types of arch pattern.*

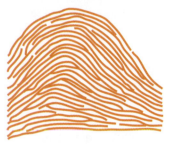

A. Plain arch

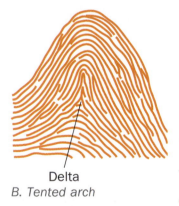

Delta
B. Tented arch

are touched or cut by the line. The double loop whorl (Figure 6-9C) has two separate loop formations and two deltas. The accidental whorl (Figure 6-9D) has two or more deltas and is a combination of two of the other patterns (but not a plain arch).

Arches may be divided into plain arches (4%) and tented arches (1%). The plain arch (Figure 6-10A) shows ridges entering one side, rising in the center, and flowing out the other side without making an angle. The plain arch has no characteristics of the loop pattern. The tented arch (Figure 6-10B) does form an angle, or it may possess some characteristic of the loop pattern, such as a delta.

While looking at the basic fingerprint patterns can quickly help eliminate a suspect, in order to positively match a print found at a crime scene to an individual, more information is needed. Every individual, including identical twins, has a unique fingerprint resulting from unique **ridge patterns** called **minutiae** (because the details are so small). Recognizing these details in the differences between ridges, their relative number, and their location on a specific fingerprint is called *fingerprint identification*. There are about 150 individual ridge characteristics on the average full fingerprint. When forensic examiners identify a fingerprint, they are in theory identifying the unique signature of a person, and they can be pretty sure they are characterizing one, and only one, particular individual in the world. To match fingerprints, a minimum number of points of comparison are needed.

In Figure 6-11, on the next page, fingerprint minutiae are described. In the lab activities, you will practice the techniques necessary to identify and match fingerprints, including analyzing these ridge characteristics.

TYPES OF FINGERPRINTS

There are three types of prints found by investigators at a crime scene. **Patent fingerprints,** or visible prints, are left on a smooth surface when blood, ink, or some other liquid comes in contact with the hands and is then transferred to that surface. **Plastic fingerprints** are actual indentations left in some soft material such as clay, putty, or wax. **Latent fingerprints,** or hidden prints, are caused by the transfer of oils and other body secretions onto a surface. They can be made visible by dusting with powders or making the fingerprints in some way more visible by using a chemical reaction.

Fingerprints of suspects are taken by rolling each of the 10 fingers in ink and then rolling them onto a ten card that presents the 10 fingerprints in a standard format. In Activity 6-5, you will learn how to roll your own fingerprints.

Figure 6-11. *Some minutiae patterns used to analyze fingerprints.*

Name	Visual Appearance
1. Ending ridge (including broken ridge)	1.
2. Fork (or bifurcation)	2.
3. Island ridge (or short ridge)	3.
4. Dot (of very short ridge)	4.
5. Bridge	5.
6. Spur (or hook)	6.
7. Eye (enclosure or island)	7.
8. Double bifurcation	8.
9. Delta	9.
10. Trifurcation	10.

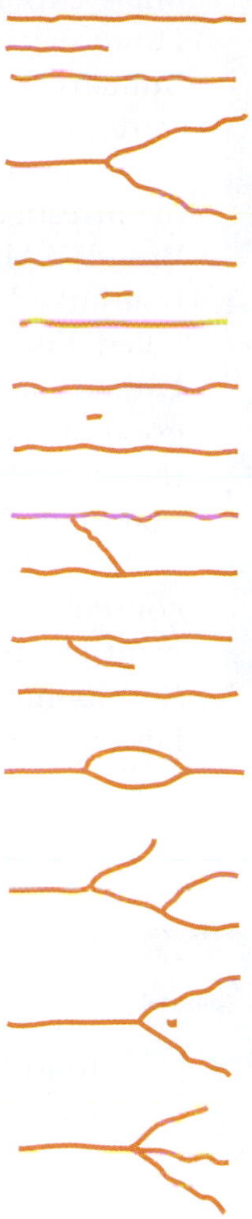

FINGERPRINT FORENSIC FAQs

Can Fingerprints Be Altered or Disguised?

As soon as fingerprints were discovered to be a reliable means of identification, criminals began to devise ways to alter them so they could avoid being identified. American Public Enemy Number One in the 1930s, John Dillinger (Figure 6-12), put acid on his fingertips to change their appearance, something he likely learned from stories of workers in the pineapple fields in Cuba who did not have fingerprints. This is because several chemical substances found in the pineapple plant, when combined with the pressure of handling the plants, dissolved the workers' fingerprints. What Dillinger did not learn is that when these workers ended their contact with the pineapples, their fingerprints grew back! Fingerprints taken from Dillinger's body in the morgue on his death were compared to known examples he left behind during his life of crime. Despite his efforts to destroy his fingerprints, they still allowed him to be identified.

Figure 6-12. *Wanted poster for John Dillinger.*

©AP Photo/Paul Walsh

Figure 6-13. *Brandon Mayfield was falsely accused of involvement in the Madrid train bombing on the basis of the mistaken analysis of fingerprint evidence.*

©Brent Wojahn/The Oregonian/Corbis

How Reliable Is Fingerprinting as a Means of Identification?

Many experts claim that fingerprint identification is flawless. However, humans input and analyze the information, and humans make mistakes. In 1995, 156 fingerprint examiners were given a test. One in five examiners made at least one false-positive identification. More recently, the Federal Bureau of Investigation (FBI) arrested and jailed Oregon lawyer Brandon Mayfield (Figure 6-13) based on fingerprint evidence that linked him to the Madrid train bombings in 2004, which killed 170 people. Mayfield, who had not traveled out of the United States for 10 years, claimed the fingerprint was not a good match. Mayfield was held in custody for two weeks, until the Spanish authorities told the FBI that the mark was, in fact, that of an Algerian citizen.

In light of the fallibility of human nature, and the serious consequences of making a mistake, it is important that fingerprint examiners be held to a high standard of performance. Results need to be checked and double-checked to prevent false convictions and to maintain the integrity of the science.

Figure 6-14. *A technician compares fingerprints in the AFIS system.*

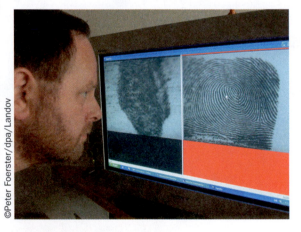

©Peter Foerster/dpa/Landov

How Are Fingerprints Analyzed?

Contrary to what we see on television, fingerprint matching is not carried out by a computer in a matter of seconds. By 1987, the FBI had 23 million criminal fingerprint cards on file, and getting a match with a fingerprint found at a crime scene and one stored on file required manual searching. It could take as long as three months to find a match. In 1999, with the cooperation of the national law enforcement community, the FBI developed the Integrated Automated Fingerprint Identification System (IAFIS or AFIS).

The IAFIS provides digital, automated fingerprint searches, latent searches, electronic storage of fingerprint photo files, and electronic exchange of fingerprints and test results. It operates 24 hours a day, 365 days a year. Now agencies submitting fingerprints electronically for matching can expect results for criminal investigations within two hours (Figure 6-14). Currently, the IAFIS maintains the Criminal Master File, which is the largest database of its kind in the world. It contains the fingerprints and criminal histories for more than 47 million people. State, local, and federal law enforcement agencies submit these data voluntarily. Federal and state fingerprinting agencies do not pool their databases.

Did You Know?

There are about 30 recorded cases of permanent, intentional fingerprint mutilations. In all of these cases, the mutilation was discovered by law enforcement.

How Are Latent Fingerprints Collected?

As mentioned earlier, latent fingerprints are not visible, but techniques can "bring them out." Dusting surfaces such as drinking glasses, the faucets on bathroom sinks, telephones, and the like with a fine carbon powder can make a fingerprint more visible. Tape is then used to lift and preserve the fingerprint. The tape with the fingerprint is then placed on an evidence card on which the date, time, location, and collector of the print is logged. Proper evidence collection techniques involve photographing the fingerprints before they are lifted. Metal or magnetic powders can also be used. They are less messy than carbon-based powders.

Figure 6-15. *Methods used for visualizing latent fingerprints.*

Chemical	Uses	Application	Safety	Chemical Reaction	Latent Print
Ninhydrin	Paper	Object dipped or sprayed in Ninhydrin Wait 24 hours	Do not inhale or get on your skin	Reacts with amino acids (proteins) found in sweat	Purple-blue print
Cyanoacrylate Vapor	Household items: plastic, metal, glass, and skin	Heat sample in a vapor tent	Do not inhale or get on your skin: irritant to mucous membranes	Reacts with amino acids	White print
Silver Nitrate	Wood Styrofoam	Object dipped or sprayed in Silver Nitrate	Wear gloves to avoid contact with skin	Chloride from salt in perspiration on the print combines with silver nitrate to form silver chloride	Black or reddish brown print under UV light
Iodine Fuming	Paper Cardboard Unpainted surfaces	In a vapor tent, heat solid iodine crystals	Toxic to inhale or ingest	Iodine combines with carbohydrates in latent print	Brownish print (fades quickly) must be photographed or sprayed with a solution of starch

To recover a print from a surface that is not smooth and hard requires the use of different chemicals. When the fingerprint residue combines with these chemicals, the fingerprint image becomes visible. Figure 6-15 summarizes some of the more common chemicals used to produce a latent print, and Figure 6-16 shows a trained officer lifting a print.

THE FUTURE OF FINGERPRINTING

Fingerprinting is not going away anytime soon. With the new scanning technology and digital systems of identifying patterns, fingerprints can be scanned at the rate of 500 to 1,000 dots per inch. This provides an image that reveals minute pore patterns on the fingerprint ridges, allowing for even better pattern matching (Figure 6-17, next page). Perhaps with time, the chances of mistakes being made will be virtually eliminated.

Entirely new uses for fingerprints are also being developed. Scientific studies have shown that much of the material we touch in our daily lives leaves trace evidence on our fingers and hands, which is in turn left behind on the objects that we touch. Dr. Sue Jickells is doing research to ask how things criminals may touch, such as explosives, cigarettes, and drugs, can leave behind traces on

Figure 6-16. *A forensic specialist lifting a print.*

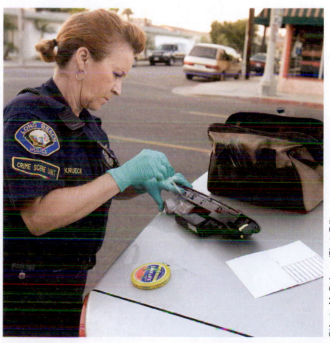

©Kayte M. Deioma/PhotoEdit

Figure 6-17. *This high-resolution fingerprint is a digital image that shows the pores along the ridges, which appear as holes in the lines.*

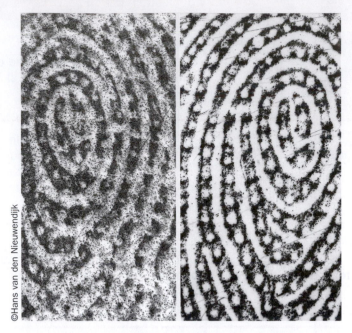

©Hans van den Nieuwendijk

the skin. When identified and studied, these trace substances could tell us much more about the lives of fingerprint donors than just their identities.

Technologies currently being developed use other physical features to identify people, including retinal patterns in the eyes, facial patterns, and the pattern of veins in the palm of the hand. Who knows what else the future holds!

SUMMARY

- Humans have noticed the patterns on their hands for thousands of years, but it was not until 1684 that these patterns were described in detail. In the mid-1800s, the idea of a fingerprint's uniqueness was studied, and the application of fingerprints to an identification system began. By the late 1800s, two effective systems were being used to identify criminals, and fingerprints were collected as evidence in crimes.

- The lines in a fingerprint are called friction ridges. Fingerprints consist of several main ridge patterns, including whorls, loops, and arches. They have a core, which is an area where ridges separate or unite after running in a parallel direction. The triangular region located near a loop pattern, or whorl, is called a delta.

- Fingerprints are formed in the womb at about week 10 of gestation. They are formed between two layers of skin, and their shape does not change during a person's lifetime. They are unique to an individual. Not even identical twins have identical fingerprints.

- Fingerprints left on an object are created by the naturally occurring ridges in the skin of fingers. Secretions from sweat glands leave small amounts of oils and salts when the ridges are pressed against an object. The residues mirror the shape of the ridges found on the finger of the donor.

- The basic types of fingerprints are patent (visible) fingerprints, plastic (indentations) fingerprints, and latent (hidden) fingerprints. They are characterized as either loops, whorls, or arches matched on the basis of minutiae.

- Criminals have sought to alter their fingerprints with chemicals, surgery, and superficial destruction. Some fingerprints can be altered by long-term contact with rough surfaces. Most attempts at fingerprint alteration have not been successful.

- Although mistakes in fingerprint analysis have led to wrongful convictions, the errors lie with the humans who are doing the analyzing. Higher standards of performance for analysts and a system of checks and balances help limit false convictions based on fingerprint evidence.

- The Integrated Automated Fingerprint Identification System (IAFIS) is a national digital database that holds about 47 million fingerprint records and operates 24 hours a day, 365 days a year.

- Fingerprints can be collected from surfaces by dusting them with certain powders and impressing them on tape, or putting them into contact with certain reactive gases to bring them out.

CASE STUDIES

Pedro Ramón Velásquez (1892)

On June 29, 1892, in the village of Necochea, Buenos Aires, two children, Ponciano Carballo Rojas, age six, and his sister Teresa, age four, were found brutally murdered in their home. Their mother, Francesca, age 27, was found with a superficial knife wound to the throat.

The police started an investigation that baffled them. Francesca told police that her neighbor, Pedro Ramón Velásquez, had committed the crime. Velasquez, a one-time suitor of Francesca's, did not confess, even after being tortured.

Inspector Commissioner Alvárez went to the crime scene to reexamine it, searching for any trace of evidence that might have been overlooked. He spotted bloody fingerprints on the doorpost of the house. Because Francesca had denied touching the bodies of her children, Alvárez believed he had found an important clue.

He took the bloody doorpost and fingerprint samples of Pedro Velásquez (Figure 6-18) to Juan Vucetich, who in late 1891 had opened the first fingerprint bureau in South America in Buenos Aires. Vucetich examined the fingerprints and found they did not match. Alvárez became suspicious of Francesca, who had been so insistent that Velásquez had committed the crime. He took a sample of her fingerprints and discovered that they matched the bloody prints found on the doorpost of the house.

When Francesca was confronted with the evidence against her, she confessed. She had murdered her own children, faked an attack on herself, and cast blame on an innocent man, intending him to die for the crime. Her reasons for the murder and for blaming Velásquez were that he had interfered in a romance between her and another suitor, and she felt she would be more appealing to the other man if she did not have children.

Francesca Rojas was the first person in the Americas to be convicted of a crime based on fingerprint evidence.

Stephen Cowans (1997)

On the afternoon of May 30, 1997, Boston police officer Gregory Gallagher was shot with his own gun in a backyard in Roxbury, Massachusetts. Still carrying the gun, the assailant ran to a nearby residence, where he received a glass of water as he wiped off the gun. Stephen Cowans was eventually identified as the shooter. Investigators found a print on the glass used by the individual. The print was matched to prints of Cowans by two fingerprint examiners with the Boston Police Department. Cowans maintained his innocence. With the compelling fingerprint evidence, Cowans was convicted of the shooting and sentenced to 30 to 45 years in the state prison.

In 2004, Cowan's defense team requested DNA testing of the glass and a baseball cap dropped at the scene of the shooting. Neither DNA sample matched Cowan's DNA, although they did match each other. The original verdict was overturned. As Suffolk County reexamined the fingerprints as it prepared to retry Cowans, the assistant district attorney discovered "conclusively and unequivocally that . . . the purported match was a mistake." Cowans was released from prison after 6½ years. As a result, Boston police and the Suffolk County District Attorney's office established new guidelines for identification and evidence handling.

 Think Critically "To get a conviction, I would rather have one good fingerprint than a pound of hair and fiber evidence." Do you agree or disagree? Support your answer.

Bibliography

Books and Journals

Cole, Simon A., *Suspect Identities: A History of Fingerprinting and Criminal Identification*. Cambridge, MA: Harvard University Press 2001.

Cummins, Harold, and Charles Midlo. *Finger Prints, Palms, and Soles: An Introduction to Dermatoglyphics*. Philadelphia, PA: The Blakiston Company, 1943; reprinted 1976.

Garfinkel, Simson. *Database Nation: The Death of Privacy in the 21st Century*. O'Reilly Media, 2001.

Jones. T. "Inherited Characteristics in Fingerprints (or Theory of Relativity)," The Print, 4(5), 1991.

Lafontaine, Ryan. "Man Found Without Fingerprints," The Sun Herald, Biloxi, Mississippi, May 31, 2006, http://www.truthinjustice.org/fingerprints.htm.

Matera, Dary. *John Dillinger: The Life and Death of America's First Celebrity Criminal*. New York: Carroll & Graf, 2004.

McCann, J. S. "A Family Fingerprint Project," Identification News, May 1975, 7–11.

Shahan, Gaye. "Heredity in Fingerprints," Identification News, XX(4): 1, 10–14, April 1970.

Web sites

Gale Forensic Sciences eCollection, school.cengage.com/forensicscience.

http://safety-identification-products.com/fingerprint-information.html

"High Resolution Scans," http://csdl2.computer.org/persagen/DLAbsToc.jsp?resourcePath=/dl/trans/tp/&toc=comp/trans/tp/2007/01/i1toc.xml&DOI=10.1109/TPAMI.2007.17

"The Wrong Man Arrested for Madrid Bombings," http://www.msnbc.msn.com/id/5092810/site/newsweek.

"Wrongly Charged with Perjury on Fingerprint 'Evidence'," http://www.shirleymckie.com.

http://www.fbi.gov/hq/cjisd/iafis.htm

http://www.fbi.gov/hq/cjisd/takingfps.html

http://www.americanmafia.com/Feature_Articles_168.html

http://www.livescience.com/humanbiology/041102_fingerprint_creation.html

http://galton.org/fingerprints/books/henry/henry-classification.pdf

Peter Paul Biro

A Hungarian immigrant currently living and working in Canada, Peter Paul Biro is an art conservator who, in 1984, was the first to make studies of the fingerprints left behind on paintings by artists. After years of careful study, he began using these marks as a means of identifying artists.

Biro's job is to discover, by the use of fingerprint comparison, who painted a work of art and to put forward evidence to support the claim. No two fingerprints are alike, so Biro's evidence is extremely valuable to those who buy and sell art, because a painting that can be positively attributed to a particular artist rises in value.

Biro only uses fingerprints on artworks that were clearly made during the original creation of the work. These include imprints left in the paint while it is still wet, or prints left as a result of the use of a fingertip to apply paint (Leonardo da Vinci often used his fingertips as paintbrushes), or a palm print that might have resulted from applying varnish by hand. The fingerprint used for comparison should come from unquestioned works of art by the artist.

In 1993, Biro examined a painting discovered in the early 1980s entitled "Landscape with Rainbow"

Courtesy, Peter Paul Biro, Forensic Studies in Art, Montreal, QC, Canada.

Peter Paul Biro.

that was thought to be painted by the famous artist J. M. W. Turner. During the restoration process of this painting, fingerprints were discovered in the paint. Even though a match was found between a fingerprint on "Landscape with Rainbow" and fingerprints photographed on another Turner painting, "Chichester Canal," art experts and scholars alike discounted the evidence. Turner, who was known to work alone with no assistants, had used his fingertip on both paintings to model the still-wet paint and was the only possible donor for both prints. When an independent fingerprint examination by John Manners of the West Yorkshire Police confirmed the conclusions that the fingerprints on both paintings were identical, the unbelievers changed their minds. This case was the first successful use of fingerprints to authenticate artwork. The newly authenticated Turner painting sold for much more money that it would have otherwise.

The fingerprint in Figure 6-18 is that of Leonardo da Vinci. Although not visible in this photograph, Biro and other experts found nine distinct characteristics to identify da Vinci's prints.

Figure 6-18. *Leonardo Da Vinci's fingerprint is in the right center of this document.*

©AP Photo/Piero Lucco

Learn More About It

To learn more about the work of a fingerprinting expert, go to school.cengage.com/forensicscience.

CHAPTER 6 REVIEW

True or False

1. Fingerprints are a result of oil and secretions from skin mixing with dirt.

2. Fingerprints are considered to be a form of class evidence.

3. It is necessary to obtain a full print from a suspect in order to match his fingerprint with a fingerprint found at the crime scene.

4. Plastic prints must be dusted or treated in order to identify the ridge patterns.

5. Loops are the most common form of fingerprints.

6. Fingerprints are formed deep within the dermis layer of the skin.

7. With the aid of IAFIS, it is possible to obtain a "match" within several hours.

8. The type of powder used to dust prints will vary depending upon the weather conditions when the print is lifted.

9. Fingerprints of the left hand are mirror images of the fingerprints on the right hand.

10. Similar print or ridge patterns can also be found on the toes.

Multiple Choice

11. Fingerprints are formed
    a) shortly after birth
    b) at about two years of age
    c) at 10 weeks' gestation
    d) at 17 weeks' pregnancy

12. Fingerprints that are actual indentations left in some soft material such as clay or putty are referred to as
    a) plastic fingerprints
    b) patent fingerprints
    c) latent fingerprints
    d) indented fingerprints

13. The use of fingerprints in identification is not perfect because
    a) The current technology depends on humans to input and analyze the information, and humans make mistakes
    b) Many people have the same exact fingerprints
    c) People can easily change their fingerprints
    d) All of the above are correct answers

14. The three main types of fingerprints are classified as
    a) loops, whorls, and deltas
    b) whorls, bifurcations, arches

c) loops, whorls, and arches
d) arches, core, and deltas

15. A small triangular region is one characteristic found in a fingerprint. This triangular region is known as a

a) spur
b) eye
c) bridge
d) delta

Short Answer

16. Describe how to take a ridge count from a fingerprint.

_____

_____

_____

17. Write a brief definition of the term fingerprint.

_____

_____

_____

18. Describe how fingerprints are formed.

_____

_____

_____

19. Is it possible to alter fingerprints? Defend your answer.

_____

_____

_____

20. Another way to make prints visible is to apply certain chemicals. What aspect of a fingerprint chemically reacts with each of the following?
a. ninhydrin

_____

_____

_____

b. cyanoacrylate

_____

_____

c. silver nitrate

_____

_____

d. iodine fuming

_____

_____

Connections

Refer to the two prints below. The first print is taken from the FBI files of a known suspect. The second print has been lifted off a glass taken from a crime scene. Determine if this is a match. Justify your answer.

- Identify the type of ridge pattern found in both prints.
- Use colored pencils to circle areas of similarity or differences.

©Cengage Learning

©Cengage Learning

ACTIVITY 6-1
STUDY YOUR FINGERPRINTS

Objectives:

By the end of this activity, you will be able to:
1. Identify your fingerprints.
2. Compare your fingerprints to those of your classmates.

Time Required to Complete Activity: 40 minutes

Materials:

clear, adhesive tape ¾ inch in width
ruler
pencil
3 × 5 card
magnifying glass

Safety Precautions:

No special precautions

Procedure:

1. On a lined 3 × 5 card, rub the end of a graphite pencil in a back-and-forth motion, creating a patch of graphite about 2 by 3 inches.
2. Rub your right index finger across the graphite patch, gently rolling from side to side so that the fingertip becomes coated with graphite from the first joint in the finger to the tip, and from fingernail edge to fingernail edge.
3. Tear off a piece of clear adhesive tape about 2 inches long. Carefully press the sticky side of the tape onto your finger from the edge of your fingernail across your finger pad to the other side of your fingernail.
4. Gently peel off the tape.
5. Press the tape, sticky side down, into the box provided on the next page.
6. Examine your fingerprint using a magnifying glass.
7. Compare your fingerprint to the pictured samples.
8. Identify whether your fingerprint pattern is a loop, arch, or whorl.

©Cengage Learning (all)

Arches – 5%   *Whorls – 30%*   *Loops – 65%*

Tape your fingerprint into the space provided and identify its pattern type.

Which hand? _____

Which finger? _____

Fingerprint pattern? _____

Data Collection from Class:

Complete the table: Count the number of students showing each of the three types of fingerprint patterns and place those numbers in the Data Table.

Data Table

	Loop	Whorl	Arch
Number of students showing trait			
Total size of class (This will be the same total for each column)			
Percentage of class showing the trait (Divide the number of students with trait by the total size of class, then multiply by 100%)			
Experts say this percentage should be	65%	30%	5%

Questions:

1. Did the class percentage agree with the value given by experts (yes or no)? Explain your answer using data for support.

2. Describe how to improve this data-collecting activity so that your results might be more reliable.

ACTIVITY 6-2
GIANT BALLOON FINGERPRINT

Objectives:

By the end of this activity, you will be able to:
Create a giant fingerprint for use in studying various ridge patterns.

Introduction:

Ridge patterns help make fingerprints unique and identifiable. By studying your own thumbprint and those of your classmates, you will be able to identify these patterns.

Time Required to Complete Activity: 20 minutes

What you will need to do this experiment: a white balloon and an inkpad.

©Cengage Learning

Materials:

1 large white balloon
fingerprinting inkpad
hand soap or moist wipes
paper towels

Safety Precautions:

Before doing this activity, determine if any students are allergic to latex. If so, you can substitute purple nitrile gloves in place of the latex balloon.

Procedures:

1. Slightly blow up a large balloon or half inflate a large balloon.
2. Ink your thumb from thumbnail to thumbnail and past the first joint.
3. Carefully roll your thumb over the balloon from nail edge to nail edge, leaving a thumbprint. Make sure your print is situated about a quarter of the way from the top, and two-thirds of the way from the bottom.
4. Fully inflate the balloon and examine your thumbprint.
5. Identify your thumb pattern as a loop, whorl, or arch.
6. Examine the balloons of your classmates and identify the ridge types you find.
7. Deflate your balloon and save it.

ACTIVITY 6-3
STUDYING LATENT FINGERPRINTS

Objectives:

By the end of this activity, you will be able to:
1. Explain the significance of fingerprint evidence.
2. Describe how to take and identify latent fingerprints.

Introduction:

Every person has a unique set of fingerprints, even identical twins. Whenever you touch a surface without gloves or other protection, you leave behind an invisible fingerprint. Law enforcement agencies use various fingerprint powders and chemicals to help visualize these telltale prints.

Time Required to Complete Activity: 40 minutes

Materials:

newspaper
black dusting powder
adhesive tape ¾ inch wide
dusting brush
cloth
magnifying glass
drinking glass, watch with glass face, other pieces of glass or Plexiglas®
soap or premoistened handwipes
paper towels

Safety Precautions:

Cover the work area with newspapers.
Handle the dusting powder with care, because it can be very messy.

Procedure:

1. Cover the worktable with newspaper.
2. Wipe off a drinking glass, watch glass, piece of window glass, or Plexiglas® with a clean cloth.
3. Take your thumb and run it along the side of your nose or the back of your neck. These areas of your body are rich in oils and will help lubricate the ridges of the thumb to produce a clearer print.
4. Choose an area on the glass object and touch the glass with your thumb. Use a paper towel or other type of cloth in your other hand to prevent leaving other fingerprints. Be careful to avoid placing any other fingerprints in this area.
5. Dip the dusting brush lightly into the fingerprint powder. Place the brush between your hands and gently twist the brush, so that the bristles spin off excess powder near the surface of the object you are dusting. A latent (hidden) fingerprint should begin to appear. Continue to dust lightly, touching the surface until you have exposed as much

of the latent print as possible. Gently blow off the excess powder. (Be prepared for dust to settle on everything in the classroom.)

6. Tear off a three-inch piece of adhesive tape and place it over the fingerprint and press down.

7. Peel off the tape and place it on the Data Table. This process is called lifting the print.

The three types of fingerprints.

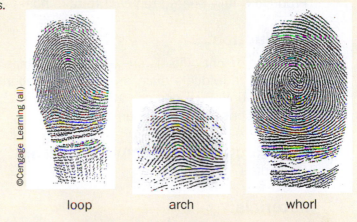

©Cengage Learning (all)

| loop | arch | whorl |

Data Table

	Thumbprint #1
Tape your latent print in the box to the right.	
Identify your print pattern as either loop, arch or whorl.	

Further Study

If time permits, clean the glass and place additional fingerprints on the surface and repeat the technique; then exchange your glass for a classmate's. Dust, lift, and identify his or her print.

ACTIVITY 6-4
HOW TO PRINT A TEN CARD

Objectives:

By the end of this activity, you will be able to:
Produce a fully printed ten card to take home with you.

Introduction:

Law enforcement officials prepare and use fingerprint cards to identify individuals such as criminals, security workers, teachers, and bus drivers, and to register children and those persons licensed to carry firearms.

Criminals are not the only people who are fingerprinted.

©Thinkstock/Jupiter Images

Time Required to Complete Activity:

50 minutes

Materials:

ten card
inking strips or an inkpad
magnifying glass
pre-moistened cleansing wipes, soap, paper towels, newspaper

Safety Precautions:

Cover the work area with newspaper.

Procedure:

1. Cover the work area with newspapers.
2. Fold the ten card along the line between the right-hand fingers and those of the left hand.
3. Place the fingerprint card on the front edge of the table.
4. Starting with your right hand, ink your thumb side to side from fingernail edge to fingernail edge and from the first joint to the tip of the finger. Try to keep your fingers parallel to the surface of the card when printing.
5. Working from the inside to the outer edge, place your thumb in the first square of the card and gently roll side to side using constant pressure. Do not rock back and forth. Make only one wipe. A good print should include from the joint of the finger to the tip.
6. Continue to ink and roll each finger, printing them in order in the squares on the ten card.
7. When you are finished fingerprinting the right hand, fold the card so that the squares for printing the left hand are closest to the edge of the table.
8. Repeat steps 5 to 6 for the left hand.
9. Re-ink the fingers of the right hand and press them gently into the box labeled "first four" for the right hand. Do the same for the right thumb and place the print in the box labeled right thumb.
10. Repeat the above process for the left hand.
11. Using the magnifying glass, examine each fingerprint and label them as a loop, arch, or whorl.

ACTIVITY 6-5
IS IT A MATCH?

Objectives:

By the end of this activity, you will be able to:
1. Describe and identify different types of fingerprint minutiae.
2. Identify different minutiae patterns found in fingerprints.

Time Required to Complete Activity: 30 minutes

Introduction:

Latent fingerprints found at crime scenes are usually incomplete (partial) prints. Investigators need to examine the characteristics of a fingerprint very carefully. The simple identification of a whorl, loop, or arch is not sufficient. Other markers (minutiae) need to be identified.

Materials:

examples of recovered latent fingerprints
red pen

Procedure:

1. Study the picture below. It shows fingerprints obtained from a suspect and a crime scene (mirror). Notice how the investigator has labeled the points of comparison with the same letter on the rolled ink print and the latent print from mirror. Use the chart of characteristics in your text to identify the specific characteristics.

   Rolled ink print taken from suspect (left) and latent fingerprint lifted from the crime scene (right).

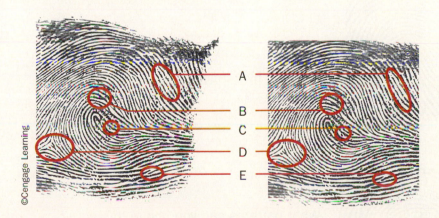

©Cengage Learning

Fingerprint Ridge Patterns

2. Identify each of the patterns labeled in the ridge pattern diagram. Refer to the chart on minutiae in the text.

A. _____

B. _____

C. _____

D. _____

E. _____

3. Examine each of the fingerprints below. Using a red pen and referring to the chart in your text, circle the minutiae pattern and then label it with the appropriate number.

Arthur

1. Bifurcation
2. Island ridge
3. Ending ridge

Doris

4. Eye
5. Spur or hook
6. Ridge ending

Alice

7. Double bifurcation
8. Island ridge

Suspect

What patterns can you find in this print?

_____

_____

_____

ACTIVITY 6-6
FINGERPRINT MATCHING

Objectives:

By the end of this activity, you will be able to:
1. Match the latent crime scene print to one of the suspect's fingerprints.
2. Justify your match by identifying the fingerprint pattern along with as many fingerprint minutiae found in both the crime scene print and the suspect's fingerprint.
3. Circle the common minutiae points on both the crime scene print and the suspect's fingerprint.

Introduction:

Using a red pen, encircle and identify as many minutiae reference points shared by the crime scene fingerprint and the suspect's fingerprint.

Time Required to Complete Activity: 30 minutes

Crime-scene print

Suspect A	Suspect B	Suspect C	Suspect D

Suspect E	Suspect F	Suspect G	Suspect H

All photos ©Cengage Learning

CHAPTER 7

DNA Fingerprinting

DNA FINGERPRINTING AND FAMILY RELATIONSHIPS

Andrew was a teenage boy living in England with his mother Christiana and her other three children. The family had originally come from Ghana. On his way back from a family visit to Ghana in 1984, immigration authorities stopped Andrew at Heathrow airport in London. They suspected that his passport had been altered or forged and that Andrew was not its true owner. Andrew was not allowed to enter the country. Because his father's whereabouts were unknown, no blood specimen could be collected from the father. Traditional biological evidence, such as blood typing, could only be used to prove that Andrew was related to Christiana, but could not prove he was her son.

©AP Photo/Rui Vieira

Dr. Alec Jeffreys.

In 1985, lawyers sought help with Andrew's case from Dr. Alec Jeffreys, a famous British geneticist. Dr. Jeffreys had developed the DNA fingerprinting technique in 1984. In this technique, DNA fingerprints appear as a pattern of bands on X-ray film. Except for identical twins, no two humans have identical DNA or DNA fingerprints. Because half of a child's DNA is inher-ited from the mother and the other half is inherited from the father, DNA fingerprints can be used to prove family relationships. Any bands in a child's DNA fingerprint that are not from the mother must be from the father.

Dr. Jeffreys collected blood samples from Andrew, Christiana, and her other three children. He isolated the DNA and ran DNA fingerprints. He used the DNA fingerprints of Christiana and her three children to reconstruct the DNA fingerprint of the miss-ing father. Each of the bands in Andrew's DNA fingerprint could be found in the DNA finger-prints of Christiana or her three children. About half of the bands in Andrew's DNA fingerprint matched Christiana's fingerprint. The remaining bands matched the reconstructed bands match-ing the fingerprint of the father of Christiana's children.

DNA fingerprinting provided the evidence needed to prove that Andrew was indeed a member of the family. The case against Andrew was dropped. He was allowed to enter the coun-try and was finally reunited with his family.

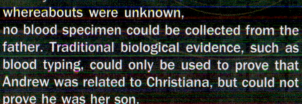

OBJECTIVES

By the end of this chapter you will be able to

✔ Explain how crime-scene evidence is collected for DNA analysis.

✔ Describe how crime-scene evidence is processed to obtain DNA.

✔ Describe how radioactive probes are used in DNA fingerprinting.

✔ Explain how DNA evidence is compared for matching.

✔ Explain how DNA fingerprinting is used to determine if specimens come from related or unrelated individuals.

✔ Explain how to use DNA fingerprinting to identify DNA from a parent, child, or relative of another person.

TOPICAL SCIENCES KEY

BIOLOGY EARTH SCIENCES CHEMISTRY

PHYSICS PSYCHOLOGY MATHEMATICS π

VOCABULARY

allele an alternate form of a gene; for example, a gene for human hair color may have alleles that cause red or brown hair

chromosome a cell structure that contains genetic information along strands of DNA

DNA fingerprint pattern of DNA fragments obtained by examining a person's unique sequence of DNA base pairs (also called DNA profiling)

DNA probe a molecule labeled with a radioactive isotope, dye, or enzyme that is used to locate a particular sequence or gene on a DNA molecule

electrophoresis a method of separating molecules, such as DNA, according to their size and electrical charge using an electric current passed through a gel containing the samples

gene segment of DNA in a chromosome that contains information used to produce a protein or an RNA molecule

PCR (polymerase chain reaction) a method used to rapidly make multiple copies of a specific segment of DNA; can be used to make millions of copies of DNA from a very small amount of DNA

restriction enzyme a molecule that cuts a DNA molecule at a specific, base sequence

STR (short tandem repeat) tandem (next to each other) repeats of short DNA sequences (two to five base pairs) with varying numbers of repeats found among individuals

VNTR (variable number of tandem repeats) tandem (next to each other) repeats of a short DNA sequence (9 to 80 base pairs) with varying numbers of repeats among individuals

INTRODUCTION

Except for identical twins, no two people on earth have the same DNA (deoxyribonucleic acid). Advances in DNA technology have allowed criminal cases to be solved that previously were thought unsolvable. Since the 1980s, DNA evidence has been used to investigate crimes, establish paternity, and identify victims of war and large-scale disasters. Because each human is unique, DNA evidence from a crime scene or from an unidentified body can be traced back to one and only one person. DNA evidence can be used to link a suspect to a crime or to eliminate a suspect. It can also be used to identify a victim, even when no body can be found. DNA evidence has been used to identify human remains of victims of large-scale disasters, such as plane crashes, tsunamis, and hurricanes.

HISTORY OF BIOLOGICAL EVIDENCE IN FORENSICS

Several types of biological evidence, such as skin, blood, saliva, urine, semen, and hair, are used in forensics for identification purposes. Biological evidence is examined for the presence of inherited traits, such as blood type or enzyme variants. Most laboratory techniques used in forensics were originally developed for other purposes, such as medical diagnosis or treatment. When human cells are present in biological evidence, their chromosomes can be examined to determine whether the evidence comes from a male or a female. The analysis of chromosomes is known as karyotyping. Blood-typing techniques, which were first developed to make transfusions safe, are now used in forensics. Blood-type information obtained from crime-scene evidence can help investigators to exclude suspects. Blood typing can also be used to determine if blood found at a crime scene comes from only one person or from multiple individuals. Because blood type is an inherited trait, blood typing is used in paternity testing.

DNA fingerprinting, also known as DNA profiling, is used in criminal and legal cases to determine identity or parentage. DNA can be extracted from relatively small amounts of biological evidence, such as a drop of blood or a single hair follicle. When DNA fingerprinting is performed and interpreted by qualified forensic scientists, the results can very accurately predict whether an individual can be linked to a crime scene or excluded as a suspect.

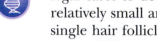

Did You Know?

If the DNA in one human chromosome were unwound, it would reach a length of 6 feet.

THE FUNCTION AND STRUCTURE OF DNA

To understand how DNA fingerprinting works, it is important to know about the function and structure of DNA. DNA is the blueprint of life and contains the genetic material of a cell. DNA holds all of the information and instructions needed for a cell to make proteins and to replicate (make copies of itself). Genetic information is stored in molecules of DNA making up structures called **chromosomes**. The structure of DNA is shown in Figure 7-1 on the next page.

If you tease apart a human chromosome by dissolving away the protective histone packaging, you will see that it is made up of two strands, and each DNA strand is tightly coiled around protein molecules and itself. Because each DNA molecule is composed of two strands, DNA is known as a double helix. James Watson and Francis Crick received the 1953 Nobel Prize for their work on describing the structure of DNA—a double helix that resembles a twisted ladder. The sides of the ladder are called the backbone of each of the two strands of DNA. The backbones consist of alternating sugar and phosphate molecules. The rungs of the ladder are made up of pairs of molecules called *nitrogenous bases* (or bases, for simplicity).

Figure 7-1. *The structure of DNA showing the double-helix shape.*

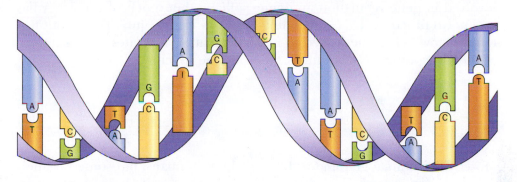

THE DIFFERENT DNA BASES

DNA has four different bases: A (adenine), C (cytosine), G (guanine), and T (thymine). Each base is bonded to the sugar molecule of the backbone on one side and can form weak hydrogen bonds with one specific base on the other strand of the DNA. The base-pairing rules are as follows: adenine (A) binds only with thymine (T), and cytosine (C) binds only with guanine (G). When the base pairs of two human DNA strands form a double helix, the DNA strands are considered to be complementary. For example, if the order of the bases in a section of one strand is CGTCTA, then the order of bases in the complementary section of DNA in the other strand is GCAGAT.

There are 23 pairs (a total of 46) of chromosomes in the nucleus of most human body cells. (As an exception, the egg and sperm cells have only one of each of the 23 chromosomes.) One chromosome in each pair is inherited from the mother; the other chromosome of the pair is inherited from the father. This means that half of an individual's nuclear genetic information comes from the mother and the other half comes from the father. DNA in chromosomes is called nuclear DNA and is virtually identical in all cells of the human body. Another type of DNA is found in the mitochondria of the cell. Mitochondrial DNA exists in the form of a circular loop and, unlike nuclear DNA, is inherited only from the mother. Mitochondrial DNA is passed onto the offspring in the cytoplasm of the egg cell; none of the mitochondria come from the sperm cell.

GENES AND ALLELES

Each chromosome contains many genes. **Genes** are DNA sequences that have instructions that determine our inherited characteristics or traits, such as blood type. Genes can also make another type of nucleic acid called RNA (ribonucleic acid). An **allele** is one of two or more alternative forms of a gene. For example, one allele of a gene might code for normal hemoglobin, while another allele codes for abnormal hemoglobin. One allele comes from the mother, and the other allele comes from the father.

Did You Know?

When a human egg and sperm combine, nearly a billion different gene combinations are possible.

The human *genome,* which is the total amount of DNA in a cell, is contained in chromosomes and mitochondria. The DNA in the chromosomes contains approximately 3 billion base pairs. Chromosomes contain DNA used to make proteins or other molecules, called encoded DNA (*exons*) and unencoded DNA (*introns*), that does not produce protein or RNA molecules but have been found to be important for gene splicing. The chromosomes in the nucleus of each human cell contain some 23,688 encoded genes, with each gene averaging 3,000 base pairs. This is less than 1.5 percent of DNA in the genome. The rest of the human genome—more than 98.5 percent—is noncoding DNA that does not make any protein. Some noncoding DNA is involved in gene regulation and gene splicing, but much of this DNA has no known function. This non-coding DNA was called "junk DNA," but this term is now being questioned as new information indicates that segments of the noncoding DNA are important for gene splicing.

DNA IDENTIFICATION

Most of the human genome is the same in all humans, but some variation exists among individuals. Scientists are able to identify individuals based on this variation. Much of this variation is found in the noncoding DNA. What is interesting about noncoding DNA is that much of it is in the form of repeated base sequences. Individuals have unique patterns of repeated base sequences in the noncoded DNA, and certain base sequences may be repeated many times. DNA sequences have different lengths and different sequences of the bases in different individuals. Within a human population, these differences in DNA sequences are called *polymorphisms.*

In 1984, Dr. Alec Jeffreys at the University of Leicester observed that DNA from different individuals contains different polymorphisms. His laboratory developed a technique for isolating and analyzing these variable areas that is known as DNA fingerprinting, or DNA profiling. A DNA fingerprint appears as a pattern of bands on X-ray film. When DNA fingerprinting is used to analyze biological evidence, variable regions appear as a pattern of bands. The unique patterns of repeated base pairs can be analyzed and used to identify an individual. Because the number and location of polymorphisms are unique in each individual, each individual's DNA has a unique band pattern. The examination of DNA profiles can help forensic scientists to decide if two or more DNA samples are from the same individual, related individuals, or unrelated individuals.

Different repeated sequences appear in different places in the genome of each individual. These repeats may be studied to aid in the identification of individuals. Forensic scientists focus on two types of repeating DNA sequences in the noncoding sections of DNA known as VNTRs (variable numbers of tandem repeats) and STRs (short tandem repeats) found in the DNA.

VNTR

Within the noncoding sections of DNA, certain short sequences of DNA are repeated multiple times. The number of copies of the same repeated base sequence in the DNA can vary among individuals. For example, if the repeated base sequence is CATACAGAC, there might be three copies (CATACAGAC CATACAGAC CATACAGAC) in the DNA of one individual, while another individual might have seven copies. Because the number of repeats varies from one person to another, these multiple tandem repeats are known as **variable number of tandem repeats (VNTR)**. The length of a VNTR can be from 9 to 80 bases in length.

STR

DNA sequences with a high degree of polymorphism are most useful for DNA analysis. A **short tandem repeat (STR)** is a short sequence of DNA, usually only two to five base pairs in length, within the noncoding DNA. An STR is much shorter than a VNTR. For example, an STR GATA repeats four times in an individual with the sequence GATAGATAGATAGATA. The polymorphisms in STRs result from the different number of copies of the repeat element that occur in a population. The use of STRs is becoming the preferred method of analysis because of its accuracy and because small and partially degraded DNA samples may be analyzed to identify individuals. Because the repeated units in VNTRs are longer, the resulting strands of DNA being compared among individuals are also much longer than with STRs. This greater length makes it more difficult to separate the VNTR sequences.

Digging Deeper
with Forensic Science e-Collection

Dr. Alec Jeffreys is the British geneticist who developed the DNA fingerprinting technique. In the late 1970s Dr. Jeffreys and his associates at the University of Leicester in England were looking for variable regions in human DNA that could be used to distinguish among individuals. They hoped their work would enable them to map genes and develop diagnoses for inherited diseases. At that time, they were looking for variable pieces of DNA in regions where the DNA code is repeated. In 1983, Dr. Jeffreys used DNA restriction enzymes to cut the DNA taken from several people. The DNA was cut into fragments of different lengths. They found that the DNA from each individual formed a unique pattern. Dr. Jeffreys called this pattern a DNA fingerprint. Go to the Gale Forensic Science eCollection on school.cengage.com/forensicscience and research how Dr. Jeffreys applied DNA fingerprinting to paternity testing and forensics. Write a short report describing a case investigated by Dr. Jeffreys. How has forensics changed as a result of the development of DNA fingerprinting?

DNA PROFILE

Because of the great variability of VNTRs and STRs in the genome, a DNA profile or DNA fingerprint of a person can be developed when several different VNTRs and STRs are examined. VNTR and STR data from DNA fingerprints are analyzed for two main purposes—tissue matching or inheritance matching. For tissue matching, two samples that have the same band pattern are from the same person. This is the case when the DNA fingerprint from crime-scene evidence matches the DNA fingerprint from a suspect. For inheritance matching, the matching bands in DNA fingerprints must follow the rules of inheritance. Each band in a child's DNA fingerprint must be present in at least one parent.

POPULATION GENETICS AND DNA DATABASES

Population genetics is the study of variation in genes among a group of individuals. The proportion of people in a population who have a particular characteristic, such as black hair or type O blood, is determined by the proportion of alleles for these traits in the population. For example, among Asian populations, blue eyes are rare. This means that few people in this population have the alleles that codes for blue eye color. However, the alleles for blue eye color occur more frequently among northern European populations. Another example of varying allele frequencies among different populations concerns ABO blood types. Among the European populations of Great Britain and the United States, 46 percent of the people have type O blood. However, among native South Americans, almost 100 percent of the population has type O blood.

To calculate the percentage of people who have a particular allele, population studies are conducted to collect data. This data is entered into a database. The greater the number of people included in the study, the more accurate the statistics will be. Therefore, to determine the probability that a suspect's blood type matches the blood type of crime-scene evidence, it is important to refer to a database with blood-type information compiled from a representative population. Calculations are made to determine the probability that a random person in the population would have the same allele as the suspect in a crime or the alleged father in a paternity case.

SOURCES OF DNA

DNA is found in the nucleus of cells in the human body. A perpetrator may leave biological evidence, such as saliva, blood, seminal fluid, skin, or hair, at a crime scene. This biological evidence contains the perpetrator's unique DNA. Because this evidence is capable of identifying a specific person, it is known as *individual evidence*. A saliva sample can be collected from an envelope, a toothbrush, or a bite wound. DNA can be isolated from a sample of biological evidence as small as a drop of blood or a hair follicle.

When the amount of evidence left at a crime scene is very small, it is considered to be *trace evidence*. One of the problems encountered in dealing with trace evidence is that the evidence may be totally consumed during forensic testing. The use of the **polymerase chain reaction (PCR)** technique helps resolve this problem. Dr. Kary Mullis invented the PCR technique, for which he shared the Nobel Prize in 1993. PCR generates multiple identical copies from trace amounts of original DNA evidence. This enables forensic scientists to make billions of DNA copies from small amounts of DNA in just a few hours. The DNA produced with PCR can be analyzed using DNA fingerprinting techniques.

COLLECTION AND PRESERVATION OF DNA EVIDENCE

Cases have been solved because of the presence of a single strand of hair that links a suspect to the crime scene. Because extremely small samples of DNA are used as evidence, attention to contamination issues is necessary when identifying, collecting, and preserving DNA evidence. Contamination of DNA evidence can occur when DNA from another source is mixed with DNA that is relevant to the case. This may happen if someone sneezes or coughs over the evidence or if the person collecting the evidence touches his or her own mouth, nose, or other part of the face before touching the area that may contain the DNA to be tested. To avoid contamination of DNA evidence, the following precautions can be taken:

Did You Know?

The polymerase chain reaction (PCR) was featured in the motion picture *Jurassic Park*. In this story, scientists find fossilized blood-sucking insects. These insects, which had bitten dinosaurs, were caught in tree sap that fossilized into amber. The scientists in the film extracted dinosaur blood from the fossilized insects and used the DNA in the blood to recreate dinosaurs through cloning.

1. Wear disposable gloves and change them often.
2. Use disposable instruments for handling each sample.
3. Avoid touching the area where you believe DNA may exist.
4. Avoid talking, sneezing, and coughing over evidence.
5. Avoid touching your face, nose, and mouth when collecting and packaging evidence.
6. Air-dry evidence thoroughly before packaging.
7. Put evidence into new paper bags or envelopes.
8. If wet evidence cannot be dried, it may be frozen.

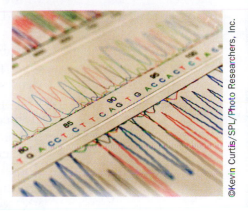

Figure 7-2. *DNA sequences differ from person to person.*

©Kevin Curtis/SPL/Photo Researchers, Inc.

It is important to keep evidence dry and cool during transportation and storage. Moisture can damage DNA evidence, because it encourages mold growth, which can damage the DNA material. Thus, avoid using plastic bags to store evidence that may contain DNA. Direct sunlight and warm conditions are also harmful to DNA, so avoid keeping evidence in places that may get hot, such as a room or a car without air-conditioning.

PREPARING DNA SAMPLES FOR FINGERPRINTING

Why do DNA fingerprints differ from person to person (Figure 7-2)? The DNA sequence of one person is different from that of another person. As a result, when DNA is mixed with special enzymes that cut the DNA in specific places, it will form different-sized fragments of DNA. Using the process of gel **electrophoresis**, these different sized DNA fragments are separated within an agarose gel. As a result, they will form different band patterns, or a fingerprint, of the different DNA fragments within the gel.

STEPS OF DNA FINGERPRINTING

Several steps are necessary before DNA samples can be analyzed and compared. These steps are summarized as follows and then expanded upon in more detail following the summary:

1. Extraction. DNA is extracted from cells.
2. Restriction fragments. In some VNTR analyses, DNA is cut by restriction enzymes. **Restriction enzymes** recognize a unique pattern of DNA bases (restriction sites) and will cut the DNA at that specific location. Restriction fragments of varying lengths are formed when the DNA is cut.
3. Amplification. In the case of other VNTR analyses analysis, specifically chosen DNA fragments are amplified using polymerase chain reaction.
4. Electrophoresis. DNA is loaded into the wells found in an agarose gel. When an electric current is passed through the gel, the negatively charged DNA fragments (pieces of DNA) migrate toward the positive end of the gel. DNA fragments are separated by size, with the smallest DNA fragments moving the fastest through the gel.

Extraction

The first step in preparing a sample for DNA fingerprinting is to extract the DNA from the cell nucleus. Cells are isolated from tissue and are then disrupted to release the DNA from the nuclear and cell membrane as well as from proteins and other cell components.

Restriction Fragments

In the case of some DNA profiles, after the DNA is extracted, the sample is mixed with a restriction enzyme to cut the long strands of DNA into smaller pieces called DNA restriction fragments. Restriction enzymes are "molecular scissors" that cut DNA at specific base sequences. Restriction enzymes are often produced and used by bacteria to defend themselves against invading viruses. There are many different restriction enzymes, and each one binds to a specific recognition site. Moreover, the enzyme cuts the DNA strand at specific locations within that restriction site. For example, the restriction enzyme *Hind III* recognizes the AACGTT base sequence. The *Hind III* restriction enzyme cuts the DNA between the two AA bases. When restriction enzymes cut DNA into pieces, fragments of many different lengths are produced. Within some of these fragments are unique sequences called VNTRs. Several different restriction enzymes may be used to cut the DNA in a sample.

Amplification

With some DNA analyses, polymerase chain reaction (PCR) can be used to amplify certain pieces of the DNA that contain the VNTRs. In STR profiles, restriction enzymes are unnecessary and PCR allows the amplification of the strands with the STR sequences.

Electrophoresis

The fragments of cut DNA are separated by gel electrophoresis. A gel is the matrix (usually agarose) used to separate DNA molecules. Electrophoresis is the method of separating the molecules within an electric field based on the size of the DNA fragments. The gel forms a solid but porous matrix for the DNA fragments to move through.

For this technique, the gel is placed into a gel electrophoresis chamber.

Figure 7-3. *Visualize fragments of DNA forming bands. The arrow shows the direction of movement of the DNA fragments from the negative to the positive end of the gel. Smaller fragments occupy the bands to the right, or positive, end of the gel.*

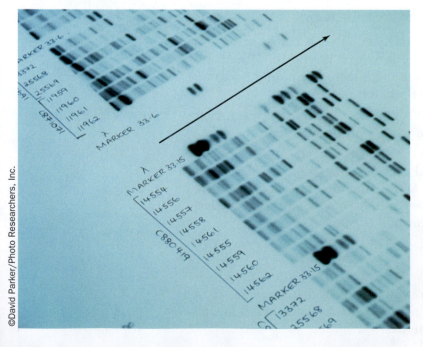

©David Parker/Photo Researchers, Inc.

Then, each DNA sample containing the amplified fragments is drawn up into a micropipette and placed into a separate well or chamber along the top of a gel. One well contains a control, a solution containing DNA fragments of known lengths called a DNA Ladder or Standard DNA.

An electric current is passed through the gel. The negatively charged fragments of DNA in the wells move toward the positively charged opposite end of the gel. DNA fragments of different sizes are separated as the smaller DNA fragments move easily from the negative end of the gel toward the positive end of the agarose gel (Figure 7-3). All of the DNA fragments line up in bands along the length of the gel, with the shortest fragments forming bands closest to the positive end of the gel and the longest fragments forming bands closest to the negatively charged end.

Figure 7-4. *A diagram of an electrophoresis apparatus running five samples of DNA. The diagram shows the direction of movement of the negatively charged DNA fragments through the field created by the battery in the gel matrix.*

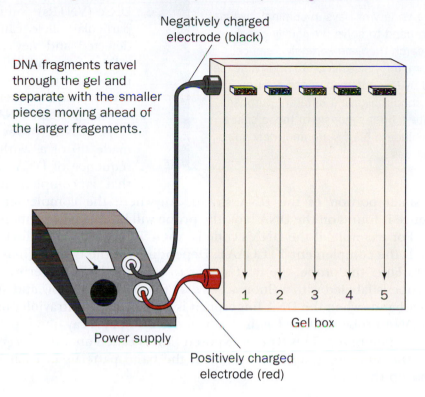

Negatively charged electrode (black)

DNA fragments travel through the gel and separate with the smaller pieces moving ahead of the larger fragements.

1 2 3 4 5

Gel box

Power supply

Positively charged electrode (red)

Electrophoresis is complete when the DNA fragments have *migrated through the gel*. When specific fragments have not been amplified by PCR, it is necessary to visualize the specific fragments of interest by using radioactive probes that bind to specific sequences in those fragments. To do this, the fragments on the gel are then transferred to a membrane in a process called *Southern blotting*. Figure 7-4 shows an electrophoresis apparatus.

DNA AND PROBES

It would be impossible to look at all the DNA restriction fragments produced from an individual's DNA. There would be so many fragments that the DNA fingerprint would not appear as separate bands, but instead it would be a smear of DNA. Therefore, when viewing a DNA fingerprint, we identify specific DNA sequences by using radioactive probes (Figures 7-5 and 7-6).

In methods that amplify specific sections of DNA, such as STR analysis, there will be no genomic DNA smear on the gel, only those fragments that were amplified. In this case, it is only necessary to make any DNA on the gel visible to see the bands of the fingerprint. One method for doing this is to include a fluorescent stain in the liquid the gel is bathed in that inserts itself into the gaps of the DNA double helix. An ultraviolet light source can then be used to cause the DNA to fluoresce so the bands of the fingerprint are visible.

Figure 7-5. *Diagram of electrophoresis gels showing DNA from two individuals analyzed side by side (a) and the same DNA fragments flourescing after probe is applied (b). Note how the probe highlights unique patterns*

a   b

Figure 7-6. *A photograph of an electrophoresis gel showing a whole genome DNA smear in two lanes.*

©Cengage Learning

Digging Deeper
with Forensic Science e-Collection

DNA evidence has been used in a variety of ways in criminal investigations. For example, it has been used to solve previously unsolved cases, so-called "cold cases." Search the Gale Forensic Science eCollection on school.cengage.com/forensicscience to find two cases in which DNA evidence was used by law enforcement to solve or shed light on a case. The cases should highlight different purposes for which the DNA was used. Write a brief synopsis of these cases, explaining what happened, and include a brief statement describing the role DNA played in solving the case.

PROBES

The areas of repeating DNA (VNTRs) within a particular allele can be detected and viewed in a DNA fingerprint. DNA probes are used to identify the unique sequences in each person's DNA.

A **DNA probe** is made up of a synthetic sequence of DNA bases that is complementary to a small portion of the DNA strand. Anywhere the complementary sequence is found on the DNA blot, the probe will bind to complementary bases. For example, if the DNA's code is AAGCTTA, then the radioactive probe is the complement TTCGAAT. Depending on the type of chemical used to label the probe, the next step is to view the DNA fingerprint. If the probe is labeled with a fluorescent dye, the probe will glow and show distinct bands when the DNA fingerprint is viewed under ultraviolet light. If the DNA probe is labeled with a radioactive isotope, X-ray film is placed over the membrane. This type of exposed film is called an *autoradiograph*. When the autoradiograph is developed, the band patterns in each lane appear on the film.

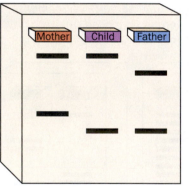

Figure 7-7. DNA bands showing genetic inheritance from parents.

PATERNITY

A child inherits two genes for each trait—one gene from the mother and one gene from the father (Figure 7-7). If the two genes from the parents are different, then two bands will appear in the child's DNA fingerprint. If the two genes from the parents are identical, then only one band would appear on the gel.

The greater the number of probes used on a DNA fingerprint, the greater the accuracy of the DNA fingerprint. In most criminal cases, six to eight different probes are used in the DNA fingerprinting process. Both the position and the width of the bands are significant in matching samples of DNA, and all of the bands have to match exactly.

ANALYSIS OF DNA FINGERPRINTS

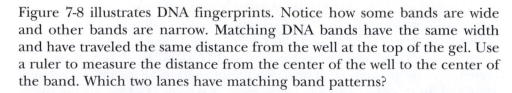

Figure 7-8 illustrates DNA fingerprints. Notice how some bands are wide and other bands are narrow. Matching DNA bands have the same width and have traveled the same distance from the well at the top of the gel. Use a ruler to measure the distance from the center of the well to the center of the band. Which two lanes have matching band patterns?

APPLICATIONS

Several countries have databases of DNA profiles. In the United States, the Combined DNA Index System (CODIS) is an electronic database of DNA profiles. It is similar to the Automated Fingerprint Identification System (AFIS) database. Every state in the country maintains a DNA index of indi-

Figure 7-8. *The results of a DNA fingerprint analysis.*

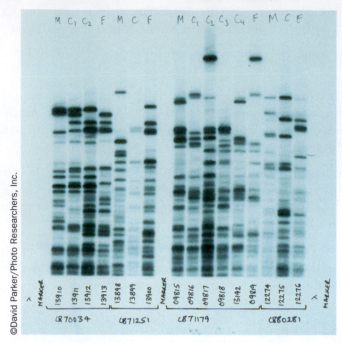

©David Parker/Photo Researchers, Inc.

viduals who have been convicted of certain crimes, such as rape, murder, or child abuse. When perpetrators are convicted, their DNA profiles are entered into the DNA database. Just as fingerprints found at a crime scene can be run through AFIS to identify a suspect or a link to another crime scene, DNA profiles from a crime scene can be entered into CODIS to identify possible suspects. DNA fingerprinting is used in criminal and legal settings. In the case studies and activities that follow, you will see how DNA fingerprinting can be used for the following:

- Crime-scene DNA matching with suspect's DNA
- Paternity and maternity determination
- Identify family members and relatives
- Suspect elimination
- Freeing those who have been falsely accused and imprisoned
- Identification of human remains

SUMMARY

- Present DNA analysis allows a tissue sample to be identified with a single individual; it has been used to solve crimes, establish paternity, and identify victims of war or large-scale disasters.

- Human biological evidence has in the past included karyotyping and blood-type determination.

- DNA fingerprinting can now use a small sample of tissue to identify an individual.

- DNA carries the genetic information, in a sequence of nitrogenous bases, required to produce an organism from a single fertilized egg.

- DNA contains, in noncoding regions, many repeated sequences that vary in number among individuals; these differences can be used to produce a DNA fingerprint for an individual.

- Past difficulties with genetic analyses of crime-scene evidence resulted from the tiny samples that were usually available. Polymerase chain reaction (PCR) for DNA amplification has largely eliminated this problem.

- DNA evidence must be collected carefully to avoid contamination with other DNA.

- DNA analysis involves extraction of the DNA from the sample, in some cases restriction enzyme digestion of the DNA (cutting), amplification using PCR, electrophoresis of DNA fragments, and visualization of the DNA fragments using probes.

- Visualized fingerprints from different tissue samples can be compared in their patterns to determine whether they came from the same person (same pattern) or show a parent—offspring relationship (offspring shares bands with both parents).

- DNA profiles are kept by police agencies of several countries in electronic databases. These databases can be used to connect crime-scene evidence to suspects who were sampled in connection with other crimes.

CASE STUDIES

Colin Pitchfork (1983)

Two schoolgirls in the small British town of Narborough in Leicestershire, United Kingdom, had been raped and murdered three years apart. The methods used were the same for both cases. Blood-type testing revealed that semen samples collected from both victims were from a person with type A blood. This blood type matched 10 percent of the adult male population in the area, but without further information, no suspects could be identified. The noted geneticist Dr. Alec Jeffreys, developer of the DNA fingerprinting technique, was consulted.

To match a suspect to the DNA found in the semen, Dr. Jeffreys suggested that police launch the first-ever DNA-based manhunt. Every young man in the entire community was asked to submit a blood or saliva sample. Blood group testing was performed and DNA fingerprinting was carried out on the 10 percent of men with type A blood. At first no DNA match to the crime-scene evidence was found among the 5,000 samples collected. Then it was discovered that Colin Pitchfork, a local bakery worker, had asked a friend to give a blood sample on his behalf. The police then forced Pitchfork to give a blood sample. His DNA matched the DNA evidence found on both victims. He confessed to the crimes and was sentenced to life in prison. This was the first time DNA fingerprinting was ever used to solve a crime. Joseph Wambaugh's book, *The Blooding*, is based on this case.

©Darrin Klimek/Getty Images

Tommie Lee Andrews (1986)

Nancy Hodge, 27, worked at Disney World in Florida. She was attacked and raped at knifepoint in her apartment. Her attacker's face was covered (concealed). A series of rapes followed, and police suspected that up to 27 attacks could be attributed to the same man. Tommie Lee Andrews was apprehended and linked to the rapes by conventional fingerprint and DNA profile evidence. He was sentenced to more than 100 years in prison. This was the first time DNA evidence was used in the United States to convict a criminal.

Ian Simms (1988)

Helen McCourt was last seen alive as she boarded a bus on her way home from work in Liverpool, England. Evidence found in the apartment of Ian Simms, a local pub owner, linked him to McCourt's disappearance. His apartment was covered with blood, and part of McCourt's earring was found there. The rest of her earring was found in the trunk of his car. Bloody clothing belonging to McCourt was found on the banks of a nearby river. Her body was never recovered. Dr. Alec Jeffreys analyzed the blood found in Simms's apartment and matched it to blood from her parents. Dr. Jeffreys determined that there was a high probability that the blood found in Simms's apartment matched that of Helen McCourt. Simms was found guilty of murder, and sentenced to life imprisonment. This was the first time DNA evidence was used to convict a murderer in a case where the victim's body was not found.

Kirk Bloodsworth (1984)

Dawn Hamilton, age nine, was found raped and beaten to death in a wooded area near her home in 1984. In 1985, Kirk Bloodsworth was accused and convicted of the crime, despite evidence supporting his alibi. Because of a legal technicality, his case was retried, and he was again found guilty in 1986. He was sentenced to three terms of lifetime imprisonment. Bloodsworth continued to maintain his innocence. In 1992, a semen sample from the victim's clothing was analyzed by both a private laboratory and the FBI laboratory. Using PCR and DNA fingerprinting, both laboratories determined that Bloodsworth's DNA did not match the DNA evidence from the crime scene. He was pardoned after spending nine years in prison.

Identification of Human Remains: Hurricane Katrina

DNA testing has been used to match the dead to the families of missing disaster victims. To identify human remains of disaster victims, epidemiologists and genetic counselors work with local officials to collect data on family history. Talking to the families about the missing family members requires skilled detective work by qualified genetic clinicians and genetic counselors. They have to ask the right questions based on what information is needed to fill in the family pedigree.

The DNA identification process following a natural disaster such as Hurricane Katrina, which destroyed New Orleans and other parts of the Gulf Coast on August 29, 2005, is a complex job. That is because many of the missing victims lost their identifying personal effects in the disaster. Items such as hairbrushes and toothbrushes, which could contain hair or saliva samples, were contaminated or destroyed by the flooding. Many medical and dental records normally used to match dead bodies with names of the missing were destroyed, leaving few resources to identify the dead. Family members whose DNA is required to make a match were also displaced, making contact difficult.

 Think Critically Should DNA evidence alone be sufficient to convict when there is no corroborative evidence? State your opinion and provide support for it.

CAREERS IN FORENSICS

Kary Banks Mullis, Nobel Prize–Winning Chemist

Kary Banks Mullis, Nobel Prize–winning chemist, developed the PCR technique.

©AP Photo/Rhonda Birndorf

In 1983, Dr. Kary Mullis developed the polymerase chain reaction (PCR) technique for amplifying DNA. The PCR technique is now used throughout the world in medical and biological research laboratories to detect hereditary diseases, diagnose infectious diseases, and clone human genes. Both commercial and forensics laboratories use the PCR technique in paternity testing and DNA fingerprinting.

Dr. Mullis received a Bachelor of Science degree in chemistry from the Georgia Institute of Technology in 1966 and earned a Ph.D. in biochemistry from the University of California at Berkeley in 1972. After several years of postdoctoral work, Dr. Mullis joined Cetus Corporation, a biotechnology company in Emeryville, California, as a DNA chemist in 1979. During his seven years there, Dr. Mullis conducted research on oligonucleotide synthesis and developed the PCR method. Although methods for DNA amplification existed before the development of the PCR method, the DNA polymerases used in those methods were destroyed when the DNA was heated and needed to be repeatedly replaced. In 1983, Dr. Mullis had the idea of using a DNA polymerase to amplify segments of DNA. He used *Taq* DNA polymerase, isolated from the bacterium *Thermophilus aquaticus*, a strain of bacteria that lives in hot springs. *Taq* polymerase is heat resistant and, when added once, makes the technique much simpler and less expensive.

In 1983, Dr. Mullis received a Nobel Prize in chemistry for inventing the PCR technique. The process, which Dr. Mullis conceptualized in 1983, is hailed as one of the monumental scientific breakthroughs of the 20th century. A method of amplifying DNA, PCR multiplies a single, microscopic strand of the genetic material billions of times within hours. The process has multiple applications in medicine, genetics, biotechnology, and forensics.

Learn More About It
To learn more about the work of a DNA expert, go to
school.cengage.com/forensicscience.

CHAPTER 7 REVIEW

True or False

1. Millions of individuals have identical DNA profiles.

2. Mitochondrial DNA is inherited from the father.

3. The two main purposes for analyzing VNTR data from DNA fingerprints are matching tissues and inheritance.

4. The technique used to amplify DNA is polymerase chain reaction (PCR).

5. In gel electrophoresis, the shortest DNA fragments travel greater distances than the longest DNA fragments.

Multiple Choice

6. Half the bands in a baby boy's DNA fingerprint are from his
   a) uncle
   b) cousin
   c) sister
   d) father

7. DNA restriction enzymes
   a) are used to repair DNA
   b) are involved in DNA synthesis
   c) prevent DNA from being amplified
   d) cut DNA into fragments of different lengths

Short Answer

8. What are the advantages of STR fingerprinting over VNTR fingerprinting?

9. What causes the difference between different STR alleles?

10. Explain the difference between using DNA profiles to determine paternity and to match a tissue sample with a suspect.

Connections

Use colored pencils to draw and color a strand of DNA with at least eight pairs of bases. Make sure that the bases are paired correctly. Then explain how the picture you drew illustrates how DNA can be unique.

Bibliography

Books and Journals

Lazer, David, ed. *DNA and the Criminal Justice System*. Cambridge, MA: The MIT Press, 2004.

Russell, Peter J., Stephen L. Wolfe, Paul E. Hertz, and Cecie Starr. *Biology: The Dynamic Science*. Belmont, CA: Thomson Higher Education, 2008.

Solomon, Eldra, Linda Berg, and Diana W. Martin. *Biology*, 8th ed. Belmont, CA: Thomson Higher Education, 2008.

Starr, Cecie, and Ralph Taggart. *Cell Biology and Genetics*. Belmont, CA: Brooks-Cole Publishing, 2005.

Strachan, Tom, and Andrew P. Read. *Human Molecular Genetics*, 2nd ed. Hoboken, NJ: Wiley-Liss, 1999.

Web sites

"Advancing Justice Through DNA Technology," U.S. Department of Justice, http://www.usdoj.gov/ag/dnapolicybook_cov.htm.

"DNA Forensics," Human Genome Project Information, http://www.ornl.gov/sci/techresources/Human_Genome/elsi/forensics.shtml.

Gale Forensic Science eCollection, school.cengage.com/forensicscience.

"Handbook of Forensic Evidence: DNA Examination," http://www.fbi.gov/hq/lab/handbook/intro6.htm#dna, revised 2003.

"How DNA Evidence Works," http://science.howstuffworks.com/dna-evidence.htm.

Innocence Project, http://www.innocenceproject.org.

March, David, ed. "Hopkins Genetics Experts Aid Efforts to Identify Hurricane Katrina Victims," Medical News Today, 2006, http://www.medicalnewstoday.com/medicalnews.php?newsid=40670.

Moenssens, Andre A. "A Mistaken DNA Identification? What Does It Mean?" http://www.forensicevidence.com/site/EVID/EL_DNAerror.html.

http://www.usatoday.com/tech/science/2005-12-11-katrina-mystery-deaths_x.htm

http://www.dnai.org/index.html (Internet Activity)

ACTIVITY 7-1
DNA FINGERPRINTING SIMULATION USING DYES

The apparatus

Objectives:

By the end of this activity, you will be able to:
1. Describe the charge on a DNA molecule.
2. Explain the process of electrophoresis.
3. Describe how DNA fragments can be visualized.
4. Compare two DNA fingerprints to determine if they match.

©Gregory G. Dimijian, M.D./Photo Researchers, Inc.

Introduction:

When an egg is fertilized by a sperm, half of the DNA is from the mother and the other half is from the father. To verify paternity, DNA fingerprints of the child, the mother, and the alleged father are compared. Paternity is established if half of the child's DNA bands match those of the alleged father.

Probability Calculation:

By using four different probes for four different genes, it is possible to calculate the probability that a DNA fingerprint belongs to a specific person. For example, suppose that:

1 in 100 people have a gene for trait A
2 in 50 people have a gene for trait B
1 in 10 people have a gene for trait C
3 in 20 people have a gene for trait D

To calculate the probability that one person possesses all four of these traits, you need to multiply the individual probabilities:
$$(1/100) \times (2/50) \times (1/10) \times (3/20) = 6/1,000,000$$

To improve the accuracy of matching bands in a child's DNA fingerprint to a parent's bands, additional probes can be used.

Scenario:

In this activity you will use dyes to perform a simulated DNA fingerprint. The purpose of this activity is to determine the paternity of a child.

A couple had been trying to conceive a child for three years. No cause for their infertility was identified after they underwent many tests. The couple continued to try to conceive a child without success. The doctor at the fertility clinic suggested using a sperm donor and artificial insemination. One month after the artificial insemination, the couple learned of the pregnancy. Eight months later, when the baby was born, the baby bore a strong resemblance to the husband. The couple wanted to know if the husband and not the sperm donor could be the biological father. They contacted the fertility clinic, and were told that a lab technician had not kept good records. Semen from two sperm donors had been used on the day of the artificial insemination. The baby could be either the husband's biological child or it could be the child of one of the two sperm donors.

The parents were anxious to know if the husband was the biological father. The doctor suggested doing a DNA fingerprint. DNA was extracted from blood samples collected from the mother, the baby, and the husband. DNA was extracted from semen specimens of the two sperm donors. The DNA samples were treated with restriction enzymes.

Your task is to load the digested DNA into the wells and produce a DNA fingerprint. Analyze the gel to determine whose sperm fertilized the mother's egg. Who is the biological father of this baby?

Time Required to Complete Activity: 40 to 90 minutes

Materials:

(student groups of five)
simulated DNA labeled #2-6
1 DNA sample
1 package of colored pencils or markers
1 gel box
prepared 0.8 M agarose gel
1 set of five prepared unknown samples of simulated DNA labeled 1 to 5
1 DNA standard sample (pre-cut DNA with loading dye already added)
6 10-microliter micropipettes (Wards' 15-3000) with plungers
300 mL TBE buffer 5%
DNA stain
power supply
plastic ruler

Safety Precautions:

Make sure all gels are loaded and connected before turning on the power supply for gel boxes.
Properly dispose of all used pipettes.

Procedure:

Omit steps 1 to 4 if the gel has been prepared by your instructor.
1. Loosen the cap before heating.
2. Warm the bottle of agarose gel in a hot-water bath or autoclave until the bottle is warm to the touch and the agar is melted. Be sure that all of the agar has melted and that there are no chunks of agar in the heated solution.
3. Insert the comb into the gel box and seal the edges before adding the warmed agar.
4. Fill the gel box with agar so that the teeth of the comb are covered to a depth of at least one-half to two-thirds of the length of the teeth. Gels can be prepared up to several days in advance.
5. Allow the gel box to cool for 15 minutes. The agarose gel will turn from clear to opaque white.
6. Cover the gels with plastic wrap to store.
7. When you are ready to use the gel, uncover and remove the comb by pulling straight up simultaneously from both ends of the comb.
8. Place the gel box containing the gel in the electrophoresis chamber.
9. Place the wells near the negative (black) end of the box.

10. Add TBE buffer. Fill one reservoir, and then fill the second reservoir in the gel box.

11. Add enough TBE buffer to cover the entire top of the gel. The TBE buffer should be at least 2 millimeters above the surface of the gel. Use a clean 10 ml pipette for each of the six DNA samples.

12. Add 10 microliters of the following samples to the wells of your gel. Wells 7 and 8 are left empty.

Sample 1	Standard	Well 1
Sample 2	Mother	Well 2
Sample 3	Child	Well 3
Sample 4	Man 1	Well 4
Sample 5	Man 2	Well 5
Sample 6	Husband	Well 6

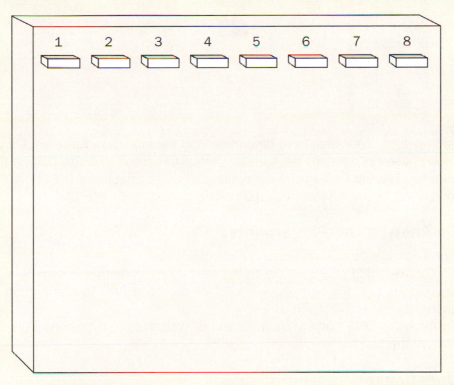

13. Set the power supply to 110 V DC. The power should remain on for approximately 45 minutes, or until the loading dyes have moved toward the positive end of the gel.

14. After the gel has run, complete the sketch found in question #1 using colored pencils or markers.

15. Measure the distance the bands traveled from the center of the well to the center of the color band. Record the measurement at the bottom of the band on the DNA Fingerprint Data figure.

Analysis:

This is a simplified simulation of how a DNA fingerprint is made using gel electrophoresis. When an actual DNA fingerprint is made, many bands are examined. In this simulation, we are assuming that the gel was washed with only one probe, resulting in just two visible bands under each well. You can verify the parentage of a child by comparing the DNA fingerprints for each of the parents with the child's DNA fingerprint. Recall that one-half of the DNA in the child is donated by

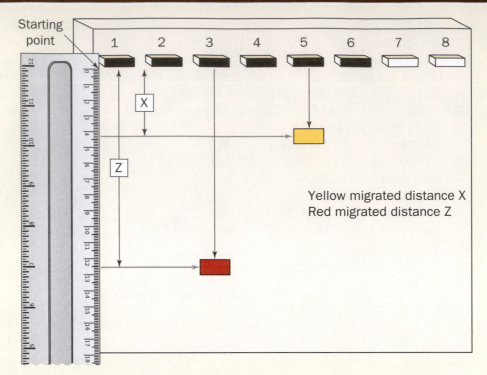

Yellow migrated distance X
Red migrated distance Z

each parent. Your task is to determine who the biological father is. On your DNA fingerprint sketch, be sure to sketch the results of the DNA gel. Be sure to align the DNA in the correct lane. To distinguish the DNA from each individual, use a different colored pencil for each person's DNA.

To Analyze the Fingerprints:

First examine the bands under lane 3, the child.
Determine which bands of the child came from the mother in lane 2.
Remember that the remaining bands of the child had to come from the biological father.
Examine the DNA from lanes 4, 5, and 6. Which of the men contributed his DNA to the child?

Questions:

1. In your sketch of the DNA fingerprint, use a pink-colored pencil or marker to circle the DNA bands shared by both mother and child. Use a blue-colored pencil or marker to circle the DNA bands shared by both the child and the biological father.

2. Who is the biological father of the child——the husband, Man #1, or Man #2? Support your answer with data from your DNA gel test.

3. A different DNA profile is performed with DNA from a father, a mother, and their child. The results are shown in the DNA fingerprint seen below. Which of the three lanes represents the child's DNA? (Hint: One-half of the child's DNA comes from each parent.) Explain your answer.

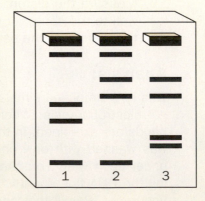

ACTIVITY 7-2
WHERE'S THE CAT? SIMULATION

Scenario:

A man was assaulted and robbed of his wallet by two men outside a bar. He was unable to describe either of his attackers, but he did manage to scratch the face of one of the assailants. Surveillance video was examined, and two suspects were identified and detained.

DNA samples from the two suspects were compared to the DNA obtained from scrapings collected from the victim's fingernails. Your task is to determine if the DNA collected from the victim matches the DNA from either of the suspects. Will DNA evidence identify either of the two suspects?

Objectives:

By the end of the activity, you will be able to:
1. Describe how restriction enzymes cut DNA.
2. Describe how to prepare and load a gel for gel electrophoresis.
3. Determine if a DNA fingerprint of a suspect matches the DNA fingerprint from the evidence obtained from an assault victim.

Time Required to Complete Activity:

Two 40-minute class periods

Background:

DNA fingerprinting may be performed on DNA extracted from relatively small samples of cells, such as a bloodstain the size of a nickel (about two drops) or a semen stain the size of a dime. When performed under properly controlled conditions and interpreted by an experienced forensic scientist, such DNA fingerprinting can link a suspect to a particular incident with compelling accuracy or completely exonerate a suspect.

Lab Procedure for DNA Fingerprinting

These steps are followed when performing DNA fingerprinting on crime-scene evidence. Isolate the DNA from a tissue sample. (Blood, semen, saliva, skin, and hair follicles are sources of DNA.) It may be necessary to amplify the DNA using a method of copying DNA called PCR (polymerase chain reaction). DNA is treated to separate the double helix into two single strands. Single-stranded DNA is cut with a restriction enzyme to produce specific fragments. Each kind of restriction enzyme recognizes a specific sequence of bases on the DNA and acts as "molecular scissors" to cut the DNA strand within the recognition sequence.

In this activity, you will use the enzyme *Hind III*. This enzyme recognizes the AAGCTT base sequence in DNA. The *Hind III* enzyme cuts the DNA between the two A bases: A/AGCTT. Within the human genome, the AAGCTT base sequence is found in multiple places. *Hind III* cuts the human genome into many pieces—perhaps as many as a million different fragments!

Loading the Gel

Once the digested DNA has been cut into fragments, these fragments are loaded into wells at the negative end of an agarose gel.

Electrophoresis of the DNA Fragments

After the DNA is loaded onto the gel, an electric current is passed through the gel. Because DNA is negatively charged, the fragments of DNA will migrate toward the positive electrode in the gel chamber. DNA **runs to red**, that is, toward the positive electrode. A control standard ladder of known sizes or known lengths of DNA fragments is run alongside the unknowns to provide size standards for comparison. After electrophoresis is complete, the DNA from the gel is transferred to a membrane in a process known as Southern blotting.

If DNA were stained at this point, there would be so many fragments that the DNA would appear to be a continuous smear. In order to distinguish between the DNA of one individual and that of another, we need to label only the particular fragments of interest. In this case, we will label only those fragments of DNA that contain the sequences.

Radioactive Probe

Complementary sequences of DNA labeled with radioactive isotopes are called probes. These probes recognize and bind to the repeating CATCAT within the DNA sequence.

Note: If the VNTR is CATCATCAT, then the radioactive probe is

     GTAGTAGTA

     (G pairs with C, A pairs with T)

A solution containing the radioactive probes is allowed to interact with the DNA bands on the Southern blot membrane. Probes that do not attach are washed away, and the radioactivity is visualized by placing X-ray film in contact with the Southern blot membrane. To repeat this process for examination of different VNTRs, the first probe is removed. Then probes for different sequences are added, and the process is repeated.

Materials:

(per group, students will work in groups of five)
1 control standard ladder of DNA (general population) in envelope 1
1 DNA sample from the Victim in envelope 2
1 DNA sample from underneath the victim's fingernails in envelope 3
   (crime-scene DNA)
1 DNA sample from Suspect 1 in envelope 4
1 DNA sample from Suspect 2 in envelope 5
1 set of GTA (radioactive) probes cut from red paper in envelope
6 colored pencils or markers
1 sheet construction paper approximately 2 feet by 3 feet
scissors
meter stick
black magic marker
tape
glue stick

Safety Precautions:

No special precautions are needed.

Procedure:

Working in a group of five students, assign a role to each person in your group:
Standard DNA
Crime-scene DNA (DNA from underneath the victim's fingernails)
DNA from the victim
DNA from suspect 1
DNA from suspect 2

Obtain your DNA samples from your instructor:
Standard DNA (envelope 1)
Crime-scene DNA (DNA from underneath the victim's fingernails) (envelope 2)
DNA from the victim (envelope 3)
DNA from suspect 1 (envelope 4)
DNA from suspect 2 (envelope 5)

Note that in this lab simulation, only a very short sequence of a person's DNA is being used.

If you are working with the Standard DNA sample (envelope 1):
1. Cut out the DNA fragments.
2. You will not need to look for any recognition sites because your section of DNA has already been broken into known sizes (lengths) of DNA (predigested).

If you are working with DNA samples from the crime scene, victim, suspect 1, or suspect 2, then you will need to do the following:
1. Remove the pieces of DNA from your envelope.
2. Attach the strips together using tape so that the asterisks (*) are taped together, making one long piece of DNA.
3. Look at your DNA samples and locate the restriction sites for the *Hind III* enzyme.
4. You will need to find all of the restriction sites on your segment of DNA. Be sure to find the entire sequence: AAGCTT.
5. Using a pencil, draw a line between the first two A/A as shown below.

Identify the sequence A/AGCTT in all of the lengths of DNA. If you did this correctly, you should find eight cuts for each DNA sample.

Restriction Enzyme Digest
1. Using scissors, cut the DNA between the A/AGCTT.
2. Count the number of restriction fragments. (You should have a total of nine fragments for each DNA sample.)

Prepare the Gel
1. Have a member of your group prepare the gel.
2. Obtain a 2' x 3' piece of construction paper.
3. Using a ruler and a black magic marker, label wells for each of the DNA samples as noted in the diagram below, with the wells at the top of the paper.

4. Label the positive and negative ends of the gel.

Electrophoresis of the DNA Fragments

1. Beginning with the standard, simulate running the DNA fragments through the gel by separating each of the fragments of DNA by size.
2. Recall that the largest DNA pieces move more slowly than the smaller DNA pieces. Move the DNA fragments, allowing a small space both at the top and at the bottom of the gel.
3. Simulate running each of the successive DNA fragments through the gel, sorting the DNA fragments according to size.
4. Use the standard DNA fragments as a size reference to determine placement of other DNA fragment samples.

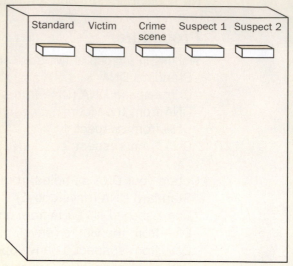

Model for 2' x 3' construction paper

If the DNA fragment contains 10 nitrogen bases and your DNA ladder has a fragment containing 10 nitrogen bases, then these fragments will move through the gel at the same time and should be parallel. If your DNA sample contains 25 nitrogen bases and your DNA ladder has one band at 20 nitrogen bases and another at 30 nitrogen bases, then you know to place your DNA fragment in a location between the 20 and 30 base bands of your DNA ladder.

When you are satisfied that all fragments have correctly moved through the gel, use a glue stick to attach each DNA fragment to the paper gel. If you have two DNA fragments that are the same length but have different bases, place them next to each other on the appropriate line.

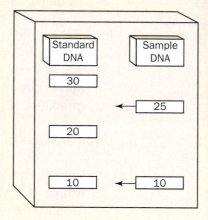

Radioactive Probes

1. Obtain the radioactive probes from your instructor (envelope 6).
2. Cut the probes so that you have fragments of GTA. "Probe" your gel with the radioactive GTA fragments.
3. Anywhere on the gel where a CAT appears, the radioactive GTA fragments will attach to the CAT.
4. Place your GTA probes on top of the CAT in the DNA fragments.

CAT → CAT →
GTA

Data Report:

1. Each team will submit a glued paper model of the DNA fingerprint (one per team).
2. Each student will complete Figure 1 like the one at the top of this page. Refer to the large DNA fingerprinting sheet as a reference.
   - When drawing in the bands for the DNA standard ladder, you will need to count the number of bases within the fragment. For example, if the fragment is GAATCGGCCA, that fragment has 10 bases. So you will align this 10-base band next to number 10. If the band is GTCTA, then you will align this 5-base band next to number 5.

- As you draw in the bands for each of the other four samples of DNA, you need to count the number of bases in each fragment and align the DNA fragment on the gel according to size. (Refer to the DNA standard on the left.)
- In Figure 1, use colored pencils or markers to shade in the area where a DNA fragment would be located. Do not copy the GT(s). Indicate the position of the radioactive probes by placing an X within the DNA fragment. Write the number of repeats in front of the X. (Remember you are probing for the VNTR of "CAT" within each of the fragments.)

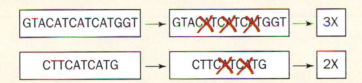

Analysis:

When you have completed "running your gel" and adding the radioactive probes, try to determine if the crime-scene DNA evidence matches any of the other samples of DNA. If the crime-scene DNA matches any of the other DNA samples, then:

- The band patterns and positions will be the same.
- The radioactive probes will be aligned in the same location.
- The number of VNTRs (repeats of CAT) will be the same within each gene.

Questions:

After completing your DNA gel and Table 1, answer each of the following questions:

1. Which DNA samples match? Explain your answer.
2. Does this evidence provide sufficient information for a conviction? Why or why not?
3. Explain why it was necessary to run a sample of the victim's own DNA.
4. Explain how the DNA fragments are separated within the gel.
5. Complete the table on the next page.

Complete the table below:

Materials Used in DNA Fingerprinting	State the Function
Gel	
Restriction Enzymes	
Electric Current	
VNTR	
Radioactive Probes	

Further Study:

1. How are restriction enzymes named? (e.g., *Hind III*, EcoRI)
2. How is the recognition sequence for EcoR1 determined?
3. Lawyers usually cite statistics when analyzing a DNA fingerprint. In the O.J. Simpson trial, the odds that the crime-scene evidence DNA sample belonged to anyone else were extremely low. This was determined by using several different probes on the DNA fingerprint. Dr. Eric Lander of MIT took issue with the method of analysis. Research how this probability is determined and identify Dr. Lander's concerns.

ACTIVITY 7-3
WARD'S DNA FINGERPRINTING SIMULATION

Scenario:

A man was convicted and sentenced to life in prison for murder. The convicted murderer continues to maintain his innocence. After the sentencing, another man confessed to the crime. The lawyer for the convicted man requested a DNA analysis of evidence found at the crime scene to compare to his client's DNA. The lawyer also requested a DNA comparison with the man who confessed to the murder. In this activity, your task will be to perform the DNA analyses and determine whose DNA matches the DNA found at the crime scene.

Objective:

By the end of this activity, you will be able to:
Determine whose DNA was found at the crime scene after performing Gel Electrophoresis

Introduction:

In this activity, a simulated DNA gel electrophoresis is run, and a suspect's DNA is compared to DNA found at a crime scene. Suspect 1 is the convicted murderer, and Suspect 2 is the person who confessed to the crime.

Time Required to Complete Activity:

40 to 90 minutes depending on preparation of group

Materials:

(class of 24, four students per group running 6 gels)
Ward's Natural Science Kit–DNA Detectives 36W6231 (contains pre-digested DNA and loading dye)
3 gel boxes (36W5160), so each gel box can run two gels
6 gel casting trays (36W5172)
12 snap-on end dams (36W5173)
6 dual-sided gel combs (36W5171)
2 power supply boxes, which will accommodate three gel boxes (36W5112)
1 package 10 microliter micropipettes (Wards; 15-3000) with plungers
TBE buffer
digital camera (optional)

Safety Precautions:

Make sure all gel boxes are loaded and covered with TBE buffer before connecting them to the power supply.
Wash your hands before and after handling gels.

Procedure:

These steps need to be revised and renumbered.
 1. Pour a gel. (This step may have already been performed by your teacher.)

2. After the gel has cooled, slowly remove the comb by lifting it straight up.
3. Remove the dams from the end of the gel.
4. Place the gel in the gel box with the wells at the negative end of the box.
5. Cover the gel with TBE buffer. Slowly fill the gel box by adding the buffer to one end of the gel box. Fill the opposite end of the gel box with buffer so that the buffer covers the top of the gel and fills the wells.
6. Using a clean pipette, add 10 microliters of each of the following DNA samples to the wells at the negative end of the gel.

Sample 1	DNA standard marker
Sample 2	Crime-scene DNA
Sample 3	Suspect 1 DNA sample
Sample 4	Suspect 2 DNA sample

7. Turn the power supply boxes to 110 V DC. Run the gels for about 45 minutes or until the blue dye migrates to within about one-half inch from the positive end of the gel.
8. Analyze your DNA gel and determine if the DNA from the crime scene matches the DNA of suspect #1 or of suspect #2.
9. Make a sketch of DNA gel on the diagram below. (If a digital camera is available, take a photograph of the gel.)

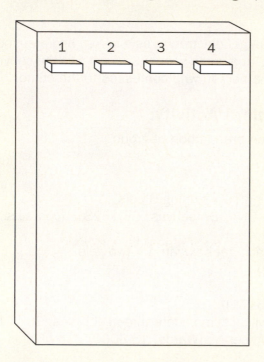

Questions:

1. Which suspect's DNA profile matches the crime-scene DNA evidence?
2. Justify your answer, referring to the DNA fingerprint.
3. Explain what other testing could be done using DNA gel electrophoresis to increase the probability that a person's DNA either matches or does not match the evidence DNA.

ACTIVITY 7-4
WHO ARE THE PARENTS?

Scenario:

Three baby boys were born on the same morning in the same hospital. That morning, the hospital had started using new identification bracelets. When the babies were bathed, the identification bracelets slipped off and the nurses thought a mix-up might have occurred. Given the information from the DNA profile in the diagram, determine which baby belongs to each set of parents.

Objective:

By the end of this activity, you will be able to:
Use DNA profiles to match a child to his parents.

Materials:

3 colored pencils or markers (red, green, blue)
ruler

Safety Precautions:

None

Time Required to Complete Activity: 10 minutes

Procedure:

1. Recall that 50 percent of a child's DNA is obtained from each parent.
2. Use a ruler to align the DNA bands of the baby with any DNA bands of the parents. Determine if any of the parents share the same band of DNA with the babies.
3. There is only one correct set of parents matching a baby.
4. Use colored pencils or markers to circle the band patterns shared by baby and parents. Use red for Baby 1 and his parents, blue for Baby 2 and his parents, and green for Baby 3 and his parents.

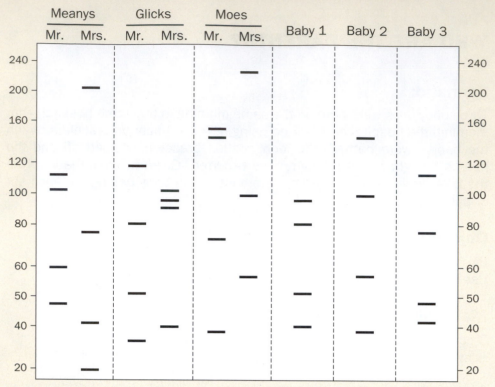

Questions:

1. Which baby belongs to the Meanys?
2. Which baby belongs to the Glicks?
3. Which baby belongs to the Moes?
4. Is it possible for a child to have a DNA band that is not found in the mother's DNA? Explain your answer.

ACTIVITY 7-5
WHICH MAN IS THE FATHER?

Scenario:

Two men are claiming to be the father of the child of a rich heiress who died suddenly without leaving a will. They are both suing for custody of the child. A DNA sample was collected from a hair found in the hairbrush of the dead heiress. Blood samples were collected from each man and the baby. Which man is the father of the child?

Objective:

By the end of this activity, you will be able to:
Use DNA fingerprinting to identify the father of a child (establish paternity) or to prove that a man is not the father of a child (exclude paternity).

Materials:

two colored pencils or markers (red and blue)
ruler

Safety Precautions:

None

Time Required to Complete Activity: 10 minutes

Procedure:

1. Refer to the DNA profiles below Recall that 50 percent of a child's DNA comes from each parent.
2. Use a ruler to help align the positions of DNA band patterns in the diagram of the baby. Determine which DNA bands of the baby were inherited from the mother. Determine if any of the baby's DNA aligns with the DNA band of the two men claiming to be the father.
3. Use colored pencils or markers to circle any band patterns shared by both child and mother. In both profiles, use red for child and mother, and use blue for any band patterns shared by the child and alleged father.
4. Remember: Any band patterns not found in the mother must come from the father.
5. After analyzing the DNA, answer the questions at the end of this activity.

Profile set #1

Ladder

Mother

Child

Alleged father #1

Profile set #2

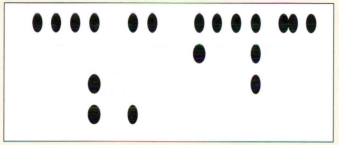

Ladder

Mother

Child

Alleged father #2

Questions:

1. Can either man be excluded as the father? Explain.
2. Which man may be the father of the child? Explain.
3. How many radioactive probes were used in this activity?
4. Is this DNA profile sufficient to establish paternity? Why or why not?

ACTIVITY 7-6
THE BREAK-IN

Scenario:

One afternoon, a break-in occurred at a high school, and several computers were stolen. At the time of the break-in, the building was empty. A motion detector tripped by movement in one of the hallways alerted police. When the police arrived to investigate, they found that one of the doors leading into the school had been propped open with paper wedged into the door-jamb. The door appeared to be locked, but it could easily be pushed open. Near the door, police found a cold soft drink can. Because of the cool temperature of the drink, police suspected that the can was left by one of the intruders.

The can was bagged as evidence, and in the forensics laboratory, a DNA sample was obtained from the lip of the can. The neighborhood was can-vassed, and a clerk in a convenience store remembered selling canned soft drinks to two young males just before the break-in occurred. The surveil-lance video in the convenience store was examined, and the clerk provided the police with the names of all males who were in the store just prior to the break-in. Three suspects were identified from the surveillance video, and blood samples and conventional fingerprints were collected from the suspects.

Introduction:

Using a DNA sample obtained from the soft drink can collected at the crime scene, a PCR was run to amplify the amount of DNA, and then a DNA profile was performed. Cheek swabs were obtained from three suspects, and their DNA was tested. The results are shown below.

Objective:

By the end of this activity, you will be able to:
1. Describe how DNA fingerprinting can be used to identify a suspect.
2. Determine if the suspect's DNA matches the DNA found at the crime scene.

Materials:

colored pencil or marker
ruler

Safety Precautions:

None

Time Required to Complete Activity: 15 minutes

Procedure:

1. Review the steps to the DNA autoradiograph shown in Figure 7-7.
2. A standard DNA ladder of known lengths of DNA has been provided for comparison.
3. Your task is to try to match the crime-scene DNA sample with a DNA samples from three suspects.
4. Use a ruler to check the positions of DNA band patterns in the autoradiograph. DNA from the same source should have band patterns that line up. Do the DNA patterns of any of the suspects' DNA match the DNA pattern of the crime-scene DNA?
5. Use a colored pencil or marker to circle the band patterns shared by the crime-scene DNA and each suspect.

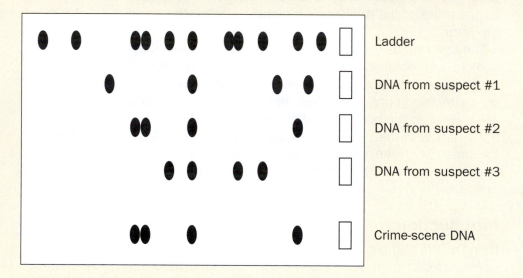

Ladder

DNA from suspect #1

DNA from suspect #2

DNA from suspect #3

Crime-scene DNA

Questions:

1. Does the crime-scene DNA match the DNA from any of the suspects? If so, which one(s)?
2. Is there more than one DNA match? Explain.
3. Is this DNA profile sufficient to convict a suspect? Explain.

ACTIVITY 7-7
INTERNET SEARCH

Background:

The Romanov family ruled Russia for 300 years. In 1918, the last ruling Romanov, Nicholas II, and his family were executed and buried in a mass grave in Siberia. When the grave was exhumed, two of the bodies were missing. In this activity, you will use forensic tools to solve the mystery of the missing Romanovs.

Go to www.dnai.org/d/index.html.
Click on "Applications."
Review Module Two: Recovering the Romanovs and the use of mitochondrial DNA.

Romanov family history:
Delve into the history of the Romanovs, the last imperial family of Tsarist Russia.

The mystery of Anna Anderson:
Meet Anna Anderson, who claimed to be the missing Anastasia Romanov, and compare her features with Anastasia's.

Science solves a mystery:
Find out how DNA science was used to determine whether Anna Anderson was the missing Anastasia Romanov. After reviewing the module, answer the following questions.

Questions:

1. How is it possible that a family with the same genotypes as the Romanovs have no children with hemophilia?
2. How can skeletons be identified?
3. How is mitochondrial DNA different in shape from nuclear DNA ?
4. From which parent is mitochondrial DNA inherited?
5. How was DNA science used to prove the identity of skeletal remains from the Yekaterinburg?
6. What did the comparison of mitochondrial DNA sequences show?
7. Do you think Anna Anderson is Anastasia? Support your answer with the evidence.

CHAPTER 8

Blood and Blood Spatter

BLOOD PAINTS A PICTURE

Concerned neighbors called the police after hearing a woman screaming in the upstairs apartment. When the police arrived, they found the woman with a split lip, bruised cheek, and several facial cuts. Her left eye was swollen shut, and it looked like a black eye was developing. Both of her upper arms were bruised. Behind the woman's chair, blood spatter was evident.

When questioned, the woman said she had been climbing the stairs and had fallen on the last step, striking her head as she fell. Her husband had a small cut on

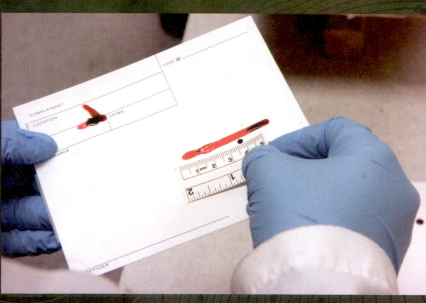

Collecting blood evidence at a crime scene.

©AP Photo/Pat Vasquez-Cunningham

his fist and some blood on the cuff of his shirtsleeve. He told the police that the blood on his shirt was his wife's blood. He explained that the blood must have gotten on his sleeve as he tried to help her wipe the blood from her wound.

The police photographed and noted all of the evidence. However, was the description of the accident consistent with the blood evidence? Did the woman fall on the steps? If the woman had fallen on the steps, then why wasn't any blood found on the steps? The blood-spatter evidence found behind the chair where the woman was sitting indicated that the blood originated from the area of the chair and moved toward the wall. Was the blood on the man's shirt his wife's blood or his own? Were the woman's injuries of the type that would occur by falling on steps, or were they more consistent with a beating from her husband? These are the types of questions a blood-spatter expert tries to answer.

OBJECTIVES

By the end of this chapter you will be able to

✔ Explain the composition of blood.

✔ Describe the functions of blood cells.

✔ Describe a brief history of the use of blood and blood-spatter analysis in forensics.

✔ Describe how to determine the blood type of a sample of blood.

✔ Describe how to screen for the presence of human blood.

✔ Calculate the probability of certain blood types within a population.

✔ Conduct a blood-spatter analysis.

✔ Examine stab wounds and describe the nature of the weapon.

✔ Use blood-spatter evidence to recreate the events at a crime scene.

TOPICAL SCIENCES KEY

BIOLOGY · EARTH SCIENCES · CHEMISTRY · PHYSICS · PSYCHOLOGY · MATHEMATICS

VOCABULARY

agglutination the clumping of molecules or cells caused by an antigen–antibody reaction

antibodies proteins secreted by white blood cells that attach to antigens

antigen–antibody response a reaction in which antibodies attach to specific antigens

antigens any foreign substance or cell in the body that reacts with antibodies

cell-surface protein proteins embedded in the cell membrane

lines of convergence a two-dimensional view of the intersection of lines formed by drawing a line through the main axis of at least two drops of blood that indicates the general area of the source of the blood spatter

point of origin a three-dimensional view formed using lines of convergence and angles of impact of at least two different drops of blood to identify the source and location of blood splatter

red blood cells donut-shaped cells that carry oxygen throughout the body

satellite drop of blood secondary drop formed when some blood breaks free from the main contact drop of blood

white blood cells cells that police the body by destroying foreign materials

INTRODUCTION

Blood left at a crime scene can be analyzed in several ways by a criminal investigator. Blood typing may provide class evidence because more than one person has the same blood type. Because white blood cells contain DNA, it is possible to determine with a high degree of certainty using DNA profiling whether evidence blood left at a crime scene matches the blood of a suspect (or victim).

Blood-spatter evidence can also be used to help recreate a crime scene to validate the information provided by a witness or suspect. By using blood spatter, it is possible to note the direction from which the blood originated, the angle of impact, and the point of origin of the blood. Further examination of the blood drops might indicate if the blood spatter resulted from a high- or low-velocity impact, indicating the type of weapon used to cause the injury.

In this chapter, we will explore how to analyze blood found at a crime scene. By examining blood types and studying blood-spatter patterns, you will learn how a criminalist is able to use this evidence to help solve crimes.

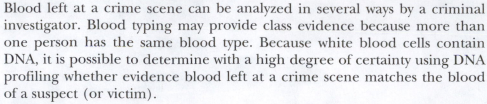

Did You Know?

In 1937, Dr. Bernard Fantus established the first Blood Bank.

BLOOD HISTORY

Blood has been studied in one way or another for thousands of years. The table shown in Figure 8-1 provides a snapshot of the gradual developments in our knowledge of blood.

Figure 8-1. *A chronological history of the study of blood.*

Date	Who	Contribution
2500 B.C.	Egyptians' bloodletting	Effort to cure disease
500 B.C.	Greeks	Distinguished between arteries and veins
175 A.D.	Galen	Established that circulatory paths exist
1628	Sir William Harvey	Noted continuous circulation within body
1659	Antony Leeuwenhoek	Viewed blood cells with microscope
1795	Philip Syng Physick	Performed first blood transfusion (not recorded)
1874	Sir William Osler	Discovered platelets
1901	Karl Landsteiner	Discovered three blood types: A, B, O
1902	Alfred on Decastello	Discovered blood type AB
1922	Percy Oliver	Established blood donor service
1925	various	Discovered that about 80% of human population "secretors"
1935	Mayo Clinic	Developed a method to store blood for transfusions

Date	Who	Contribution
1937	Dr. Bernard Fantus	Established Blood Bank
1900s	Kastle	Developed first presumptive blood test
1940	Karl Landsteiner	Discovered Rh protein
1940	Edwin Cohn	Devised a method to fractionate blood for transfusions
1941	American Red Cross	Organized civilian blood banks for WWII
1943	Dr. Paul Beeson	Described transfusion-transmitted hepatitis
1959	Dr. Max Perutz	Described structure of hemoglobin
1959	Belgian Congo	Recorded first case of AIDS
Late 1960s	Dr. Kenneth Brinkhaus	Produced large quantities of Factor VIII
1971	Dr. Blumberg	Developed method of antibody detection
1984	Dr. Robert Gallo	Identified virus causing AIDS
1985	Various	Developed an ELISA test
1987–2002	Various	Development of blood-screening tests for infectious disease

COMPOSITION OF BLOOD

Blood is a circulating tissue consisting of three types of cells: **red blood cells**, **white blood cells**, and platelets (Figure 8-2). These cells are suspended in a liquid known as plasma. Plasma is similar to salt water in composition. It carries dissolved proteins, such as antibodies, hormones, and clotting factors, and nutrients such as glucose, amino acids, salts, and minerals.

BLOOD CELLS

Each blood cell performs a different function. Red blood cells (erythrocytes) carry respiratory gases, mainly oxygen and carbon dioxide. The hemoglobin in red blood cells is an iron-containing protein that binds to oxygen in the lungs and transports the oxygen to cells in all the tissues in the body. Hemoglobin in red blood cells is also responsible for the red color in blood. White blood cells (leukocytes) fight disease and foreign invaders. Platelets (thrombocytes) aid in blood clotting and are involved in repairing damaged blood vessels.

Our bodies have the ability to discriminate between their own cells and molecules (self) and foreign invaders (non-self). The immune system functions to protect our bodies by identifying cells or molecules that are foreign, such as viruses, bacteria, and other parasites. When the immune system recognizes

Figure 8-2. Pie chart showing the distribution of the various components that make up blood.

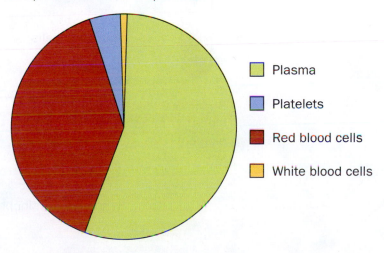

- Plasma
- Platelets
- Red blood cells
- White blood cells

the presence of invading foreign molecules, white blood cells, which migrate throughout the body, concentrate in the location of the invading material—whether it is a virus, bacteria, or protein. The white blood cells engulf and digest the invader. Other white blood cells secrete proteins, known as **antibodies,** which assist in the immune response. Because white blood cells are the only type of blood cell that contains a nucleus, they are the only blood cells that can be used as a source of DNA for DNA profiling. Figure 8-3 shows various types of blood cells.

HISTORY OF DNA PROFILING

In 1982, white blood cells were used as a source of DNA by Dr. Alec Jeffreys to produce the first DNA profile. The first legal case involving DNA evidence is described in a novel entitled *The Blooding* by Joseph Wambaugh. Today, DNA profiling or DNA fingerprinting is widely accepted and is used by such programs as The Innocence Project to help free inmates who have been falsely convicted of crimes (see Chapter 1). The cellular components of blood are shown in Figure 8-4.

Figure 8-3. *Types of blood cells.*

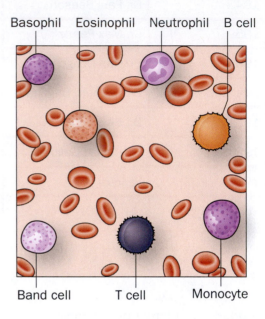

Figure 8-4. *The cellular components of blood. (Cells not drawn to scale.)*

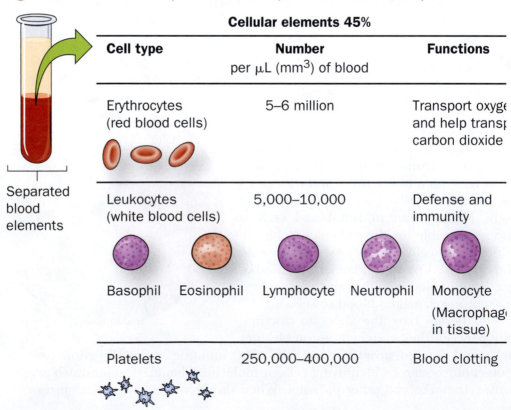

Cellular elements 45%		
Cell type	Number per μL (mm³) of blood	Functions
Erythrocytes (red blood cells)	5–6 million	Transport oxyge and help transp carbon dioxide
Leukocytes (white blood cells)	5,000–10,000	Defense and immunity
Platelets	250,000–400,000	Blood clotting

Separated blood elements

Basophil Eosinophil Lymphocyte Neutrophil Monocyte (Macrophag in tissue)

BLOOD TYPING

Blood typing is less expensive and quicker for analyzing blood evidence than is DNA profiling. Because many different people share the same type, this blood evidence is considered to be class evidence. By typing the blood found at a crime scene, it is possible to link a suspect to a crime scene or to exclude a suspect. However, matching blood types does not prove guilt because many people share the same blood type.

Discovery of Blood Types

In 1900, Karl Landsteiner found that blood from one person did not always freely mix with blood from another person. Instead, clumping might occur, which could result in death. The presence or absence of particular proteins found embedded within the cell or plasma membranes of red blood cells determine a person's blood type. In 1901, Landsteiner described the A and B proteins found on the surface of red blood cells. Other red blood cell proteins, such as the Rh factor, were later identified. The presence or absence of these **cell-surface proteins** gives rise to our present system of blood typing. An antibody reaction test is used to identify each blood type.

A and B Proteins

A and B proteins are found on the surface of some red blood cells (Figure 8-5). If a person's blood contains only protein A, then he or she has type A blood. If the blood has only protein B, then the person has type B blood. If the person's blood has both the A and the B proteins, then he or she has type AB blood. The blood of some people lacks both the A and the B proteins. This blood type is designated as type O blood.

Four main human blood types using the ABO system and the percentage of the U.S. population with that particular blood type include:

Type O (43%)
Type A (42%)
Type B (12%)
Type AB (3%)

Figure 8-5. *A diagrammatic representation of the cell-surface proteins for the different human ABO blood types.*

 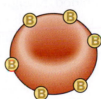 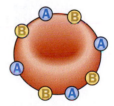 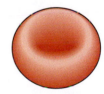

Type A   Type B   Type AB   Type O

Rh Factor

In 1940, Alexander Weiner, working with *Rhesus* monkeys, noticed another type of red cell protein. Eighty-five percent of the human population has a protein called Rh factor on their red blood cells. Blood that has the Rh factor is designated Rh+ (positive), while blood that does not have this factor is designated Rh− (negative) (Figure 8-6).

Naming of Blood Types

A person's blood type is based on the presence or absence of the AB and Rh proteins.

The earliest performed blood transfusions sometimes helped the recipient of the blood, but many times the transfusion killed the person. In the early days of transfusion, doctors were not aware that people have different blood types. The presence of different blood types was not discovered until 1901. When a person receives a blood protein that is foreign to him or her, antibodies will cause the blood to clump and may cause death.

Figure 8-6. *Rh factor and ABO blood type examples.*

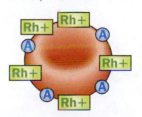 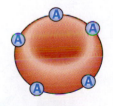

Protein A present
Rh protein present
Type A+

Protein A present
Rh protein absent
Type A−

Antibodies

To help white blood cells identify foreign proteins, B-lymphocytes, specialized type of white blood cells, secrete antibodies. The antibodies are Y-shaped protein molecules that bind to the molecular shape of an **antigen**, fitting like two complementary puzzle pieces. The binding site of the antibody is located on the tip of the Y-shaped molecule (Figures 8-7 and 8-8). The antibody recognizes a foreign substance as an invader and attaches to it.

Figure 8-7. *The general structure of an antibody.*

Figure 8-8. *The shape of the antibody fits the bonding site on the foreign protein or antigen.*

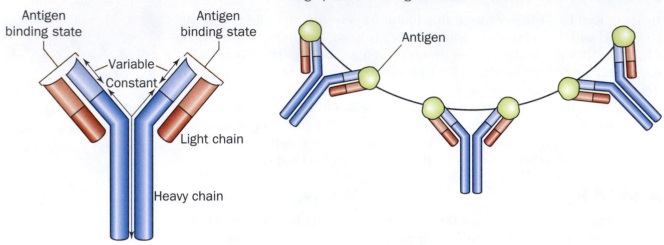

Antigen binding state

Antigen binding state

Variable
Constant

Light chain

Heavy chain

Antigen

Antigen–Antibody Response

When a foreign invader is recognized by the immune system, an attack is launched against that invader. This is called the **antigen–antibody response.** White blood cells recognize a substance as foreign and try to destroy it. The foreign invaders may be viruses, bacteria, or even the red blood cell proteins from a person with a different blood type.

For example, if a person with type A blood receives a transfusion containing type B blood, foreign type B antigens alert the immune system to the presence of an invader. The body responds in two ways: (1) B lymphocyte makes specific antibodies against that protein; these antibodies are Y-shaped and have specific binding sites that attach to foreign proteins; and (2) phagocytes, a type of white blood cells, engulf the invader.

Agglutination

There are more than 300 known blood group proteins and more than 1 million different protein-binding sites on each red blood cell. Many of these proteins are on the surface of the cell membrane. When one arm of the Y-shaped antibody attaches to the red blood cell, the second arm of the Y attaches to another red blood cell. The result is **agglutination**, or clumping, of the red blood cells (Figure 8-9).

If this clumping occurs within the circulatory system of a person receiving a blood transfusion, blood could cease to flow. The blood vessels will become obstructed by clumped red blood cells. Without blood circulation,

Figure 8-9. *An antibody reaction to surface proteins on red blood cells causes agglutination, or clumping, of the cells.*

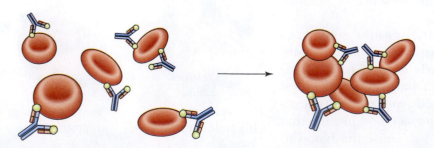

the cells of the body cannot receive oxygen or eliminate carbon dioxide, and the person will die.

Blood Typing Tests

Blood typing is a way to identify and match blood samples. When a patient needs a blood transfusion, his or her blood needs to be typed to ensure that the transfused blood does not contain any foreign blood proteins. The person's blood is tested for the presence of three red blood cell proteins: A, B, and Rh.

Three separate tests are performed. The patient's blood is mixed with antibodies that bind to the A protein. If the patient's blood clumps or agglutinates, that means that the person's blood contains protein A. If the blood being tested does not react with the antibodies that bind the A protein, then that person's blood does not contain protein A.

Similar testing is done with antibodies to protein B and the Rh factor. If the blood clumps or agglutinates in the presence of these antibodies, it means that those proteins are present in the blood. In the absence of these blood proteins, no clumping will occur (Figure 8-10).

Figure 8-10. Notice that agglutination occurs in only two of the blood tests, indicating that the person has blood type B+ (clumping with anti-B and with anti-Rh antibodies).

| Antibodies A | Antibodies B | Antibodies C |

Additional Blood Proteins and Probability

Many other blood proteins have been identified. The presence or absence of these proteins is inherited from our parents. In addition to A, B, and Rh protein, people can have M proteins and N proteins.

Approximately 30 percent of the population is pure MM, inheriting a gene for the M protein from each parent. Twenty-two percent of the population is pure NN, inheriting a gene for the N protein from each parent. The remaining 48 percent of the population have MN, inheriting a gene for M from one parent and a gene for N from the other parent (Figure 8-11).

Figure 8-11. A summary of the population percentages having different blood proteins and enzymes in the United States.

ABO

Type	Percent	Fraction
A	42%	42/100
B	12%	12/100
AB	3%	3/100
O	43%	43/100

MN

Type	Percent	Fraction
MM	30%	30/100
MN	48%	48/100
NN	22%	22/100

Rh

Type	Percent	Fraction
Rh +	85%	85/100
Rh–	15%	15/100

Additional enzymes and proteins have been found in the blood, which are important for identification purposes. They include:

1. Phosphoglucomutase (PGM)
2. Adenylate kinase (AK)
3. Adenosine deaminase (ADA)
4. Esterase D (EsD)
5. Glucose-6-phosphate dehydrogenase (G-6-PD)
6. Polymorphic proteins: Group-specific Components (Gc) and haptoglobins (Hp)

PROBABILITY AND BLOOD TYPES

Given the frequency of different genes within a population, it is possible to determine the probability or chance that a particular blood type will appear within a particular population. To determine the probability of two separate events, it is necessary to multiply their individual probabilities.

Example 1: What is the chance of throwing dice and getting two sixes?

The probability of one die showing a six is $\frac{1}{6}$ (one side out of six).

The probability of the other die showing a six is $\frac{1}{6}$ (one side out of six).

So, the chance of throwing two dice and getting both showing a six is calculated by multiplying the individual probabilities:

$\frac{1}{6} \times \frac{1}{6} = \frac{1}{36}$, or one chance of throwing two sixes out of 36 possible combinations!

Try to estimate the probability that an individual will have a particular blood type. Refer to the preceding charts to determine the individual probability of a particular blood protein. Here are some examples to guide you.

Example 2: What percentage of the population would have A+ blood?

Type A blood = 42% of the population

Rh+ = 85% of the population

Step 1. Convert the percentages to decimals.

Type A blood = 42% = 0.42 of the population

Rh+ = 85% = 0.85 of the population

Step 2. Multiply the decimals.

$0.42 \times 0.85 = 0.357$ of the population should be both A and Rh+.

Step 3. Multiply by 100 to convert the decimal to a percentage.

$0.357 \times 100\% = 35.7\%$ of the population should be both A and Rh+.

Therefore, about 36 out of every 100 people would have Type A+ blood.

Example 3: What percentage of the population would have the following combination of blood-type proteins? Type O–, MN

Step 1. Convert the percentages to decimals.

Type O = 43% = 0.43

Rh– = 15% = 0.15

MN = 48% = 0.48

Step 2. Multiply the decimals.

$0.43 \times 0.15 \times 0.48 = 0.031$

Step 3. Multiply by 100 to convert the decimal to a percentage.

$0.031 \times 100\% = 3.1\%$ of the population are O– MN. Therefore, only 3 out of every 100 people would have Type O– MN blood. This would make the suspect population very small.

By identifying the additional proteins in the blood, we can limit the size of our suspect population and help identify a suspect. For example, analysis of

Digging Deeper
with Forensic Science e-Collection

How are new blood-typing tests developed? What types of new blood-typing tests are being used? How is the accuracy of new blood-typing tests checked? Go to the Gale Forensic Science eCollection on school.cengage.com/forensicscience and enter the search terms "blood typing" and then click on Academic Journals and read about some of the problems and successes associated with new blood-typing tests. Write a brief essay describing your findings, and be sure to cite your resources.

blood found at a crime scene resulted in data with the following characteristics: Type A, N, Hp-1, Rh−, PGM-2. What is the probability of someone having all of these blood proteins?

The probability of finding another person in the population with the same blood type can be calculated by finding the product of the individual probabilities:

Type A × N × Hp-1 × Rh− × PGM-2 = occurrence of this combination in the population. Substituting, 0.42 × 0.22 × 0.14 × 0.15 × 0.06 = 0.000116, or about 1 in every 8,600 people should have this combination of blood-type proteins. By testing for more blood-type proteins, the probability for uniqueness continues to increase, and the number of other people with the same combination as our suspect decreases.

BLOOD SPATTER

When a wound is inflicted and blood leaves the body, a blood-spatter pattern may be created. A single stain or drop of blood does not constitute a spatter. Instead, a grouping of bloodstains composes a blood-spatter pattern. This pattern can help reconstruct the series of events surrounding a shooting, stabbing, or beating.

HISTORY OF BLOOD-SPATTER ANALYSIS

In 1894, Pitoroski wrote the earliest reference to blood-spatter analysis. In 1939, Balthazard was the first researcher to analyze the meaning of the spatter pattern. In 1955, blood-spatter evidence was used by the defense in the Sam Shepard case, helping to exonerate him. In 1971, Dr. Herbert MacDonnell used blood-spatter analysis as a tool in modern forensic examinations. Today, blood-spatter evidence is used to explain events at a violent scene.

BLOOD-SPATTER ANALYSIS

In the laboratory activities in this chapter, you will study how blood-spatter patterns can be used to recreate a crime scene. Given blood-spatter patterns, it is possible to determine the direction the blood was traveling, the angle of impact, and the point of origin of the blood. Blood-spatter patterns can help determine the manner of death, based on the blood velocity. Instructions on blood-spatter analysis are provided within each activity.

Did you ever wonder why blood forms droplets as it falls from a wound? If blood is a mixture, then why doesn't it separate in the air before it hits the ground or an object? Why does a drop of blood have a curved surface when it lands on a flat surface instead of spreading out flat? The answers to these questions have to do with what happens when the forces of gravity, cohesion, adhesion, and surface tension act on blood.

Recall that blood is a thick mixture of blood cells and plasma. When a person is injured and is bleeding, gravity acts on blood, pulling it downward toward the ground (Figure 8-12). The blood droplet has a tendency to become longer than it is wide as a result of gravity (Figure 8-13). Blood is cohesive. This means that the blood mixture is attracted to similar blood mixtures and tends to stick together and not separate as it falls (Figure 8-14).

Figure 8-12. A falling droplet of blood.

Figure 8-13. The effect of gravity on blood.

Figure 8-14. The cohesive forces in a blood droplet.

Figure 8-15. *Cohesive forces resist droplet flattening.*

The effect of the downward force of gravity combined with the cohesive force of the blood results in a net effect on the blood droplet as it falls. Thus, the blood maintains a circular or round appearance.

When a drop of blood falls on a flat surface, the blood drop will have a curved surface. The blood drop does not totally flatten out (Figure 8-15). The reason for this shape is the cohesive nature of blood causing the blood to pull together and resist flattening out on a surface. The result is that the surface of the blood is elastic, giving the top of the blood spatter a spherical appearance.

If any of the blood does overcome cohesion and separate from the main droplet of blood, it will form small secondary droplets known as **satellites** (Figure 8-16).

Figure 8-16.

Note the smaller satellites are not attached to the main drop of blood but have broken free

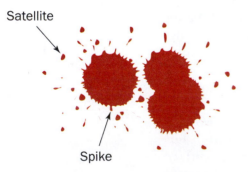

If blood is dropped onto a smooth surface, such as glass or marble, the edge of the blood drop appears smooth and circular. However, if the blood lands on a porous surface, such as wood or ceiling tile, then the edge of the drop of blood may form small spikes or extensions (Figure 8-17). Notice that spikes are still connected to the main droplet of blood, whereas satellites are totally separated.

In 1902, Dr. John Glaister first described the six patterns into which blood spatters could be classified. They include:

1. Blood falling directly to the floor at a 90-degree angle will produce circular drops, with secondary satellites being more produced if the surface hit is textured. This is known as a passive fall.

2. Arterial spurts or gushes typically found on walls or ceilings are caused by the pumping action of the heart.

3. Splashes are shaped like exclamation points. The shape and position of the spatter pattern can help locate the position of the victim at the time of the attack.

4. Smears are left by a bleeding victim depositing blood as he or she touches or brushes against a wall or furniture.

5. Trails of blood can be left by a bleeding victim as he or she moves from one location to another. The droplets could be round or smeared or even appear as spurts.

6. Pools of blood form around a victim who is bleeding heavily and remains in one place. If the bleeding victim moves to another location, there may appear to be droplets or smearing connecting the first location with a second.

Figure 8-17. *Satellites and spikes in a blood drop pattern.*

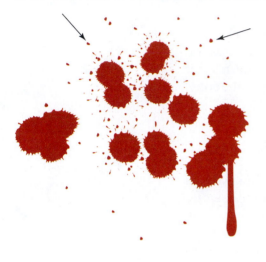

Satellite

Spike

The size and shape of blood droplets help identify the direction from which the blood originated. Round droplets, for example, are caused by blood dripping downward at a 90-degree angle. Blood droplets with tails or satellite droplets help us determine the direction from which the blood originated.

Spatter patterns can help the investigator determine the type of wound. A fine-mist spatter pattern is produced by a high-velocity impact, such as a gunshot wound. A beating with a pipe will produce blood cast off with a lower-velocity pattern. Voids (empty spaces) in the spatter pattern could help determine the presence of a person or object moved after the attack.

By using the spatter pattern to determine the angle of impact of various blood droplets, the examiner can determine the point of impact or convergence, a two-dimensional representation of the location of the victim at the time of the injury.

The **point of origin** can next be determined by the mathematical relationship between the width and length of the blood droplets. These relationships will be addressed in the activities section.

EXAMINATION OF DIRECTIONALITY OF BLOOD

The shape of an individual drop of blood provides clues to the direction from where the blood originated. A circular drop of blood (width and length are equal) indicates that the blood fell straight down (90-degree angle of impact). This is typical if the blood was passively produced (without any force). This would be typical of blood dripping from a wound.

When a blood drop is elongated (longer than it is wide), it is possible to determine the direction the blood was traveling when it struck a surface.

As moving blood strikes a surface, several forces affect the droplet of blood. These forces are cohesion, adhesion, and surface tension. *Cohesion* is a force between two similar substances. *Adhesion* is a force between two unlike surfaces, such as blood and the surface of a wall. *Surface tension* is an elastic characteristic along the outer edge of a liquid caused by the attraction of like molecules.

When blood comes into contact with another surface, the blood tends to adhere or stick to it. As a result, the point of impact may appear to be darker and wider than the rest of the drop of blood spatter (Figure 8-18).

Figure 8-18. *Pattern of cast-off blood.*

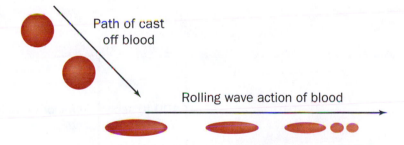

Path of cast off blood

Rolling wave action of blood

Momentum tends to keep the blood moving in the direction it was traveling. As it travels, some of the blood adheres to the new surface. However, because of cohesion, most of the blood tends to remain as one drop. As blood droplets move away from their source, the blood droplet elongates and may produce a thinner tail-like appearance. The tail points in the direction of blood's movement. Smaller satellite or secondary droplets may break away from the main drop of blood. These satellites will appear in front of the moving droplet of blood (Figure 8-19). Note that satellites are not connected to the main drop of blood.

Figure 8-19. *Satellite drops.*

Direction blood is traveling

LINES OF CONVERGENCE

The location of the source of blood can be determined if there are at least two drops of blood spatter. By drawing straight lines down the long axis of the blood spatter and noting where the lines intersect, this will indicate the **lines of convergence** (Figure 8-20). When there are numerous blood spatters, the area where the lines of convergence meet is where the source of blood originated. One can draw a small circle around this intersecting area to note the area of convergence. The circle locates the area of convergence and identifies in a two-dimensional view the location of the source of the blood.

Figure 8-20. Lines of convergence.

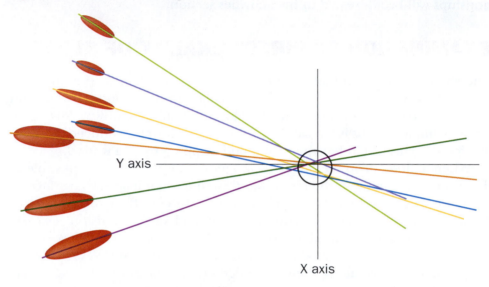

Y axis

X axis

BLOOD SPATTER TYPES

Blood spatter may also be classified by its speed or velocity on impacting a surface (Figure 8-21).

Figure 8-21. Table of blood-spatter parameters.

Velocity	Size of Droplets (mm)	Visual Image	Velocity of Blood	Examples of Injuries
High	Less than 1		100 ft/sec.	Gunshot wounds
Medium	1–4		25 ft/sec.	Beating, stabbing
Low	4–6		5 ft/sec.	Blunt object impact

All photos ©Cengage Learning

CRIME-SCENE INVESTIGATION OF BLOOD

In an attempt to hide evidence, a perpetrator may try to remove blood evidence by cleaning the area. Although a room may look perfectly clean and totally free of blood after a thorough washing of the walls and floor, blood evidence still remains. Red blood cells contain hemoglobin, the iron-bearing protein that carries oxygen. To detect hemoglobin, an investigator mixes Luminol powder with hydrogen peroxide in a spray bottle. The mixture is then sprayed on the area to be examined for blood. The iron from the hemoglobin, acting as a catalyst, speeds up the reaction between the peroxide and the Luminol. As the reaction progresses, light is generated for about 30 seconds on the surface of the blood sample.

Luminol can also be used to detect blood in darkened rooms or at night. Even with the most thorough of cleaning, blood residue is difficult to remove. Traces of blood may adhere to surfaces for years, even after cleaning. Luminol works best in an area where an attempt has been made to clean up the blood.

Once found, there are several steps used in processing a bloodstain, and each can provide a different kind of critical information:

1. **Confirm the stain is blood.**

At a crime scene, clothing is found with a red stain. It appears to be blood, but is it? Could ketchup, ink, or any other red substance cause the red stain? Before trying to collect the blood, it is first necessary to confirm that the evidence is blood. There are two tests:

- Kastle-Meyer test. If blood is present, a dark pink color is produced.
- Leukomalachite green. This chemical undergoes a color change, producing a green color in the presence of blood.

2. **Confirm the blood is human.**

Someone is accused of leaving the scene of a crime. The police later question the owner of a car that fits the description provided by witnesses. Blood is located under the bumper of the car. When asked about this blood, the owner of the car said he hit a dog a few days ago. To confirm his story, the crime-scene investigators are able to perform tests to determine if the blood came from a human.

One test known as an ELISA test (Enzyme Linked Immunosorbent Assay) involves an antibody–protein reaction. This test is similar to one used in blood typing but uses different antibodies. Human blood is injected into a rabbit or some other animal to produce antibodies against human blood. These antibodies are isolated and stored. When a sample of human blood is mixed with some of these anti-human antibodies produced by the rabbit or other animal, a positive reaction will occur, and the presence of human blood is confirmed.

Digging Deeper
with Forensic Science e-Collection

In what kinds of cases are blood-spatter patterns used? Search the Gale Forensic Science eCollection on school.cengage.com/forensicscience using the terms "spatter" and "case" or "evidence." Find articles about specific cases in which blood spatter played a role in the evidence used to solve the case. Write a brief summary of two of the cases you find, pointing out how the blood-spatter evidence was used and what it showed.

3. Determine blood type.

Blood collected from a crime scene is tested using specific antibodies. The person's blood type is determined by examining antigen–antibody reactions. Remember, the resulting match is considered class evidence. However, if the blood does not match, then a particular person may be excluded as a suspect. Depending on the circumstances, blood typing may not be done at all, just DNA analysis.

SUMMARY

- Blood consists of cellular components and plasma containing dissolved ions, proteins, and other molecules.

- Blood types result from the presence of proteins on the surface of red blood cells and vary among individual humans. This variation can be used to exclude blood samples from belonging to a group of individuals.

- Other proteins in the blood, such as enzymes, also show variation and can be used as class evidence.

- Combinations of specific blood and enzyme types can narrow the possible source of a blood sample to a fairly small group of individuals.

- Blood-spatter analysis can be used to recreate a crime scene.

- The angle of impact is calculated as the arc sin of width/length.

- The characteristics of blood drops on surfaces can show how the blood was deposited, at what rate the blood was moving, the location of the origin or source of blood, and the direction in which the blood was moving when it struck the surface.

- Crime-scene investigators use several tests to locate and identify blood at a crime scene, including visualization with Luminol and identification with the Kastle-Meyer test.

CASE STUDIES

Ludwig Tessnow (1901)

The dismembered bodies of eight-year-old Hermann and his six-year-old brother, Peter, were found in the woods near their home. Ludwig Tessnow, a local carpenter, was a suspect. Both his clothing and boots had dark stains on them. Tessnow told officials the stains were from wood dye, which he used in his carpentry work.

Three years earlier in Griefswald, similar murders had taken place. Two young girls, ages seven and eight, had been murdered in a similar manner. Their bodily remains had been found in a nearby wooded area. Ludwig Tessnow had been detained for questioning in that murder as well.

A German biologist, Paul Uhlenhuth, developed a test that could be used to differentiate between blood and other types of stains. His test could also discriminate between human and animal blood. The test he developed used an antigen–antibody reaction to test for the presence of human blood.

Uhlenhuth examined the boots and clothing belonging to Tessnow and concluded that the clothing did contain wood dye as Tessnow has claimed. Seventeen spots of human blood were identified, as well as several stains of sheep's blood. Based on this evidence, Tessnow was found guilty and executed at Griefswald Prison.

Thomas Zeigler (1975)

On Christmas Eve in 1975, Tommy Zeigler was found shot in his store. His wife, her parents, and a citrus worker were also found dead. Tommy claimed they were all victims of a gang that had attempted to rob the store. Blood evidence was found throughout the store. Herbert MacDonnell, a blood-spatter expert, was called to examine the crime scene. He spent many hours reconstructing the events of the crime scene. Blood-spatter patterns and trajectory velocities helped explain the chain of events leading to each murder. Zeigler was given a life sentence. Zeigler has since filed a request for DNA testing. Judge Whitehead rejected his plea, stating that DNA evidence would not exonerate him from any of the shootings.

Graham Backhouse (1985)

Graham Backhouse was accused of killing his neighbor and attempting to kill his wife to collect his wife's life insurance. Backhouse's wife had been injured in an explosion of a homemade car bomb. Backhouse claimed his neighbor had a grudge against him, and the bomb was really intended for him.

When police arrived at the Backhouse home, the neighbor, Colyn Bedale-Taylor, was found dead from shotgun wounds. Backhouse had sustained wounds to his chest and face. He claimed self-defense. The blood-spatter evidence contradicted Backhouse's statement. The blood spots on the kitchen floor were those made by dripping blood, circular in appearance and not what would have been produced by a violent encounter as Backhouse related. His wounds appeared to be self-inflicted. Chairs and furniture had been overturned on top of the blood spots, indicating a staged crime scene. Backhouse was convicted of the murder of his neighbor and the attempted murder of his wife.

Think Critically Select one of the Case Studies and imagine you can interview the forensic scientist who studied the blood evidence. Write the questions and answers from the interview. Be sure your questions demonstrate what you have learned about blood and blood-spatter evidence.

CAREERS IN FORENSICS

Michael Baden

Dr. Michael M. Baden is a Board-certified forensic pathologist and medical doctor. After earning a Bachelor of Science degree from the City College of New York, he attended New York University School of Medicine, where he was awarded a medical degree in 1959. He trained in internal medicine and pathology at Bellevue Hospital, and then worked in the Office of the Chief Medical Examiner in New York City from 1961 to 1986, serving as the Chief Medical Examiner from 1978 to 1979. He was also the Deputy Chief Medical Examiner for Suffolk County in New York from 1981 to 1983. Dr. Baden has held professorial appointments at Albert Einstein Medical School, Albany Medical College, New York Law School, and John Jay College of Criminal Justice.

Dr. Michael Baden, forensic pathologist and medical doctor.

©AP Photo/Robyn Beck

Dr. Baden has been a practicing medical examiner for 45 years and has performed more than 20,000 autopsies. He was the chairperson of the Forensic Pathology Panel of the U.S. Congress Select Committee on Assassinations that investigated the deaths of President John F. Kennedy and Dr. Martin Luther King, Jr.

During his career, Dr. Baden has published articles on aspects of forensic medicine in several national and international medical journals. He has also written two nonfiction books, *Unnatural Death, Confessions of a Medical Examiner* and *Dead Reckoning, the New Science of Catching Killers.*

Dr. Baden has appeared as an expert in forensic pathology in high-profile cases of national and international interest. He served as an expert witness in the O.J. Simpson trial. Dr. Baden was the forensic pathologist member of a team of U.S. forensic scientists asked by the Russian government to examine the remains of Tsar Nicholas II of Russia and other members of the Romanov family found in Siberia in the 1990s.

In addition to practicing forensic pathology, Dr. Baden has served on the boards of directors of several drug abuse and alcohol abuse treatment programs. Dr. Baden calls himself "a witness for the dead.'' Finding spirituality on the autopsy table, he says, "It's a sacred place, and I always treat a body with the utmost care. I'm not religious, but when I'm looking inside a person's body, which is the first time a human being has done that, it's a wondrous thing. And it never fails to convince me that each one of us is a miracle.

Learn More About It

To learn more about the work of blood-spatter experts, go to school.cengage.com/forensicscience.

Multiple Choice

1. Blood types are determined by the presence of protein located on
   a) all of the blood cells
   b) only the white blood cells
   c) only on the T-helper cells
   d) only on the red blood cells

2. Blood proteins that determine blood types are
   a) on the surface of the cell membrane
   b) inside of the cytoplasm
   c) both on the cell membrane and inside of the cytoplasm
   d) found in the bone marrow

3. If a person has type A– blood, then they have
   a) only the A protein
   b) both the A and the Rh proteins
   c) all three blood proteins
   d) It is impossible to tell what proteins they have

4. If a person has type O+ blood, then they have
   a) the A and the B protein, but lack the Rh protein
   b) an O protein but not the Rh protein
   c) none of the ABO nor the Rh proteins
   d) the Rh protein but not the A or the B proteins

5. To determine one's blood type, what is added to the slide?
   a) blood and antibodies
   b) only the antibodies
   c) luminol
   d) none of the above

Short Answer

For the following questions, determine the blood type being tested. State
if the person is type A, B, AB, or O. Be sure to indicate if the person is
Rh+ or Rh– for each blood test shown below.

1. Type _____

Antibodies A   Antibodies B   Anti-Rh antibodies

2. Type _____

Antibodies A   Antibodies B   Anti-Rh antibodies

Blood spatter

Refer to the pictures of blood spatter below:

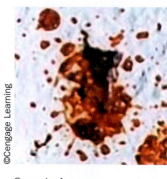

©Cengage Learning

Sample A

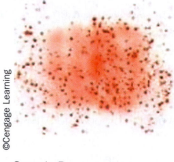

©Cengage Learning

Sample B

3. Which pattern most likely came from a gunshot and which one was produced from a blow on the head? Explain your answer.

_____

_____

4. Blood was dropped from a distance of 30 cm. Although the height was the same, two different patterns emerged. How do you account for the difference?

_____

_____

5. The following bloodstains were produced by dropping the blood from heights of 6 inches and 12 inches.

Drop 1   Drop 2

- Which one was dropped from 12 inches: drop 1 or drop 2? _____
- How do you account for the different patterns?

_____

_____

6. Explain how it is possible to tell that this blood was emitted from an artery and not just a cut.

_____

_____

7. How is it possible to determine the direction of blood flow from blood spatter?

_____

_____

8. Determine the point of origin for the blood-spatter stains.
   - Use a pencil and a straight edge.
   - Draw a circle around the point of origin.

9. A crime has been committed. From the blood spatter, it is possible to determine:

- Angle of impact for a drop of blood.

- Point of convergence for several drops of blood.

- Distance from the point of convergence and a blood drop (X).

- Point of origin, the source of the blood.

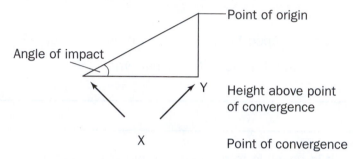

10. A prosecuting attorney was given the forensic lab results from the blood-typing activity. She found that the crime-scene blood matched the blood type of one of the suspects' blood. The prosecutor told the jurors that the crime-scene evidence blood matched the defendant's blood type. She further stated that this evidence proved that the suspect was indeed the perpetrator of the crime.

You are the defendant's attorney. How would you respond to the prosecutor's argument that because the blood types matched, then the defendant is proven to be guilty of the crime?

_____

_____

_____

_____

_____

_____

_____

Bibliography

Books and Journals

Bevel, Tom, and Ross M. Gardner. *Bloodstain Pattern Analysis*, 2nd ed. Boca Raton, FL: CPC Press, 2002.

James, Stuart H., and William G. Eckert. *Interpretation of Bloodstain Evidence at Crime Scenes*, 2nd ed. Boca Raton, FL: CFC Press, 1999.

Laber, Terry L., and Barton P. Epstein. *Bloodstain Pattern Analysis*. Minneapolis, MN: Callan Publishing, 1983.

Lewis, Alfred Allen. *The Evidence Never Lies*. New York: Dell Publishing, 1989.

MacDonnell, H. L. *Bloodstain Pattern Interpretation*. Corning, NY: Laboratory of Forensic Science, 1983.

Russell, Peter J., Stephen L. Wolfe, Paul E. Hertz, and Cecie Starr. *Biology: The Dynamic Science*. Belmont, CA: Thomson Higher Education, 2008.

Solomon, Eldra, Linda Berg, and Diana W. Martin. *Biology*, 8th ed. Belmont, CA: Thomson Higher Education, 2008.

Wonder, A. Y. *Blood Dynamics*. New York: Academic Press, 2002.

Web sites

Gale Forensic Sciences eCollection, school.cengage.com/forensicscience.

Schiro, George. "Collection and Preservation of Blood Evidence from Crime Scenes," http://www.crime-scene-investigator.net/blood.html.

http://www.fbi.gov/publications/leb/2005/feb2005/feb2005.htm

http://www.bloodspatter.com/bloodspatter.pdf

http://www.crimelibrary.com/criminal_mind/forensics/serology/8.html

http://www.albany.edu/writers-inst/turoach_baden.html

http://www.practicalhomicide.com/bio/bioBADEN.htm

http://www.shsu.edu/~chm_tgc/JPPdir/JPP1999

http://www.bobaugust.com/answers.htm#no11

http://www.nifs.com.au/FactFiles/dynamicblood/case.asp?page=case

ACTIVITY 8-1
A PRESUMPTIVE TEST FOR BLOOD

Objective:

By the end of this activity, you will be able to:
Use the Kastle-Meyer Presumptive Blood Test to determine if a given stain contains blood.

Scenario:

At dusk, a young man was seen riding his bicycle along the narrow, winding road that encircled a lake. On that same quiet road, an animated young couple was seen speeding around the many turns in the road. They were in a hurry to arrive at a friend's party at the lake. In his haste, the driver of the car lost control of the car while trying to swerve to avoid the young man on the bike. Unfortunately, the biker was hit from behind, causing him to be knocked off his bike and to fall down the hill toward the lake. The young driver of the car panicked and, without looking back, fled the scene of the crime. Although somewhat injured, the biker was able to provide a description of the car to the police.

Later that evening, the police arrived at the party on the lake to question the owner of the car. Although there was no visible blood on the car, the police did find a red stain on the car's bumper. At first the young driver said it was red paint or perhaps the blood from a squirrel he had struck the day before. Is the stain blood or paint? If it is blood, is it possible to distinguish human blood from animal blood? In this activity you will perform tests that confirm the presence of blood.

Time Required to Complete Activity: 45 minutes

Materials:

Activity Sheets for Activity 8-1
ketchup (1 oz or 10 ml)
blood from animal source (1 oz or 10 ml)
cloth or shirt with dime-sized blood stain
cloth or shirt with dime-sized ketchup stain
20 mL 3% hydrogen peroxide solution in dropper bottle
20 mL 95% ethyl alcohol in dropper bottle
20 mL distilled water in dropper bottle
20 mL 2% phenolphthalein solution in dropper bottle
biohazard container
latex or nitrile gloves

Safety Precautions:

Wear protective gloves.
Dispose of all samples in a biohazard container provided.
Assume that all red solutions are blood and handle according to safety regulations.

All procedures should be done while wearing protective gloves.

Do not contaminate any of the reagents.

Be sure to drop the solutions onto the cotton swab without touching the cotton swab.

Return the caps of all reagent bottles to the correct reagent bottle. Do not switch the caps from one bottle to the other.

Background:

Dried drops of red fluid found on a murder weapon, clothing, or automobile are noted, photographed, and analyzed. Did blood cause the red stain? If the red stain is not blood, then valuable time and money can be saved by not sending the red stain in to the laboratory for further testing. If it is determined that the red stain is blood, then further testing needs to be ordered to identify the source of the blood as human or animal.

A sample of the stain is tested using a presumptive chemical reagent. The Kastle-Meyer test is a catalytic color test that will produce a color change in the presence of blood. When phenolphthalein and hydrogen peroxide react with heme (iron) molecules in hemogloblin, the presence of blood is indicated by a pink color. A negative Kastle-Meyer test indicates the absence of blood. Because animal blood also contains heme molecules, it will also give a positive result. If animal blood is present and pertinent to the case, then additional tests can be performed to determine what type of animal blood is present.

Procedure:

1. Obtain a section of cloth that contains a known bloodstain (positive control). Before testing any unknown stains, it is important to check all reagents on a known sample of blood. If you do not get the expected results on blood, then you know that your reagents were malfunctioning and you need to replace them.

2. Wet a cotton swab with four drops of distilled water and gently rub the wet swab on the known bloodstain.

3. Drop two drops of ethyl alcohol onto the swab (don't allow the dropper to touch the swab).

4. Drop two drops of the phenolphthalein solution onto the swab (don't allow the dropper to touch the swab).

5. Drop two drops of the hydrogen peroxide onto the swab (don't allow the dropper to touch the swab).

6. A positive pink color will appear within seconds if blood is present. Record your results in Data Table 1. (This is your positive control demonstrating what happens when blood is present.)

7. Using a permanent marker, record your initials next to the stain that was just tested.

8. Dispose of all used cotton swabs and bloodstain samples into the Biohazard Waste container.

9. Using clean cotton swabs, repeat steps 1 to 8 using the shirt containing the ketchup stain. (This is your negative control since ketchup is not blood.)

10. Record the appearance of the ketchup test in Data Table 1.

11. With fresh cotton swabs, repeat steps 1 to 8 on the section of the shirt containing the unknown stain 1.
12. Using fresh cotton swabs, repeat steps 1 to 8 on the section of the shirt containing the unknown stain 2.

Data Table 1: Table of Test Results

Stains	Color Pink or not pink?	Describe Your Observations Is it blood or not blood?
Blood stain (positive control)		
Ketchup (negative control)		
Unknown 1		
Unknown 2		

Questions:

1. Complete the following Data Table indicating the role or function of each of the chemical reagents used in this experiment.

Data Table 2: The Role of Chemical Reagents in Blood Sample Analysis

Chemical	Function
1. Distilled water	
2. Ethyl alcohol	
3. Phenolphthalein	
4. Hydrogen peroxide	

2. Explain why you need to use both a positive and negative control before testing the unknown stains:
   a. Positive control
   b. Negative control
3. Should the pink color first be evident:
   a. when applying the phenolphthalein to the cotton swab?
   b. when applying the hydrogen peroxide to the cotton swab?
   Explain your answers.

4. If animal blood is different from human blood, how is it possible to get a positive reaction with the Kastle-Meyer test using dog blood?

5. List two types of substances that might produce a false-positive test when performing the Kastle-Meyer test for the presence of blood.
   a. Substance 1 _____
   b. Substance 2 _____

6. Why is it important to use a cotton swab when doing this test?

7. Why aren't the reagents applied directly to the original bloodstain?

8. Suppose that a red stain was found in a bathtub along with some bathwater. Would it be possible to detect the blood since it might have been diluted? Explain your answer.

Further Research:

1. Research the history of using this method as a presumptive blood test. Investigate the role of each of these scientists:
   a. Louis-Jacques Thenard
   b. Christian Freidrich Schonbein
   c. Dr. Kastle
   d. Dr. Meyer

2. Research how to distinguish dog blood from human blood.

3. Investigate the presumptive test for the presence of semen. Describe how this test is performed.

4. If a man has had a vasectomy, how will this affect the results of tests to detect semen? Explain.

5. Research each of the following cases. Explain the role of blood analysis in helping to solve the crime.
   a. Peter Porco case (2006)
   b. O.J. Simpson case (1994)

ACTIVITY 8-2
BLOOD TYPING

Objectives:

By the end of this activity, you will be able to:
1. Perform a simulated blood test.
2. Describe the procedure for testing blood.
3. Analyze different blood tests to determine the blood type of a suspect.
4. Describe how blood test results are used to determine if an individual is linked to crime-scene blood evidence.
5. Describe agglutination of red blood cells.
6. Describe the protein–antibody reaction that occurs when typing blood.
7. Describe the role of blood typing in forensics.

Time Required to Complete Activity: 45 minutes

Materials:

(per group of three students)
Activity Sheets for Activity 8-2
Ward's Natural Science Kit 360021 or similar artificial blood-typing kit that includes:

    similated human blood types
    antibodies for testing type A, B and Rh blood
    plastic blood testing slides each containing three wells
    plastic or paper cup labeled "Biological Waste"
    10% bleach solution in spray bottle
    paper towels
    latex or nitrile gloves
    marking pen
    red pencil or marker
    toothpicks

Safety Precautions:

Handle artificial blood as if it were actual human blood to practice lab safety techniques. Any blood spills must be cleaned using a 10-percent bleach solution. Dispose of all waste in a container labeled "Biological Waste." Wear disposable gloves. If a student is allergic to latex, substitute a different type of glove, such as nitrile. Students should be careful to avoid spilling any bleach on themselves or on their clothing. Bleach can cause skin and eye irritation and can remove color from fabric immediately.

Background:

Blood typing is a common tool used to solve crimes. It may allow the examiner to match or exclude a suspect from a crime scene. To detect the presence of blood proteins, you will add specific antibodies to individual drops of blood and determine whether clumping (agglutination) occurs.

Procedure:

1. Obtain six clean plastic three-well slides.
2. Place the slides on a clean, white sheet of paper.
3. Write the name of the blood donor in the top left-hand corner of the plastic slide. On the clean white paper, write the name of the blood donor above the slide.
4. Put on your gloves for the rest of the procedure.
5. Add two drops of blood to each of the three wells of the slide labeled Suspect 1.
6. Repeat the process for each of the Suspects 2, 3, 4, Crime Scene, and the Victim slide.
7. Add two drops of Anti-A serum (blue bottle) to each of the six wells labeled A.
8. Add two drops of Anti-B serum (yellow bottle) to each of the six wells labeled B.
9. Add two drops of Anti-Rh serum to each of the six wells labeled Rh.
10. Gently rock each slide back and forth and up and down. (Do not let the blood from one well contaminate the blood from another well!) You can also use a toothpick to help mix the contents. A new toothpick must be used in each of the wells. You will need 18 toothpicks. Stir gently to avoid scratching the plastic wells.
11. Wait five minutes to allow reactions to occur.
12. Observe the blood samples. When using this artificial blood:
    a. A cloudy, opaque, or gooey mixture is a positive reaction indicating the presence of a blood-type protein.
    b. A clear mixture is a negative reaction indicating the absence of a blood-type protein.
    Note: Placing each slide on an overhead projector may help in examining the reactions.

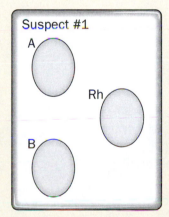

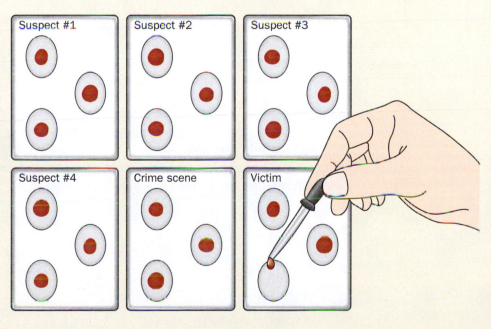

13. Record your data on Figure 1. Include the suspect's name and blood type.

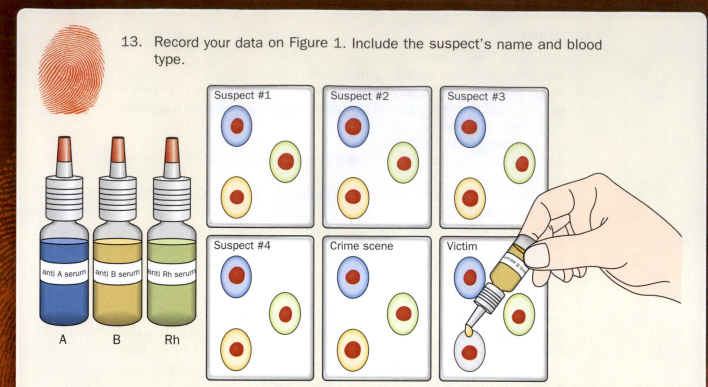

14. For a positive reaction (indicating the presence of the blood-type protein), shade in the circle using a red-colored pencil or marker. If the blood test is negative, indicating the absence of blood-type protein, leave the circle empty.

15. Dispose of all typing materials in the "Biological Waste" container provided by your instructor.

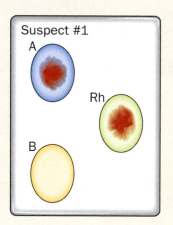

Blood Results:
1. To indicate a positive reaction (protein present), use a red pencil or marker to shade in the well.
2. To indicate a negative reaction (no protein), leave the well blank.
3. Label each reaction slide with the blood type.

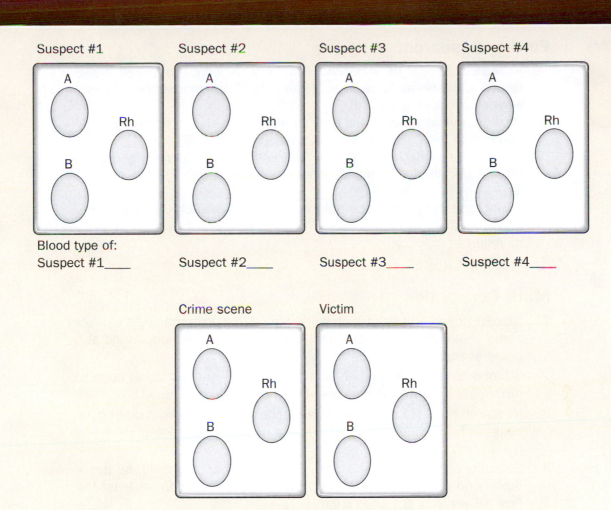

Suspect #1　　　Suspect #2　　　Suspect #3　　　Suspect #4

Blood type of:

Suspect #1____　　Suspect #2____　　Suspect #3____　　Suspect #4____

Crime scene　　　Victim

Crime scene____　　Victim____

Questions:

1. Why is simulated (man-made) blood instead of real human blood used in this activity?
2. Explain why it is necessary to type the victim's blood when trying to determine if any of the blood found at the crime scene belongs to a particular suspect.
3. In this lab activity, how many different blood-type proteins were examined?
4. List all of the blood-type proteins examined in this activity.
5. Is it possible to exclude any of the suspects based on blood types? Explain your answer.
6. Based on your results, does the crime-scene blood match the blood type of any of the four suspects? Explain your answer.
7. If blood from one of the suspects matches the crime-scene blood, does that prove that the suspect is guilty? Explain your answer.
8. Poor laboratory techniques may lead to erroneous results that could impact the outcome of a trial. Describe some examples of poor laboratory techniques involved in a blood-typing analysis that might produce erroneous results.
9. Explain why identifying the blood type found in both the suspect and at a crime scene as AB– provides a higher degree of probability of a match than if the blood type found in the suspect and at the crime scene is O+.
10. Explain why white blood cells are not used in blood typing.

Further Research:

1. Research cases of innocent people who were convicted as a result of laboratory errors. (Refer to The Innocence Project, www.innocenceproject.org)
2. Research what other blood-type proteins are used in blood typing in addition to the A, B, and Rh proteins.
3. Research information on protein and antibodies. Explain why the red blood cells clump, or agglutinate, when mixed with certain antibodies.
4. Investigate the predominant blood type found in each ethnic group:
   a. European
   b. Asian
   c. African

Math Connection:

1. Explain how the laws of probability are used in determining the probability that a particular person's blood will match the blood found at a crime scene.
2. If blood is found at a crime scene, describe what testing is done to determine if the blood is human blood or not human blood.
3. Research how it is possible to determine blood types from other body cells such as cheek cells or skin cells. (Hint: research secretors and non-secretors)
4. Paternity cases: If a man is the biological father of a child, he has a moral and a legal obligation to provide child support for at least the first 18 years of the child's life.

   Explain how blood typing can exonerate a man from paternity, but explain why it alone cannot determine paternity.
5. Investigate the role Rhesus monkeys played in the research involving the Rh protein.

ACTIVITY 8-3
BLOOD-SPATTER ANALYSIS: EFFECT OF HEIGHT ON BLOOD DROPS

Scenario:

The police examined the blood spatter at a crime scene. From the size of the droplets, it appeared that the blood had passively dripped as the injured person walked across the floor. The person may have experienced a second injury, because two different patterns of blood spatter appeared halfway across the room. The second injury seemed to be from a source higher up on the person's body.

By examining the size and shape of blood spatter, forensic scientists are able to reconstruct a crime. A partial story of the crime emerges as the blood-spatter analysis starts to "tell the story."

In this activity, you will experiment with dropping artificial blood from different heights, and you will make observations about the effect of height on blood spatter.

Objectives:

By the end of this activity, you will be able to:

1. Prepare reference cards of blood spatter produced from varying heights.
2. Compare and contrast the blood spatter produced from different heights.
3. Distinguish between the blood spatter formed at the point of contact with satellite blood droplets.
4. Distinguish between satellite droplets and spike-like formations of blood droplets.
5. Form a hypothesis about the effect of height on the size and shape of blood-spatter droplets.

Time Required to Complete Activity:

Two 45-minute class periods (one period to do the blood drop, the second period to measure the blood drops)

Materials:

(per group of four students)
2 dropper bottles of simulated blood
12 five-by-eight-inch index cards
4 meter sticks
4 six-inch rulers showing cm or four calipers
newspapers

Safety Precautions:

Cover the floor in the work area with newspaper.
Simulated blood may stain clothing.

Background:

A blood-spatter pattern is created when a wound is inflicted and blood leaves the body. This pattern can help reconstruct the series of crime-scene events surrounding a shooting, stabbing, or beating. Recall that blood forms droplets as it falls from a wound. A drop of blood that falls on a flat surface will not totally flatten out—the blood drop will have a curved surface. The reason for this shape is the cohesive nature of blood. Blood tends to pull together because of cohesion and resist flattening out on a surface. The result is that the surface of the blood is elastic, giving the top of the blood spatter a spherical appearance.

Cohesive forces keep a spherical shape to blood droplet

If any of the blood does overcome cohesion and separates from the main droplet of blood, it will form small secondary droplets known as **satellites**.

If blood is dropped onto a smooth surface, such as glass or marble, the edge of the drop of blood appears smooth and circular. However, if the blood lands on a porous surface, such as wood or ceiling tile, then the edge of the drop of blood may form small **spikes** or extensions. Notice that spikes are still connected to the main droplet of blood, whereas satellites are totally separated.

As you compare the blood dropped from various heights, note which height causes blood to form more satellites.

Procedure:

Part A: Preparation of blood-spatter reference cards for blood dropped from different heights

1. Spread newspaper on the floor of the work area.
2. You will prepare two 5 × 8 cards for each height used in the blood drop.
3. Label the top-right corner of each card with the height of the blood drop and your initials.
4. Place 12 labeled 5 × 8 index cards on the newspaper.
5. Use a meter stick to help you measure the distance above the card.
6. With the help of your partners holding a meter stick vertically for measurement, squeeze out one drop of simulated blood from a height of 25 cm onto one of the index cards. Hold your hand steady and slowly release the one drop of blood. Aim toward the top of the card.
7. Repeat this process preparing a second card held at 25 cm.
8. Repeat this process for dropping blood for heights of 50,

Your initials and height of blood drop

PB
25 cm

100, 150, 200, and 250 cm. Remember to prepare two cards for each height.

9. Allow index cards to dry. Do not move the cards until they are dry (at least 20 minutes). When you do move the cards, do not turn them on their sides, because the blood will be affected by gravity.

10. Measure the diameter of each of the spatter patterns and record the data in Table 1. Take your measurements at the widest part of the main drop. Do not include the satellites or spikes within your measurement.

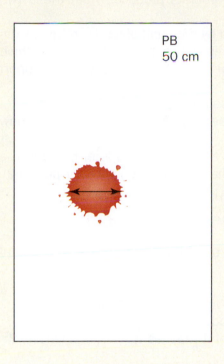

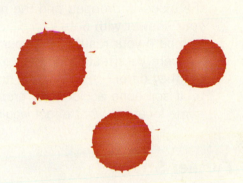

Blood dropped from various heights

11. Determine the average diameter for the blood spatter for each height, and record it in the Data Table.

12. Prepare a bar graph or histogram comparing the effect of height on the average diameter of the blood drop. Your graph should contain title, labeled x-axis and y-axis and an appropriate scale.

Data Table: *Effect of Height on Diameter of Blood Drop*

Height of drop (cm)	Diameter of drop (cm)	Diameter of drop (cm)	Average diameter (cm)
25			
50			
100			
150			
200			
250			

Questions:

1. Is there a relationship between the height from which the blood is dropped and the size of the blood-spatter droplets? Support your answer with data.
2. True or False: As the height from which the blood is dropped increases, the size of the blood spatter continues to increase. Support your answer with data.
3. Blood is dropped from heights of 25 cm and 250 cm. Compare and contrast the outer edges of blood droplets produced from these two heights.
4. Examine the blood spatter produced by dropping blood from the six different heights. Is there a relationship between the height from which the blood is dropped and the number of satellites produced? Support your answer with data.
5. Compare your results with your classmates.
   a. Were your results similar to those of your classmates? If not, how did they differ?
   b. If someone accidentally dropped two or more drops of blood in the same location, what effect would it have on the blood-spatter pattern?

Further Study:

A drop of blood will continue to pick up speed until it reaches its terminal velocity.
1. What is terminal velocity?
2. What factors affect the terminal velocity of a substance?
3. What is the terminal velocity of blood?
4. How far does blood need to fall until it reaches its terminal velocity?

You may also research blood-spatter patterns at these Web sites:
www.bloodspatter.com/bloodspatter.pdf
www.crimelibrary.com/criminal_mind/forensics/serology/8.html

ACTIVITY 8-4
BLOOD-SPATTER IMPACT ANGLE

Scenario:

Two police officers walk into a neighborhood convenience store. They soon realize that no one is inside the store. They discover a series of blood-spatter patterns on the walls and ceiling. "What happened here?" says one officer as she walks around the store. The officers call in the situation, and a forensics team is dispatched to the scene. Forensic scientists will investigate the crime scene and seek answers for the questions listed below:

- Whose blood is this?
- Does it belong to just one or several people?
- How many people were injured?
- If more than one person was injured, is it possible to tell who was injured first?
- What type of injury caused the blood loss?
- What type of weapon caused the injury?
- If the weapon was a gun, from which direction was the bullet fired? Did the shooter point the gun upward, downward, or straight ahead?
- In what direction(s) did the injured person move?

Objectives:

By the end of this activity, you will be able to:
1. Create blood-spatter patterns from different angles of impact.
2. Examine the relationship between angle of impact and blood-spatter patterns.
3. Calculate the angle of impact from blood-spatter patterns.

Time Required to Complete Activity:

Two 45-minute class periods
First period: create blood-spatter patterns from different angles
Second period: measure blood spatter and calculate angle of impact

Materials:

(per group of four students)
Activity Sheets for Activity 8-4
1 dropper bottle of simulated blood
4 five-by-eight-inch index cards
2 meter sticks
newspapers
2 clipboards
1 protractor
1 roll masking or drafting tape

Safety Precautions:

Cover the floor in the work area with newspaper.
Simulated blood may stain clothing and furniture, so care should be taken to avoid spilling blood.

Background:

Blood-spatter analysis is a powerful forensic tool. Spatter patterns allow investigators to reconstruct what happened at a crime scene. The blood-spatter patterns "tell a story" of the crime and help the investigators determine if eyewitness accounts are consistent with the evidence. To study impact angle, you will need to use trigonometry math skills.

Use trigonometric functions to determine the impact angle for any given blood droplet.

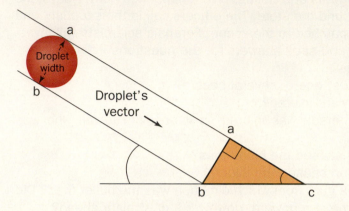

By accurately measuring the length and width of a bloodstain, you can calculate the impact angle using the following sine (abbreviation *sin*) formula:

By accurately measuring the length and width of a bloodstain, the impact angle can be calculated using the sin formula below.

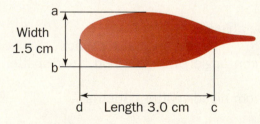

$$(c) = \frac{\text{opposite}}{\text{hypotenuse}} = \frac{\text{width (a–b)}}{\text{length (b–c)}}$$

$$(c) = \frac{1.5\ \text{cm}}{3.0\ \text{cm}}$$

$$(c) = 0.5$$

$$(c) = 30°$$

To determine the angle of impact, take the inverse *sin* of 0.5, which is 30 degrees.

Procedure:

Part B: Creating blood spatter from different angles of impact

In this activity, you will drop blood onto 5 × 8 index cards set at various angles. You will drop the blood from 30 cm from the point of impact on the 5 × 8 card. You will observe how the angle of impact affects the size and shape of the blood spatter.

You will drop simulated blood onto 5 × 8 cards that are set to represent impact angles of:

10 degrees	20 degrees	30 degrees
40 degrees	50 degrees	60 degrees
70 degrees	80 degrees	

You will be working in groups. Each group will prepare blood spatter from two different angles of impact. Divide the work as follows:

Group 1: 10 degrees and 50 degrees
Group 2: 20 degrees and 60 degrees
Group 3: 30 degrees and 70 degrees
Group 4: 40 degrees and 80 degrees

To simulate blood being cast off during bleeding, the following process is used.

1. Turn a 5 × 8 index card over so that no lines are visible.
2. Tape two 5 × 8 index cards on the clipboard as shown.
3. Label the cards with your initials on the top-right corner, along with the angle of impact that you will be using.
4. Working with your partners and a roll of masking tape, locate an area along a wall and set up the clipboard as pictured.

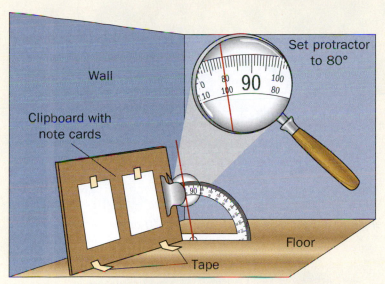

10° angle of impact

a. Place the clipboard on the floor. Turn the clipboard so that the metal clasp of the clipboard is on its side. You will move the clipboard up against the wall as indicated in the following diagrams.
b. Set your protractor at the zero mark at the end of the clipboard in contact with the floor. (See other examples on the next page.)
c. Note: To calculate the desired impact angle, set the protractor reading for 90 degrees minus the desired angle (90 − 10 = 80). Tape the end of the clipboard to the floor to keep it at this position.

5. Start with the first angle assigned. Calculate the protractor setting.
6. From a height of 30 cm drop two drops of blood onto each index card. (See examples of cards on page 233.)
7. While the first card is drying, prepare the second clipboard and card and repeat steps 1 to 6 for the second angle you were assigned.
8. Allow cards to dry completely. Do not move or pick up the cards for at least 30 minutes!
9. Measure the length and width of each droplet in millimeters as indicated in the diagram on page 230. Disregard elongated tails of blood. Measure the main football-shaped or Q-tip rounded area only. If more than one group is assigned the same angle of impact, average your readings for length and width of the drops for the same angle of impact. Record this information in the Data Table on page 233.

60 degree angle of impact and set protractor for 30 degrees

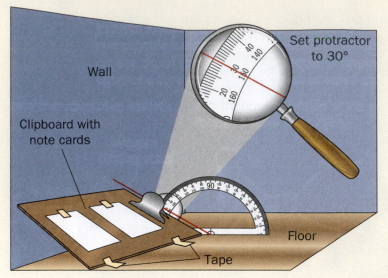

60° angle of impact

50 degree angle of impact and set protractor for 40 degrees

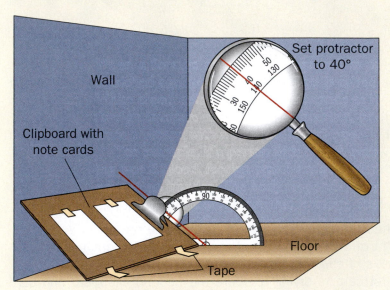

50° angle of impact

10. Determine the R value by dividing the length and width of your blood droplets.
11. Using a calculator and the Law of Sines, determine the actual angle of impact based on your blood-spatter marks.
12. Record all information in the Data Table on page 233.

Examples of blood-spatter patterns 90- to 10-degree angles of impact.

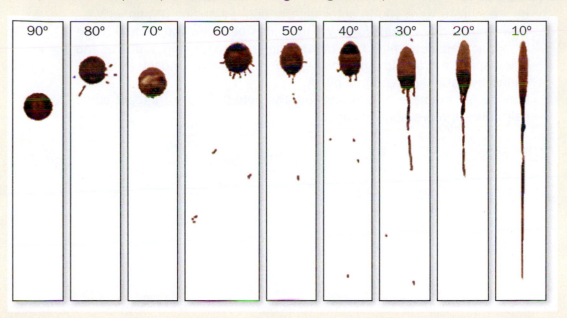

Data Table

Expected Impact Angle	Length (mm)			Width (mm)			R=W/L	Actual Impact Angle
	1st Trial	2nd Trial	Average Length	1st Trial	2nd Trial	Average Width	Average W/L	
10°								
20°								
30°								
40°								
50°								
60°								
70°								
80°								
90°								

Questions:

1. How accurate were you in obtaining the desired angles of impact?
2. How would you account for any differences between your actual angle of impact as determined by measuring the length and width of the blood-spatter droplets and your expected angle of impact as determined by your clipboard setup?
3. Provide an example of how knowing the actual angle of impact could help investigators solve crimes.

ACTIVITY 8-5
AREA OF CONVERGENCE

Scenario:

When the police arrived at a crime scene, both the victim and the attackers had already fled. Two areas of blood spatter were the only evidence that an assault had occurred. After drawing lines from the blood spatter, the crime-scene investigator determined not only the direction the blood was traveling and the approximate speed the blood was traveling but also the approximate location where the person was standing when the injury occurred.

Blood-spatter analysis can help investigators reconstruct what happened at a crime scene. When you enter a crime scene, there is always a story waiting to be discovered. An observant crime-scene investigator doesn't always need eyewitnesses to describe what happened, because the scene always tells the story. You can use blood-spatter analysis to reconstruct what happened at a crime scene.

In this activity, you will analyze blood spatter. By noting the direction of the droplet of blood, you will be able to note the direction in which the blood was moving. The size of the blood spatter will provide some indication of the velocity of the blood when it hit a surface. By examining at least two drops of blood spatter, you will be able to determine where the injured person was located when the injury occurred. When blood-spatter analysis is completed and these factors are determined, it may be possible to reconstruct what happened at the crime scene.

Objectives:

By the end of this activity, you will be able to:
1. Distinguish between blood-spatter droplets and blood-spatter satellites.
2. Distinguish between passive blood spatter and blood spatter that was emitted due to some type of force.
3. Use the shape of the blood droplet to determine the direction in which a drop of blood was moving.
4. Use the position of satellites to determine the direction in which a drop of blood was moving.
5. Use blood spatter to draw the lines of convergence to indicate the position where a person was located when bleeding occurred.

Time Required to Complete Activity: 45 minutes

Materials:

1 ruler
1 colored pencil or marker
1 pencil

Safety Precautions:

None

Background:

The shape of an individual drop of blood provides clues to the direction from which the blood originated. A drop of blood that has a circular shape (equal width and length) indicates that the blood fell straight down. When blood falls straight down, such as when it drips from a wound, the angle of impact is 90 degrees. This type of blood spatter is known as passively produced, because no applied force caused the spatter. When a drop of blood is elongated (longer than it is wide), it is possible to determine the direction the blood was traveling when it struck a surface.

The location of the source of blood can be determined if there are at least two drops of blood spatter. By drawing straight lines down the long axis of the blood spatter and noting where the lines intersect, this will indicate the lines of convergence. To determine where the source of the blood originated, draw a small circle around all of the intersecting lines. The intersection of the lines of convergence will indicate in a two-dimensional view the location of the source of the blood.

Procedure:

1. For each of the four different blood-spatter patterns pictured (Samples A–D on the next page), you will draw lines of convergence to determine the source of the blood.
2. Determine the direction in which each blood spatter is moving by locating the tail of the blood spatter and any satellites. The satellites will be found ahead of the blood spatter.
3. Draw a line through the middle of the long axis of each of the major drops of blood. Do not draw lines through the satellites.
4. Note: Begin your lines at the leading edge of the drop of blood, and draw the line in the opposite direction from the direction in which the blood was traveling. This will make your diagram easier to read.
5. Draw a small circle around the point where all of the lines intersect using a colored pencil or marker. This is the source of the blood or area of convergence.
6. For each of the four samples, determine how many incidences occurred. See the following example.

Example

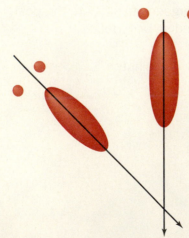

Drawn correctly!

Lines are started at the leading edge of the spatter.

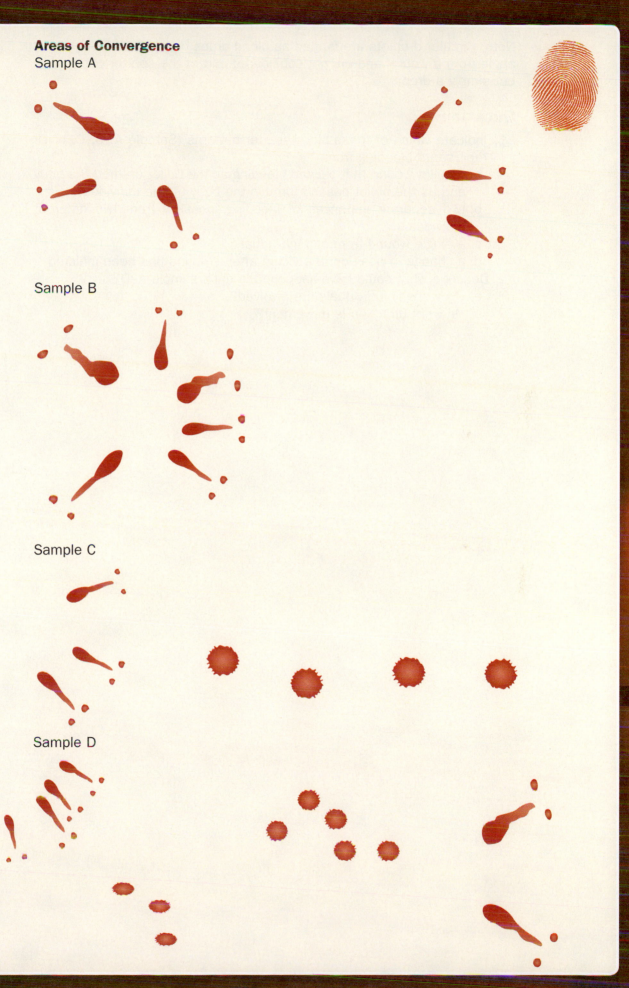

Areas of Convergence

Sample A

Sample B

Sample C

Sample D

Note: Circular droplets are formed as blood drops down at a 90-degree angle from a wound and are not considered part of the spatter pattern, but simply a drop.

Questions:

1. Indicate which of these blood-spatter patterns (Sample A, B, C, or D) represents bleeding from:
   a. a bullet wound that caused bleeding as the bullet entered the body and as the bullet passed through the body of one individual
   b. two separate instances of bleeding, possibly from two different individuals
   c. a single wound from one individual
   d. a change in position of a victim after a wound has been inflicted
2. Describe what could have happened in each sample A–D?
   a. How many individuals are involved?
   b. In what direction is movement?

ACTIVITY 8-6
POINT OF ORIGIN

Objectives:

By the end of this activity, you will be able to:

1. Determine the direction of blood flow based on the shape of the droplet.
2. Use lines of convergence to help determine the position of the victim when the wound was inflicted.
3. Calculate the angle of impact for individual drops of blood spatter.
4. Use the Law of Tangents to calculate the height above floor level where the wound was inflicted.

Time Required to Complete Activity: 45 minutes

Materials:

Activity Sheet for Activity 8-6
1 metric ruler
1 colored pencil or marker
1 pencil
calculator with tangent function
tangent tables (optional)

Safety Precautions:

None

Introduction:

In previous activities, you determined the angle of impact and area of convergence for blood spatter. Now you can use that information and the Law of Tangents to calculate the height (position) of the wound, the point of origin, on the individual.

Background:

Blood-spatter analysis helps crime-scene investigators reconstruct what happened at the crime scene. Using only blood-spatter analysis, you may be able to recognize the events leading up to the crime. Although crime-scene investigators may arrive at the crime scene after the victim and witnesses are no longer present, they still need to determine what happened. Often several witnesses give different accounts of the crime. Which witness is providing an accurate description of what really happened?

During the investigation, the crime-scene investigators need to determine if the evidence, in this case the blood spatter, matches the description given by the witnesses, the suspect(s), and the victim(s). In domestic abuse cases, the victim of domestic abuse may tell a false story to try to protect the abusing partner. A victim may state that a head injury occurred as a result of falling down stairs. However, if the blood-spatter patterns are inconsistent with this type of injury, then what type of injury did cause the blood spatter? What actually happened? Is a witness lying? Further investigation is required

when the blood-spatter evidence tells a different story than the witness's account of the incident.

In this activity, you will analyze blood spatter in three dimensions. By noting the shape of the droplet of blood, you will be able to note the direction in which the blood was moving. The size of the blood spatter will provide some indication of the velocity of the blood when it hit a surface. By examining at least two drops of blood spatter, you will be able to determine where the injured person was located when the injury occurred in two dimensions (lines of convergence). You can easily measure the distance from the area of convergence to the drop of blood. If you want to determine the point of origin, or height from the impact surface, you will need to make some calculations. By measuring the width and length of a single drop of blood, you can determine the angle of impact. By using the Law of Tangents, you can calculate the height from which the blood fell, or the point of origin for the blood.

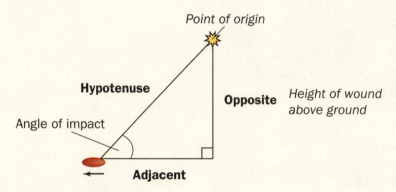

Math Review:

Right triangle
- Contains one 90-degree angle.
- The hypotenuse is the longest side of a triangle, opposite the 90-degree angle (right angle).
- The opposite side to an angle is the side directly opposite the angle of interest.
- The adjacent side to an angle is the side closest to the angle that is not the hypotenuse.

How to use a right triangle and the Law of Tangents in recreating a crime scene:

Procedure:

To recreate a crime scene from several drops of blood, you will need to perform several steps.
1. Determine the direction of blood flow in the drops that follow with an arrow next to the droplet. If the blood drop is circular, then the blood fell at a 90-degree angle. If it is not circular, then the angle of impact was less than 90 degrees. The elongated end of a drop of blood points to the direction in which the blood was moving.

2. From several drops of blood, determine the area of convergence by drawing lines through each of the blood droplets and noting where the lines intersect.
   a. Determine the direction of the blood when it struck an object.
   b. Draw your line in the direction opposite to the direction in which the blood was moving.
   c. The area where the lines intersect represents the area of convergence or the approximate location where the person was located when the blood droplets formed

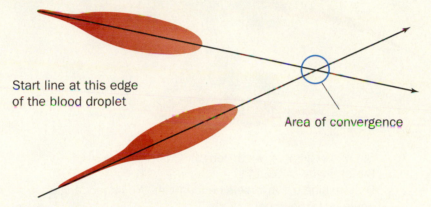

Start line at this edge of the blood droplet

Area of convergence

3. Once you have determined the area of convergence, you will measure the distance from the area of convergence to the edge of the drop of blood when it first impacted a surface. This distance is indicated in green.

Recall the diagram of a right triangle. This green line next to the angle of impact is known as the adjacent side.

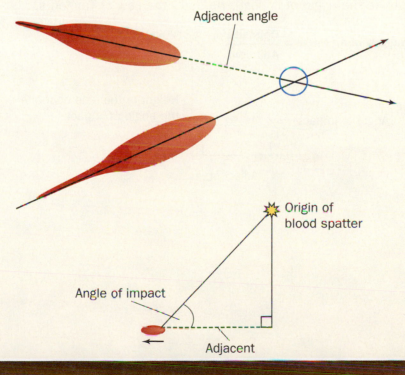

Adjacent angle

Origin of blood spatter

Angle of impact

Adjacent

4. Next determine the angle of impact for each droplet of blood. Select one of the blood droplets and determine the angle of impact for that drop of blood. To calculate the angle of impact, you will need to use the Law of Sines. Remember, when you measure the length of the blood droplet, do not include the thin extension of the leading edge.

Sin of the impact angle = width of the blood drop/length of the blood drop

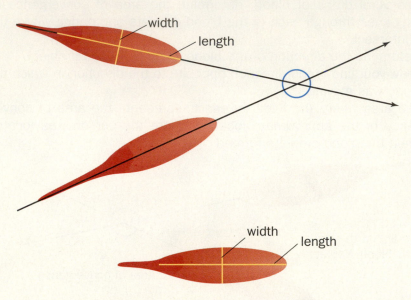

Sin of the angle = width/length = 14/45 = 0.3111
Sin of angle = 0.3111
Determining the inverse sine identifies the impact angle
Angle is 18 degrees

5. Using the Law of Tangents to solve for height. Going back to the right triangle and adding the angle of impact, we can determine the height from where the blood originated. The height of the source of blood is the side opposite the angle of impact. To solve for the height (or the side opposite the angle of impact), we apply the Law of Tangents.

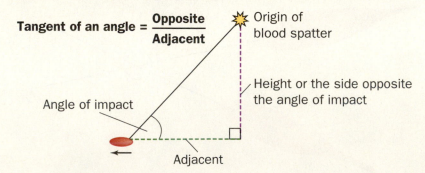

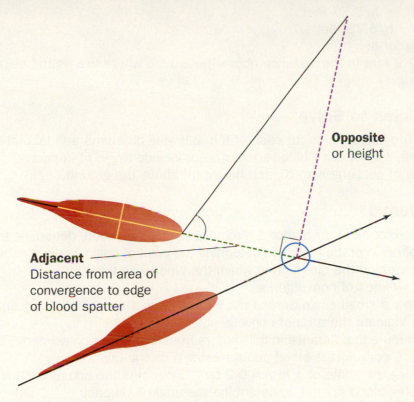

Opposite
or height

Adjacent
Distance from area of
convergence to edge
of blood spatter

Tangent of angle of impact = Opposite/Adjacent = Height/Distance
Solving for Height: Tangent of Angle × Distance = Height

Example:

Crime-scene investigators noted blood spatter on the floor of the kitchen.
The investigators drew lines of convergence and measured the distance
from the area of convergence to the front edge of a drop of blood. That
distance was recorded as 5.75 feet. After measuring the length and width
of the blood droplet and using the Law of Sines, it was determined that
the angle of impact was 27 degrees. The police wanted to determine the
point of origin, or the height from the floor where the person was bleeding.

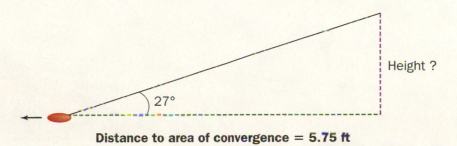

Height ?

27°

Distance to area of convergence = 5.75 ft

Solution:

Tan = Opposite/Adjacent = Height/Distance
Tangent of the blood-spatter angle = Height of the wound/Distance from
                                     blood to area of convergence
Substituting values in the equation
    tan 27° = Height of wound/distance
    tan 27° = height/5.75 ft
Consult your calculator or tan chart

tan 27° = h/5.75 feet
.5095 = h/5.75 feet
Solving for h:
h = ~2.9 feet is the distance above the ground where the wound began bleeding

Problems to Solve:

Make the calculations for each of the following problems and label the right triangle for each blood-spatter drop. Include angle of impact, distance to area of convergence (d), and height (h) above the ground.

Problem 1:

Refer to Blood-spatter Sketch 1. From these drops of blood, determine the point of origin of the blood. To determine the point of origin, you will need to:

1. Determine the direction in which the blood was traveling.
2. Draw lines of convergence.
3. Draw a small circle around the intersection of the lines of convergence to indicate the area of convergence.
4. Measure the distance in millimeters from the area of convergence to the front edge of the blood spatter using a metric ruler.
5. Using the scale of 1 mm = 0.2 feet, determine the actual distance.
6. Using blood droplet 1, determine the angle of impact:
   a. Measure the width and the length of the blood droplet.
   b. Divide the width/length ratio for the blood droplet.
   c. Using a calculator and the inverse sine function, determine the angle of impact for that blood droplet.

Blood droplets are drawn to scale but length of adjacent side is scaled to 1 mm = .2 feet

1

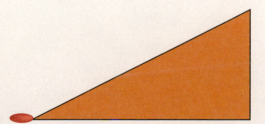

7. Using the Law of Tangents, determine the point of origin or the height of the source of blood for droplet 1.

Problem 2:

A 30-year-old man was found shot in the head in his garage. The suspect claims he was being attacked and shot the victim in self-defense. Refer to Blood-spatter Sketch 2 on the next page. From these drops of blood, determine the point of origin of the blood. To determine the point of origin, you will need to:

1. Determine the direction in which the blood was traveling.
2. Draw lines of convergence.
3. Draw a small circle around the intersection of the lines of convergence to indicate the area of convergence.
4. Measure the distance in millimeters from the area of convergence to the front edge of the blood spatter.
5. Use the scale of 1 mm = 0.3 feet to determine the actual distance.
6. Use blood droplet 1 to determine the angle of impact:
   a. Measure the width and the length of the blood droplet.
   b. Divide the width/length ratio for the blood droplet.
   c. Using a calculator and the inverse sine function, determine the angle of impact for that blood droplet.
7. Use the Law of Tangents to determine the point of origin or the height of the source for blood droplet 1.

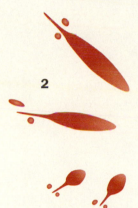

Blood droplets are drawn to scale but length of adjacent side is scaled to 1 mm = .3 feet

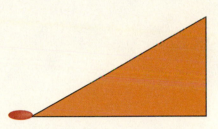

Problem 3:

A victim was found at the foot of a ladder with a chest wound. What is the approximate height of his wound when he was shot? Refer to the blood-spatter sketch below. From these drops of blood, determine the point of origin of the blood. To determine the point of origin, you will need to:

1. Determine the direction in which the blood was traveling.
2. Draw lines of convergence.
3. Draw a small circle around the intersection of the lines of convergence to indicate the area of convergence.
4. Measure the distance from the area of convergence to the front edge of the blood spatter (droplet #3) using a millimeter ruler.
5. Use the scale of 1 mm = 1.5 feet to determine the actual distance.

Blood droplets are drawn to scale but length of adjacent side is scaled to 1 mm = 1.5 feet

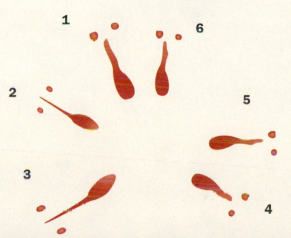

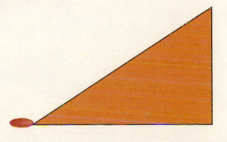

6. Use blood droplet 3 to determine the angle of impact.
   a. Measure the width and the length of the blood droplet.
   b. Divide the width/length ratio for the blood droplet.
   c. Using a calculator and the inverse sine function, determine the angle of impact for that blood droplet.
7. Use the Law of Tangents to determine the point of origin or the height of the source of blood for droplet 3.

ACTIVITY 8-7
CRIME-SCENE INVESTIGATION

Objective:

By the end of this activity, you will be able to:

1. Use information collected from a crime scene and your knowledge of spatter analysis to develop a hypothesis to describe the events that occurred at a crime scene.

Time Required to Complete Activity: 40 minutes

Materials:

ruler
pencil
calculator

Safety Precautions:

None

Procedure:

1. Examine Crime Scene Diagram on page 249 and complete the lines of convergence.
   a. Determine the position of each man at the time of the shootings. Label the position for Man 1 in the diagram. What evidence supports your answer?
   b. Label the position for Man 2 in the diagram. What evidence supports your answer?
2. Both men died. Man 1 was shot through the forehead and died instantly. Man 2 was shot in the stomach and was found dead at the scene as well. Who was shot first? Support your answer with evidence from the crime scene.
3. Data Table 1 contains some of the measurements for the bloodstains found at each position. Complete the table by filling in the blanks.
4. Did your results agree with statements made in question 2? Explain your conclusions.

Questions:

1. Based on your calculations, which man was most likely standing when he was shot? Support your answer with evidence from the crime scene.
2. In position one, there are four bloodstains in front and one bloodstain behind the victim. How do you account for this?
3. Based on the blood-spatter evidence, describe the series of events resulting in the death of these two men. Support your theory with evidence obtained from the blood-spatter analysis.

Data Table 1

Stain #	Length of Stain (L) (mm)	Width of Stain (W)(mm)	W/L Ratio (Sine Value)	Angle of Impact (nearest degree)	Distance from Near Edge of Stain (feet)	Tan Value of Angle of Impact (to four decimal places)	Height (h) of Wound above Floor (feet) h = Tan × distance
1	18.1	9.6			4.0		
2	18.6	9.0			4.4		
3	17.8	13.2			2.2		
4	18.9	12.8			2.8		
5	19.2	13.2			2.5		
6	9.0	4.5			10.1		
7	10.6	5.1			10.3		
8	8.4	3.9			10.4		
9	8.1	3.6			10.3		

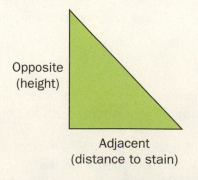

Opposite (height)

Adjacent (distance to stain)

Tan = opposite/adjacent

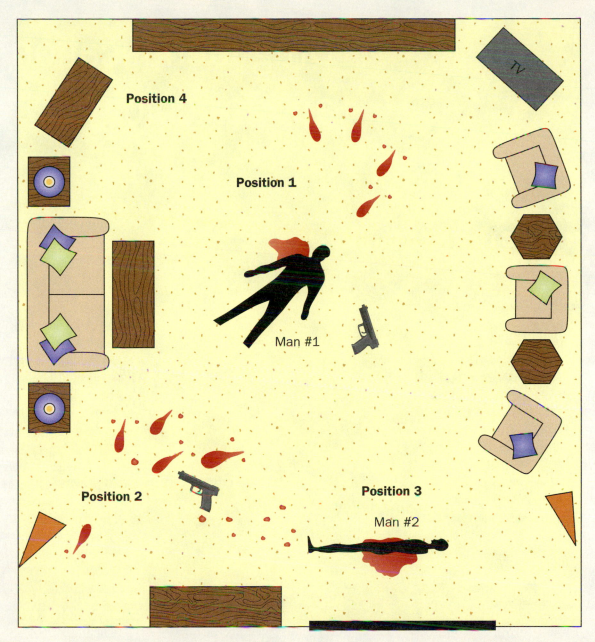

CHAPTER 9

Drug Identification and Toxicology

AN ACCIDENTAL OVERDOSE

Within weeks of her death on February 9, 2007, it was determined that model and tabloid celebrity Anna Nicole Smith had died from an accidental overdose of prescription drugs, rather than as the result of foul play or illegal drug use. But how did the medical examiner know which drugs Smith had taken? And how was it determined that the overdose was accidental rather than intentional? Specific forensic analyses by toxicologists helped determine the drugs Smith used and how they caused her death.

The Broward County Medical Examiner in Florida reported that nine drugs and a few drug metabolites were found in Smith's blood. The drugs included antianxiety and anti-depression prescriptions, such as Valium, pain and allergy medications, such as Benadryl, the antibiotic Ciprofloxacin, as well as human growth hormones. All drugs were found at therapeutic levels. Tests also found the presence of chloral hydrate, a sedative and sleeping medication.

©AP Photo/Manuel Balce Ceneta

Anna Nicole Smith died from an accidental overdose.

Specialists determined that when this sedative combined with the other drugs in her system, it led to Smith's accidental death. The combined drug effect acted on her respiratory and circulatory systems, causing them to stop working. The medical examiner ruled out that this was a suicide, as some suggested, because of the large amount of chloral hydrate remaining in the bottle and the normal levels of the other medications in her system.

Drug interactions can cause death in even small doses. Anna Nicole Smith's 20-year-old son Daniel also appears to have died from a lethal combination of drugs. In Daniel's case, it was a lethal combination of antidepressants Lexapro and Zoloft, and the drug methadone. This drug cocktail appears to have affected his central nervous system and heart, leading to his sudden death. Forensic investigations of both tragic deaths found no evidence to indicate foul play. Both overdoses were accidental.

OBJECTIVES

By the end of this chapter, you will be able to

✔ Identify the five types of controlled substances.

✔ Relate signs and symptoms of overdose with a specific class of drugs or toxins.

✔ Describe the role of various types of toxins in causing death.

✔ Discuss agents that may be used in bioterrorism.

✔ Define and describe the goals and practice of toxicology.

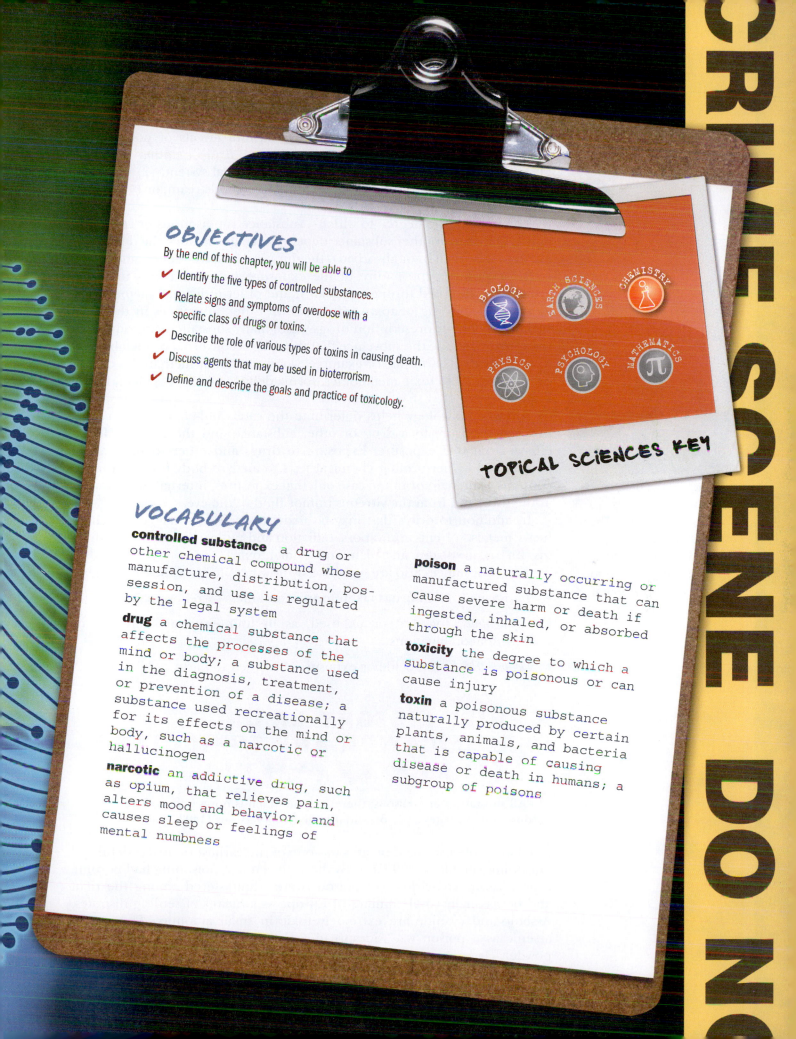

TOPICAL SCiENCES KEY

VOCABULARY

controlled substance a drug or other chemical compound whose manufacture, distribution, possession, and use is regulated by the legal system

drug a chemical substance that affects the processes of the mind or body; a substance used in the diagnosis, treatment, or prevention of a disease; a substance used recreationally for its effects on the mind or body, such as a narcotic or hallucinogen

narcotic an addictive drug, such as opium, that relieves pain, alters mood and behavior, and causes sleep or feelings of mental numbness

poison a naturally occurring or manufactured substance that can cause severe harm or death if ingested, inhaled, or absorbed through the skin

toxicity the degree to which a substance is poisonous or can cause injury

toxin a poisonous substance naturally produced by certain plants, animals, and bacteria that is capable of causing disease or death in humans; a subgroup of poisons

INTRODUCTION

Toxicology is the study of **poisons** and the identification of **drugs** and other substances a person may have used for medicinal, recreational, or criminal purposes. Toxicology also examines the harmful effects of poisons and drugs on the body. Most people are exposed to drugs or other **toxins** by (1) ingesting them so they enter the gastrointestinal system, (2) inhaling them into the lungs, (3) injecting them into the bloodstream, or (4) absorbing them through the skin.

The **toxicity**, the degree to which a substance is poisonous or can cause injury, of a drug or other substance depends on many factors: the dose (how much of it is taken in or absorbed), the duration (the frequency and length of the exposure), the nature of the exposure (whether it was ingested, inhaled, or absorbed through the skin), and other individual factors, such as whether the drug or toxin interacts with other substances in the body such as alcohol or prescription drugs. Also, many substances are only indirectly toxic because the substance that the drug is converted or metabolized to in the body is harmful. For example, wood alcohol or methanol is chemically converted to toxic metabolites, formaldehyde and formic acid, in the human liver.

Forensic toxicology helps determine the cause-and-effect relationships between exposure to a drug or other substance and the toxic or lethal effects from that exposure. Exposure to drugs and other toxins may be determined by performing chemical tests to analyze body fluids, stomach contents, skin, hair, or in the case of lethal exposures, internal organs, such as the liver, and from the vitreous humor fluid of the eye.

In addition to drugs that may be toxic, toxic agents may also include heavy metals, solvents and vapors, radiation and radioactive materials, dioxins/furans, pesticides, and plant and animal toxins.

Toxic substances also are classified by how people are exposed to them:

- *Intentionally.* As in drugs taken to treat an illness or relieve pain

- *Accidentally.* Ingested or exposed, as in unintentional overdoses or harmful combinations

- *Deliberately.* As in suicide or exposures intended to harm or kill others

Did You Know?

Saliva-based drug tests are as accurate as urine-based tests. The saliva test can generally detect illegal drugs immediately on use for up to about 72 hours.

A BRIEF HISTORY OF DRUG IDENTIFICATION AND TOXICOLOGY

"All substances are poisons; there is none which is not a poison. The right dose differentiates a poison and a remedy." Paracelsus (1493–1541)

The Greek philosopher Socrates was one of the earliest reported victims of poisoning (hemlock, 399 B.C.). By the 17th century, poisoning had become a profession. Toxic doses of poisons were administered among the rich, and occasionally royal, families of Europe as a means of settling disputes. Arsenic and cyanide are extremely toxic in small amounts. The use of arsenic as a poison was widespread and became known as "inheritance powder." It was not until the 1800s that methods of chemical analysis were developed to identify arsenic and other toxins in human tissue. The first forensic toxicologists to popularize these new methods were physicians Mathieu Orfila (1787–1853) and Robert Christison (1797–1882).

MURDER BY POISON

Although poisoning is popular in murder mysteries and detective stories, in reality, it is not a common form of murder. Less than one-half of 1 percent of all homicides result from poisoning. Throughout history, some notable individuals have died from poisoning: Nazi leaders Heinrich Himmler and Hermann Goering ingested cyanide capsules in 1945; Jonestown cult members consumed cyanide-laced punch in 1978, killing approximately 900 people; Bulgarian dissident Georgi Markov was killed by ricin in 1978; and most recently, Russian ex-spy Alexander Litvinenko was exposed to radiation in 2006. Today, the commonly used poisons include arsenic, cyanide, and strychnine, as well as an assortment of industrial chemicals that were created for other uses, such as fertilizers.

Testing for a vast array of possible toxins can be a challenge to the toxicologist. Toxicologists must distinguish between acute poisoning and chronic poisoning. *Acute poisoning* is caused by a high dose over a short period of time, such as cyanide ingestion or inhalation, which immediately produces symptoms. *Chronic poisoning* is caused by lower doses over long periods of time, which produces symptoms gradually. Mercury and lead poisoning are examples of chronic poisoning in which symptoms develop as the metal concentrations slowly rise and accumulate to toxic levels in the victims' bodies over a long period of exposure.

ACCIDENTAL DRUG OVERDOSES

Accidental deaths from drug overdoses are more common than deaths from poisoning. The deaths of comedians John Belushi and Chris Farley, actor River Phoenix, and musicians Steve Clark, Janis Joplin, Jim Morrison, and Jimi Hendrix were all linked to lethal drug combinations or overdoses.

DRUGS AND CRIME

Illegal drugs, such as heroin and lysergic acid diethylamide (LSD), are drugs with no currently accepted medical use in the United States. **Controlled substances** are defined as legal drugs whose sale, possession, and use are restricted because of the effect of the drugs and the potential for abuse. These drugs are medications, such as certain **narcotics**, depressants, and stimulants, that physicians prescribe for various conditions.

Arrests for drug abuse violations have increased steadily since the early 1990s. Drug abuse violations topped the list of the seven leading arrest offenses in

Digging Deeper
with Forensic Science e-Collection

Do research on modern detection methods and the techniques of forensic toxicology or drug-testing work, such as the different chromatography and spectrometry methods. Go to the Gale Forensic Sciences eCollection on school.cengage.com/forensicscience and research the various methods. Determine which methods are more appropriate for the major types of controlled substances. Relate your findings to the chemical properties of the major controlled substances. Cite any limitations or concerns in using any of these methods for drug testing.

the United States in 2005. Drug offenders make up more than half of the federal prison system population and about 20 percent of the state prison population.

CONTROLLED SUBSTANCES

There are five classes of controlled substances: (1) hallucinogens, (2) narcotics, (3) stimulants, (4) anabolic steroids, and (5) depressants.

Hallucinogens

Hallucinogens are often derived from plants and affect the user's perceptions, thinking, self-awareness, and emotions. Hallucinogens derived from plants include mescaline from a cactus (peyote), marijuana, and extracts from certain mushrooms. Hallucinogens, such as LSD, MDMA (the amphetamine ecstasy), and PCP (angel dust), are chemically manufactured. The effect and intensity of response to the drug varies from person to person.

LSD was originally found in 1938 in a fungus that grows on rye and other grains and is one of the most potent mood-changing chemicals. It is odorless, colorless, and tasteless and is sold in tablets or on absorbent paper divided into small decorative squares. PCP was first developed as an anesthetic, but it is no longer used because it induces hallucinations. In the illicit drug market, PCP is available in a number of forms. It may be a pure, white, crystal-like powder, a tablet, or a capsule. It can be sniffed, swallowed, smoked, or injected. Mescaline is smoked or swallowed in the form of capsules or tablets. Marijuana leaves (cannabis) may be smoked or refined, concentrated, and sold as hashish. Hashish is made from resin found on ripe flowers, which are rolled into balls and smoked. Figure 9-1 shows hallucinogenic drugs and the characteristic symptoms of an overdose.

Figure 9-1. *Table of hallucinogenic drugs and the characteristic symptoms of an overdose.*

Drug	Characteristics of Drug Overdose
MDMA (ecstasy)	Increased heart rate and blood pressure, muscle cramps, panic attacks, seizures, loss of consciousness, stroke, kidney failure, death
Mescaline	Hallucinations, euphoria, dizziness, vomiting, increased heart rate, dilated pupils, diarrhea, headaches, anxiety, irrationality of thoughts
LSD	Dilated pupils, loss of appetite, sleeplessness, increase in body temperature, increased heart rate and blood pressure, sweating, dry mouth, tremors, confusion, distortion of reality, and hallucinations
PCP	Increased heart rate and blood pressure, convulsions, sweating, dizziness, numbness, and possibly death from heart failure. Drowsiness, which can lead to accidents. Users sometimes exhibit psychosis (completely losing touch with reality) that can last for weeks.

Narcotics

Narcotics act to reduce pain by suppressing the central nervous system's ability to relay pain messages to the brain. Narcotics include opium and its derivatives—heroin and codeine. These painkillers are very habit forming. Hydrocodone (Vicodin, Lortab), methadone (Dolophine), morphine (MS Contin), oxycodone (Percocet, OxyContin), and codeine-containing pain relievers, such as Tylenol 3 (acetaminophen and codeine), are man-made narcotic painkillers that are often abused. See Figure 9-2 for a summary of narcotic drugs and the characteristic symptoms of an overdose.

Figure 9-2. Table of narcotic drugs and the characteristic symptoms of an overdose.

Drug	Characteristics of Drug Overdose
Opium	Difficulty breathing, low blood pressure, weakness, dizziness, confusion, loss of consciousness, coma, cold clammy skin, small pupils
Heroin Codeine Morphine	Difficulty breathing, low blood pressure, coma, spasms of the stomach or intestines, constipation, nausea, vomiting, sleepiness, blue fingernails and lips, death
Methadone	Difficulty breathing, drowsiness, coma, low blood pressure, muscle twitches, blue fingernails and lips
Oxycodone	Slow, difficult breathing, seizures, dizziness, weakness, loss of consciousness, coma, confusion, tiredness, cold clammy skin, and small pupils

Stimulants

Stimulants increase feelings of energy and alertness while suppressing appetite. Depression often results as the effect of the drug wears off. They are also used and sometimes abused to boost endurance and productivity. Examples of stimulants include amphetamines, methamphetamines, and cocaine (including crack), and are highly addictive. The key difference between methamphetamines and amphetamines is that methamphetamines are more potent than amphetamines. Figure 9-3 shows characteristic symptoms of an overdose with stimulant drugs.

Figure 9-3. Table of stimulant drugs and characteristic symptoms of an overdose.

Drug	Characteristics of Drug Overdose
Amphetamines (Speed)	High blood pressure, rapid heart rate, agitation, irregular heartbeats, stroke, seizures, coma, death
Cocaine/crack cocaine	Dangerous rise in body temperature, sweating, tremors, seizures, irregular heartbeats, stroke, confusion, heart attack, bleeding in the brain, death
Methamphetamines	Dangerous rise in body temperature, profuse sweating, confusion, rapid breathing, increased heart rate, dilated pupils, high blood pressure, kidney failure, bleeding in the brain, death

Anabolic Steroids

Anabolic steroids promote cell and tissue growth and division. These drugs are produced in the laboratory and have a chemical structure similar to testosterone, the male sex hormone. Anabolic steroids were originally used to treat hypogonadism, a condition in which the testes produce abnormally low levels of testosterone. Today, they are used to treat some cases of delayed puberty, impotence, and muscle wasting caused by HIV infection. In the 1930s, they gained popularity with weightlifters and bodybuilders because they act to increase body muscle and bone mass. The negative side effects of anabolic steroids range from mild side effects, such as acne, increased body hair, and baldness, to severe side effects, such as high blood pressure and cholesterol levels, impaired fertility in males, blood clotting, kidney and liver cancers, and heart attacks.

Depressants

Depressants are drugs, such as barbiturates and benzodiazepines, that relieve anxiety and produce sleep. Depressants reduce body functions, such as heart rate, by acting on the central nervous system and increasing the activity of a neurotransmitter called GABA. The result of increased GABA production is drowsiness and slowed brain activity. The user becomes very calm, which is why these drugs are used to relieve tension and promote sleep. Side effects of depressants include slurred speech, loss of coordination, and a state of intoxication similar to that of alcohol. An overdose may slow heart rate and breathing and cause coma and death. Mixing depressants with alcohol and other drugs increases their effects and health risks.

OTHER ORGANIC TOXINS

Organic toxins are poisonous substances produced by living organisms. They are usually proteins that can be absorbed by another living creature and interfere with that organism's metabolism. Poisons are generally absorbed into an organism through the intestine or the skin. A bee sting or snakebite is an example of *venom*, a toxin secreted by an animal that can be transferred to a human (Figure 9-4).

Figure 9-4. *Snake venom can be a deadly organic toxin.*

©Tom McHugh/Photo Researchers, Inc.

ALCOHOLS

All alcohols are toxic to the body. Methanol is not directly poisonous, but when it is converted by the liver to formaldehyde, it becomes very toxic. Ethanol, the alcohol found in many beverages, is called grain alcohol. It is produced by the fermentation of sugar in fruits, grains, and vegetables. Pure ethanol is tasteless, but it can damage human tissue.

The body converts ethanol to acetaldehyde and then acetic acid. When too much acetaldehyde accumulates in the blood, it may produce dehydration and the classic symptoms of a hangover, headache, nausea, and weakness. Chronic abuse of alcohol can cause liver damage as well as disturbed, dangerous behavior. Consumption of alcohol can depress the central nervous system as well.

BACTERIAL TOXINS

Botulism is the most poisonous biological substance known to humans. It is produced by the bacterium *Clostridium botulinum* and acts as a neurotoxin, paralyzing muscles by blocking the release of the neurotransmitter acetyl-choline. If the condition is diagnosed early, then an antitoxin made from horse serum may be given. Because damage caused by the toxin is irrevers-ible, acetylcholine release and muscle strength may take months to return, and recovery depends on how quickly the nerves sprout new endings.

This bacterial toxin is extremely deadly in very small amounts and causes painful spasms before death. The toxin may be ingested from contaminated food, such as canned vegetables, cured pork and ham, smoked or raw fish, and honey or corn syrup. People also become infected with bacterial spores that produce and release the toxin in the body. The spores that contain the toxin are sensitive to heat and may be destroyed by cooking and heating thoroughly at 80 degrees Celsius (176 degrees Fahrenheit) for 10 minutes or longer. Purified botulinum toxin (sometimes called "botox") has been safely used in medicine to treat muscle spasms, eye conditions, excessive sweating, and headaches, as well as to stimulate wound healing and as a cosmetic treatment.

Clostridium tetani is the bacteria that produce tetanus, a potentially deadly nervous system disease (Figure 9-5). The bacteria release tetanospasmin, a poison that blocks nerve signals from the spinal cord to the muscles, caus-ing muscle spasms so severe that they can tear muscles and fracture bones. Tetanus is sometimes called "lockjaw" because spasms often begin in the jaw and may interfere with breathing. Worldwide, tetanus causes approximately 1 million deaths per year. In the United States, tetanus accounts for about five deaths per year, primarily in persons who have not been vaccinated against the disease.

Figure 9-5. *Clostridium botulinum.*

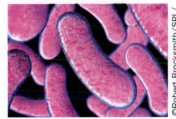

©Robert Brocksmith/SPL/ Photo Researchers

HEAVY METALS AND PESTICIDES

Applications of pesticides have been used primarily for controlling insects, mice, weeds, fungi, bacteria, and viruses that threaten plants or food crops. Pesticides are, by definition, toxic and can cause severe illness and death. Because one of the measures of toxicity of an exposure is its duration, time is of the essence in recognizing pesticide poisoning.

Metal compounds, such as arsenic, lead, and mercury, are very poi-sonous and have also been used for suicide and homicide. Metals may enter the body by ingestion and inhalation or by absorption through the skin or mucous mem-branes. Metals are stored in the soft tissues of the body and can damage many organs throughout the body. Figure 9-6 on the next page lists heavy met-als and pesticides with characteristic symp-toms of an overdose.

Other lethal agents include gases, such as hydrogen cyanide (used in gas chambers), car-bon monoxide (non-ventilated car exhausts), and potassium chloride

Digging Deeper
with Forensic Science e-Collection

Do research on how forensic toxicologists take samples from living or deceased human bodies. Go to the Gale Forensic Sciences eCol-lection on school.cengage.com/forensicscience and investigate the body organs and body fluids that can provide samples for toxicologi-cal testing. Such organs and fluids may include brain, liver, spleen, urine, blood, hair, stomach contents, and the vitreous humor from the eye. Determine which samples are often used in drug testing for employees and athletes. Find out which sample type is the preferred technique for measuring alcohol content in drunk driving cases. Investigate whether there any time restrictions for detecting drugs in the various kinds of samples.

Figure 9-6. *Table of heavy metals and pesticides, with characteristic symptoms of an overdose.*

Drug	Characteristics of Drug Overdose
Pesticides (e.g., DDT, aldrin, dieldrin)	Phosphate-containing pesticides that accumulate in fatty tissue inhibit cholinesterase, leading to excess acetylcholine, which interferes with the movement of nerve impulses and muscular contractions. Anxiety, seizures, twitching, rapid heartbeat, muscle weakness, sweating, salivation, diarrhea, tearing, coma, and death
Lead	Nausea, abdominal pain, insomnia, headache, weight loss, constipation, anemia, kidney problem, vomiting, seizure, coma, and death. Blue discoloration appears along the gumline in the mouth.
Mercury	The Mad Hatter's Disease (hat-makers in England used a mercury compound) is a progressive disorder as mercury is absorbed into the skin or lungs. Acute poisoning from inhalation causes flu-like symptoms such as muscle aches and stomach upset. Chronic poisoning causes irritability, personality changes, headache, memory and balance problems, abdominal pain, nausea, and vomiting, as well as excessive salivation and damage to the gums, mouth, and teeth. Long-term exposure can cause death.
Arsenic	Within 30 minutes of ingestion it produces abdominal pain, severe nausea, vomiting and diarrhea, dryness of the throat, difficulty speaking, muscle cramps, convulsions, kidney failure, delirium, and death. Chronic exposure produces skin lesions and changes in pigment, headache, personality changes, nausea, vomiting, diarrhea, convulsions, and coma.
Cyanide	Cyanide overdose can be fatal six to eight minutes after ingestion. Rapidly causes weakness, confusion, coma, and pink skin from high blood oxygen saturation. Produces an almond-like odor.
Strychnine	Enters the body by inhalation or absorption through eyes or mouth. Body spasms, temperature rises, violent convulsions, and rigor mortis (stiffness after death) occurs within minutes.

or sodium pentothal (used in lethal injections). These poisons produce death by inhibiting enzyme activity, interfering with production of adenosine triphosphate (ATP), which is required to provide energy for cellular function, or, in the case of lethal injections, stopping the heart by destroying the cell's potential for transmitting electrical impulses.

BIOTERRORISM AGENTS

Ricin is a component of the waste product of the manufacture of castor oil from castor beans. It is lethal in humans in quantities as small as 500 micrograms—a dose the size of the head of a pin! Ricin poisoning can be induced in various forms. It can be inhaled as a mist or a powder, ingested in food or drink, or even injected into the body. It acts by entering the cells of the body and preventing them from making necessary proteins, causing cell death. When enough cells die, death may occur. See Figure 9-7 for methods of ricin poisoning and the characteristic symptoms.

Figure 9-7. *Table of methods of ricin poisoning and the characteristic symptoms.*

Exposure	Symptoms
Inhalation	Within eight hours of exposure, difficulty breathing. Within a few hours, fever, cough, nausea, sweating, tightness in the chest, low blood pressure, excess fluid in lungs, and death
Ingestion	Within six hours of exposure, vomiting, diarrhea, bloody urine, dehydration, low blood pressure, hallucinations, seizures, and death
Skin and eye	Redness and pain

Anthrax is caused by a bacterium, *Bacillus anthracis*, that forms endospores (Figure 9-8). A spore is a thick-walled inactive cell that can later grow under favorable conditions. Infected animals can transmit the disease through spores to humans, but human-to-human transmission has not been reported. Anthrax can enter the body by inhalation, ingestion, or skin absorption. Figure 9-9 shows characteristic symptoms of anthrax exposure.

Figure 9-8. *Microscopic view of anthrax organisms.*

©NIBSC/Photo Researchers, Inc.

Figure 9-9. *Table of methods of anthrax exposure and characteristic symptoms.*

Exposure	Symptoms
Inhalation	Initially produces flu-like symptoms, such as sore throat, cough, fever, and muscle aches. Symptoms become progressively worse to include breathing problems and usually results in death.
Ingestion	Nausea, vomiting, fever, abdominal pain, and severe diarrhea. Intestinal anthrax is fatal in 25 to 60 percent of cases.
Skin absorption	Raised, itchy bumps that resemble an insect bite develop into a painless sore with a black area in the center. About 20 percent of untreated cases of cutaneous anthrax result in death. Deaths are rare with appropriate treatment.

In 2001, anthrax spread through the U.S. postal system in letter-sized envelopes caused 22 cases of anthrax infection, half of which resulted in death.

SUMMARY

- Forensic toxicology seeks to identify poisons or drugs in criminals and victims and their likely effects on those people.

- The history of intentional poisoning goes back to ancient Greece. The chemical analysis of poisons in the body began in the 19th century.

- Poisoning is rare as a form of murder, but toxicology is important in studying cases of drug overdoses and sporting violations.

- Controlled substances fall into five groups: hallucinogens, narcotics, stimulants, steroids, and depressants.

- Poisons produced by living organisms include alcohol and bacterial toxins.

- Heavy metals and pesticides are also common poisons found in humans.

- Bioterrorism agents include ricin, a poisonous compound produced by the castor bean plant, and anthrax, a bacterium that produces potent toxins.

CASE STUDIES

Mary Ansell (1899)

Mary Ansell, an English housemaid, poisoned her sister Caroline to obtain an insurance settlement. Mary sent Caroline a cake tainted with phosphorous. Caroline died after eating the poisoned cake. Evidence of Mary's recent purchases of phosphorus and a life insurance policy in her sister's name was provided at her trial. Based on this evidence, Mary was quickly convicted and executed.

Eva Rablen (1929)

Eva Rablen loved to dance. On several occasions, her husband Carroll drove her to the schoolhouse, where weekly dances were held. The First World War had left Carroll wounded and deaf. He often remained in the car while his wife danced in the schoolhouse. Eva would frequently bring Carroll coffee and sandwiches while he waited in the car. On one such evening, Carroll was found dead after consuming his food and coffee. Initially, the death was attributed to natural causes, but later a bottle of strychnine was found below the floorboards of the schoolhouse. Eva was identified by a druggist as the person who purchased the poison a few days before the death of her husband.

When Dr. Edward Heinrich examined Carroll's body, traces of strychnine were found in his stomach, in the coffee cup, and on the seat of the car. On the way to the car, Eva bumped into a woman and spilled some of the poisoned coffee on the woman's dress. Dr. Heinrich examined several drops of coffee left on that woman's dress and found strychnine. In the face of the mounting evidence, Eva changed her plea from not guilty to guilty to avoid the death penalty.

The Death of Georgi Markov (1978) and the Attack on Vladimir Kostov (1978)

After defecting from Bulgaria, Georgi Markov moved to London. While walking one day, he was injected in the leg with ricin. The delivery method used a specially constructed umbrella with a modified tip for injection. He became gravely ill, and on the third day after the attack was vomiting blood. He suffered a complete heart blockage and died. The autopsy revealed a platinum-iridium pellet the size of the head of a pin in his leg. It had been cross-drilled with 0.016-inch holes to contain the toxin. The amount of ricin in the pellet, only two milligrams of the poison, was sufficient to cause his death.

Ten days earlier, a similar assassination attempt was made against Vladimir Kostov in Paris. Kostov's heavy clothing prevented an identical projectile from entering a major blood vessel. Instead, the pellet lodged in muscle tissue, preventing the poison from circulating as it had in Markov's body. This saved Kostov's life. On hearing of Markov's death, Kostov underwent a surgical examination, and the pellet was found before sufficient toxin could be absorbed to cause his death.

Tylenol Tampering (1982)

Extra Strength Tylenol tablets dosed with cyanide claimed seven lives. The person(s) responsible have never been caught. It is believed that cyanide was added to the Tylenol and that the tainted bottles were placed on the shelves of several supermarkets and pharmacies in the Chicago area. In addition to the five bottles responsible for the seven deaths, three poisoned bottles were found on the shelves. Because they were from different production locations, investigators believed the tampering occurred after the product was shipped, rather than in the factory. This was the first documented example of random drug poisoning. The $100,000 reward posted by the drug manufacturer, Johnson and Johnson, has never been claimed. This incident led to the development of tamper-resistant packaging and caplets designed to protect the public.

In 1986, Stella Nickell, a Seattle woman, laced some Excedrin with cyanide and killed her husband for his life insurance. She placed three other poisoned bottles of Excedrin in the store to make it look like a random killing and killed another woman, Susan Snow, in the process. In 1988, Stella was sentenced to 99 years in prison.

Think Critically You are an advertising executive. Select a category of controlled substance. Using your expertise, create a message to communicate the dangers of that substance to the public.

CAREERS IN FORENSICS

Dr. Don Catlin, Pharmacologist and Drug Testing Expert

Dr. Don Catlin recently left his position as head of the UCLA School of Medicine laboratory for a new research position. The UCLA laboratory, with more than 40 researchers, helped expose many drug-related sports scandals, by identifying players who were using performance-enhancing drugs. Catlin is one of the most respected sports and antidoping drug testers in the world, and he plans to remain active in the field of research.

Catlin became a professor in the Department of Pharmacology of the UCLA School of Medicine in 1972. In 1982, his interest in substance abuse led him to found the UCLA Olympic Laboratory to do the drug tests for the 1984 Los Angeles Summer Olympics. He also ran the drug testing for the 1996 Atlanta Summer Olympics and the 2002 Salt Lake Winter Games. His job has included testifying and defending his drug-testing methods in court.

The UCLA laboratory has provided drug education and urine tests to a growing number of sports organizations, including the U.S. Olympic Committee, NCAA, NFL, and Minor League Baseball. The lab has developed novel drug tests, such as the one used to distinguish between naturally produced

Courtesy, USA Today

Dr. Don Catlin, respected sports drug tester.

and artificially taken testosterone. The laboratory is one of the world's premier places for analyzing samples from athletes to detect the use of illegal substances such as anabolic steroids, the blood-oxygen booster erythropoetin, and many other performance-enhancing drugs. It is the busiest lab of its kind in the world, with about 40,000 samples analyzed each year.

What kept Don Catlin so dedicated to the field of sports drug testing? Catlin says, "You should care about preserving something natural and beautiful. I can't think of anything more exciting than the Olympic model, where 220 countries in the world participate, and every four years they send their best to compete against the best from other countries and the best man or woman wins."

To be in the field of pharmacology, one needs a science education with graduate studies that include courses in analytical chemistry, drug metabolism, and drug pharmacokinetics. The drug-testing field requires special knowledge of legal and ethical issues. Pharmacologists can work in universities, hospitals, governmental organizations, nonprofit organizations, or pharmaceutical or related industries.

Learn More About It
To learn more about the work of a pharmacologist, go to
school.cengage.com/forensicscience.

CHAPTER 9 REVIEW

True or False

1. Toxins are poisons manufactured in laboratories.

2. The major ways people are exposed to toxins are by ingesting them, inhaling them, injecting them, or absorbing them through the skin.

3. Today, poisoning is a very common form of murder.

4. There are six basic types of controlled substances.

5. Accidental deaths from drug overdoses are more common than deaths from poisoning.

6. Anabolic steroids increase muscle mass and have no harmful effects.

7. *Clostridium botulinum* causes lockjaw.

8. All alcohols are toxic to the body.

9. Mercury can cause the symptoms of acute poisoning.

10. Some poisons, like potassium chloride, interfere with enzyme activity.

Short Answer

Choose a part of the body or a product from the body and describe what type of drug testing could be performed on that part of the body.

Urine: testing for steroid, narcotics

Hair: testing for alcohol and drug use

Breath: testing for alcohol

Muscle: testing for anabolic steroids

11. How is the test performed?

_____

_____

_____

_____

12. How expensive is the testing?

_____

_____

_____

_____

13. Is the test invasive?

_____

_____

_____

14. Can the test be easily performed on a living person?

_____

_____

_____

15. Is a skilled technician required to perform the test and to read the results?

_____

_____

_____

16. Will the test demonstrate drug usage or toxin exposure:
    a. During the past hour?
    b. During the past several hours?
    c. During the past 24 hours?
    d. During the past few months?

17. How reliable is the drug testing? What variables may affect the results?

_____

_____

_____

Bibliography

Books and Journals

Baden, Dr. Michael. *Unnatural Death Confessions of a Medical Examiner,* New York: Ballantine Books, 1989.

Benjamin, D. "Forensic Pharmacology" in *Forensic Science Handbook*, R. Saferstein ed., Upper Saddle River, NJ: Prentice Hall, 1993.

Chen, Albert. "A Scary Little Pill: A powerful medicine for pain, OxyContin has quickly become a dangerous street drug." *Sports Illustrated* 101.24 (Dec 20, 2004).

"Facing the Big Test," *The Fayetteville Observer*, Fayetteville, NC, Dec. 12, 2006.

Web sites

Gale Forensic Sciences eCollection, school.cengage.com/forensicscience.

http://abcnews.go.com/US/story?id=2861902&page=1

http://www.thesmokinggun.com/archive/years/2007/0326071anna1.html

http://www.cnn.com/2007/SHOWBIZ/TV/03/26/smith.autopsy/index.html

http://faculty.pharmacology.ucla.edu/institution/personnel?personnel_id=45462

http://outside.away.com/outside/features/200507/drugs-in-sports-2.html

http://www.usatoday.com/sports/olympics/2007-02-28-catlin-timeline_x.htm

http://www.washingtonpost.com/wp-dyn/content/article/2007/03/12/AR2007031200804.html

http://www.fdaa.com//forensicdrugabuseadvisor

ACTIVITY 9-1
DRUG ANALYSIS

Introduction:

It is essential that drug samples obtained from suspects are identified conclusively. Positive identification of a drug requires matching the unknown sample with a known sample of the drug. In this activity, students will prepare known samples of the drug to be tested (positive control) and a blank sample containing no drug (negative control). The positive and negative controls will be used for drug comparison and identification. The "drug" we will be testing is Bertinol®, a dangerous and addicting drug that, over time, destroys the liver and intestines. The test for the identification of a drug employs a chemical indicator that changes color in the presence of the drug.

Scenario:

Because of a recent incident involving the sale of the illegal drug Bertinol to junior high students, a "drug dog" was used to detect drugs in the lockers of four suspects. The police dog did detect the presence of white powders in the lockers of the four suspects. Did this white powder contain the drug Bertinol? The drugs were confiscated and sealed in a plastic vial and wrapped in evidence envelopes. The evidence envelopes were sent to the lab for positive identification.

   Your task is to perform a drug test using a chemical indicator for the drug Bertinol. You will need to report your findings to the police. If any of the white powders test positive for Bertinol, the police will have reason to bring in the suspect(s) for further questioning.

Objectives:

Upon completion of Activities A and B, students will be able to:
1. Construct a positive control for drug testing.
2. Construct a negative control.
3. Describe the importance of both types of controls.
4. Demonstrate the role of a positive and negative control in drug testing.
5. Perform a simulated drug test on four white powders.
6. Determine if any of the white powders contain the drug Bertinol.

Safety Precautions:

A carefully maintained clean area should be set aside for testing of drugs. All materials used in this activity are harmless, but it is essential to maintain appropriate techniques in handling all samples. *Treat all samples as if they were actual samples of the drug.* Maintain the chain of custody. Wear safety goggles and dispose of all materials in the manner described by your instructor.

Vocabulary:

Positive control A known sample of the material tested with the chemical indicator used to show a reaction of the known material. A positive control reaction is used to compare with any unknown sample reactions.

Negative control (blank) A sample that does not contain the drug to be tested and should therefore yield a negative test.

Time Required to Complete Activity:

45 minutes to complete both Activities A and B if working in groups of two

Materials:

(per group of two students)
6 empty clean vials with caps
marking pen
positive control envelope containing the drug Bertinol
negative control envelope containing a white powder that does not contain Bertinol
4 evidence envelopes containing white powder residues obtained from each of the four suspects
50 mL rubbing alcohol (70 percent propyl alcohol by volume) or ethyl alcohol
10 mL graduated cylinder or 5 mL pipette
flat wooden toothpicks
25 mL of Bertinol drug test solution in dropper bottles
tape

Procedure:

Part A: Creating the Positive and Negative Controls

1. Label one empty vial Negative Control.
2. Label the second vial Positive Control.
3. Into each vial, add 5 mL of rubbing alcohol.
4. Using the broad, flat side of a toothpick, remove a pinhead-sized amount of Bertinol from the envelope labeled Positive Control. Add this pinhead-sized amount of Bertinol to the vial labeled Positive Control.
5. Using the broad, flat side of a toothpick, remove a pinhead-sized amount of the white powder from the envelope labeled Negative Control. Add this pinhead-sized amount of white powder to the vial labeled Negative Control.
6. Add three drops of Bertinol drug test solution to the Negative Control vial.
7. Add three drops of Bertinol drug test solution to the Positive Control vial.
8. Observe and record the color changes in the Data Table.
9. Save these vials for comparison with the suspects' samples in Procedure B.

Part B: Comparing Samples

1. Label the four vials as follows: Suspect 1, Suspect 2, Suspect 3, and Suspect 4.
2. Using the graduated cylinder or pipette, add 5 mL of rubbing alcohol to each vial.
3. Using a clean, flat toothpick, transfer a pinhead-sized amount of the white powder from Evidence Envelope 1 to your vial labeled Suspect 1. Leave the toothpick in the Suspect 1 vial. It will be used later for stirring.

4. Reseal the Evidence Envelope properly and sign your name to maintain the chain of custody.
5. Repeat the procedure for each of the other Evidence Envelopes (i.e., Suspects 2, 3, and 4).
6. Leave the toothpicks in the suspect vials to stir the contents of each vial until dissolved. Be careful not to mix up the toothpicks.
7. Add three drops of Bertinol drug test solution to each of the four vials and stir with the individual toothpicks.
8. Observe any color changes. Record your results in the Data Table.
9. Compare test vials with the Positive Control and Negative Control vials. Do any of the evidence powders obtained from the four suspects contain the drug Bertinol?
10. Discard all liquids as described by your instructor except the two control vials.

Data Table: Drug Analysis

Sample	Appearance of Solution
Positive Control	
Negative Control	
Suspect 1	
Suspect 2	
Suspect 3	
Suspect 4	

Questions:

1. Explain the role of the positive and negative controls.
2. What measures were taken to avoid contamination of the drug samples?
3. Did any of the four suspect's white powder test positive for the presence of the drug? Explain your answer.
4. When all students in a class compared their results, they found all but one group had identical results. Determine three possible sources of error in technique that might have produced the difference in results.
5. Describe three ways to increase the reliability of this lab.
6. A student noted that when class results were compared, not every group had the same shade of color in their vials. What might account for the differences in color intensity?

ACTIVITY 9-2
URINE ANALYSIS

Introduction:

A student suddenly becomes ill during class. She demonstrates many of the symptoms of having used the drug Bertinol. When questioned, she says she had spent the previous night with three of her friends, none of whom used drugs or became ill. All four girls were asked and agreed to give a urine sample.

Objectives:

By the completion of this activity, students will be able to:
1. Prepare positive and negative controls for testing the drug Bertinol.
2. Perform a urinalysis on the four different students' urine.
3. Determine if any of the students' urine contains the drug Bertinol.

Safety Precautions:

A carefully maintained clean area should be set aside for testing of drugs. All materials used in this activity are harmless, but it is essential to maintain appropriate techniques in handling all samples. *Treat all samples as if they were actual drug samples.* Maintain the chain of custody where directed. Wear safety goggles and dispose of all materials in the manner described by your instructor.

Time Required to Complete Activity:

45 minutes working in groups of two

Materials:

(per group of two students)
Activity Sheets for Activity 9-2
6 empty vials
positive "urine" sample with the drug Bertinol for the positive control
negative "urine" sample without the drug Bertinol for the negative control
"urine" samples from four students
marking pen
Bertinol drug indicator solution
six 10 mL graduated cylinders or six 10 mL pipettes for "urine" samples
2 droppers

Procedure:

Part A: Preparation of the Positive and Negative Control Vials
1. Label one vial as the negative control.
2. Add 5 mL of negative urine to the negative urine vial.
3. Add five drops of the Bertinol drug indicator solution to the negative urine vial and swirl gently.
4. Observe your results and record them in the data table.

5. Repeat the process for steps 1 to 4 for the positive control, except this time add five ml of "urine" from the positive urine sample and add 5 drops of Bertinol drug indicator. Record your results in the data table.

Part B: Urinalysis Testing of the Students' Urine
1. Label four vials: Student 1, Student 2, Student 3, and Student 4.
2. Transfer 5 mL of urine from Student 1 to your vial labeled Student 1.
3. Add five drops of the Bertinol drug indicator solution to each of the students' vials.
4. Repeat the procedure for each of the other samples from Students 2 through 4. Use a sterile pipette each time you transfer the urine from the student urine samples to the vials to avoid contamination.
5. Observe the results.
6. Record your results in the Data Table, comparing student urine samples with the positive and negative urine test vials.
7. Discard all liquids as described by your instructor.

Data Table: Urinalysis

Urine Sample	Appearance of Solution
Positive Urine	
Negative Urine	
Student 1	
Student 2	
Student 3	
Student 4	

Questions:

1. Based on your test results, did any of the girls test positive for the drug or the drug metabolites? Justify your answer.
2. Why should the tests be conducted using designations Student 1, Student 2, and so forth rather than using the student's name?
3. The reliability of urinalysis testing has sometimes been questioned, because there is a possibility of someone altering the test results. Insurance companies frequently will request a urine sample from a prospective client. As a bodily fluid, the urine is used to detect drug use and health conditions such as diabetes. A person using drugs or who has sugar in their urine may be assessed a higher insurance

premium than a person who is drug and disease free. Other than actually watching someone produce the urine sample, the agent from the life insurance company gives the person requesting life insurance a cup and requests a urine sample. When the urine sample is given to the insurance agent, a thermometer is inserted into the urine. Why would the insurance agent want to know the temperature of the urine?

Further Study:

1. To help reduce student drug use, some schools have decided to test the students' urine for the presence of drugs. Research the following information about drug testing in junior high and senior high schools
   a. High school coaches are asked to note symptoms of steroid abuse among their athletes. What would be some warning signs that the athletes were abusing steroids?
   b. List the drugs commonly tested for in a high school urinalysis.
   c. Investigate companies that sell urinalysis kits to high schools:
      · How expensive are the drug testing kits?
      · Do these kits require a trained lab technician to read the results?
2. Other methods exist for testing someone for the presence of drugs other than a urinalysis.
   a. What are the other methods?
   b. Compare and contrast the different types of methods used to detect the presence of a drug. Include in your answer:
      · Which methods detect the current use of drugs?
      · Which method is best used to detect drug usage over a period of two months?
      · Which methods are less invasive?
      · Which methods are less likely to be tampered with or altered?
3. The U.S. Supreme Court has ruled that mandatory drug testing for all students violates the Fourth Amendment of the U.S. Constitution.
   a. Summarize the contents of the Fourth Amendment.
   b. Explain how this amendment applies to mandatory drug testing of all students.
   c. Research when the Supreme Court rendered this decision.

ACTIVITY 9-3
DRUG IDENTIFICATION

Scenario:

Neighbors at the College Apartments complained that the person in room 202 had his television continually running with the volume turned up too loud. The people in rooms 201 and 203 said the sound kept them awake all night. When the neighbors tried knocking on the door of room 202, no one answered. They became concerned and called the police.

When the police arrived, they discovered that the young man had apparently died while sitting in front of the television. While working the crime scene, the police discovered 15 identical white pills on the table next to the victim. Did a doctor prescribe the drugs for medication? Are these over-the-counter drugs purchased without a prescription? Did the person use illegal drugs? How many of these drugs did the victim take? Was this death accidental or a suicide?

The first step is to determine what drugs the 15 pills contain. In this activity, you will perform preliminary tests on "drugs" to help determine their identity.

Background:

When samples suspected of being illegal drugs are brought into a laboratory, spot tests are often performed. These tests rapidly show results and are used to identify some of the most common drugs. The tests include:

Name of the Drug Test	Drug Identified	Positive Reaction
Marquis	Opium alkaloids such as heroin, morphine, codeine, or ecstasy Amphetamines Speed OxyContin	purple orange orange-brown gray
Cobalt thiocyanate	Cocaine	blue flaky precipitate
p-Dimethlyamino-benzaldehyde (p-DMAB)	LSD	blue
Duquenois	Marijuana	purple
Cobalt acetate/iospropylamine test	Barbiturates	red-violet

Because most of these drugs to be tested are controlled substances, we will substitute similar tests, which would parallel real testing situations. The drugs in question for this case include aspirin, acetaminophen (Tylenol), naproxen (Aleve), and ibuprofen (Motrin).

Safety Precautions:

1. Anyone working at or near the testing station MUST wear safety goggles and gloves. The chemicals used are hazardous. They will be used in minute amounts.
2. Place newspapers on the desktops where testing is to be conducted.
3. Wash your hands thoroughly after testing is complete.
4. Discard all chemicals as directed by your instructor.
5. Thoroughly clean all counters and desk surfaces where testing has been completed.

Materials:

(per group of three to four students)
1 plastic well tray (24 wells per tray)
drug testing reagents in dropper bottles
Marquis, tannic acid, ferric chloride, nitric acid
samples of aspirin, Tylenol, Motrin, and Aleve
colored pencils
1 (5 × 8) index card
1 pair of scissors
1 pair of safety goggles per person
1 pair gloves per person

Procedure:

Part A: Preparing the Wells

1. Prior to testing, your instructor has prepared the unknown samples. Please treat these samples with care to avoid contamination. Wear safety goggles and gloves when performing this lab or standing in the vicinity of its performance.
2. Obtain a 5 × 8 card and trace and cut out the outline of the plastic mini-well tray. Place the plastic mini well tray on top of the 5 × 8 card. (See the example on the next page.)
3. Cut out a horizontal section of the 5 × 8 card the width of four rows of wells as indicated in the diagram on the next page to form a slotted card. This 5 × 8 card will be used to prevent contamination of one drug with the other while filling your wells. Put the slotted card aside for later use when adding your drugs to the mini well tray.
4. Obtain a second 5 × 8 card. Turn it so that the blank side is up. Place a mini 24-well tray on top of the card.
   a. Trace an outline of the mini tray on top of the 5 × 8 card.
   b. On the four-row side of the plastic mini tray, write the first letter (M, T, F, and N) for each of the chemical reagents.
   c. On the six-row side of the plastic mini tray, write the name of the "drug" to be tested in the first four rows.
   d. Leave the fifth row blank.
   e. Write "unknown" in the sixth row
   f. Record your initials in the lower right-hand corner of the 5 × 8 card.

Part B: Adding the Drugs to the Test Wells

1. The drug samples have been placed at different stations in the room. You are to take your plastic mini well tray with its lid and your labeled 5 × 8 card to each of the stations. Place the labeled 5 × 8 card behind

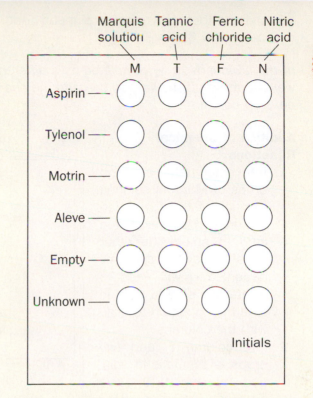

| | Marquis solution | Tannic acid | Ferric chloride | Nitric acid |
	M	T	F	N
Aspirin	○	○	○	○
Tylenol	○	○	○	○
Motrin	○	○	○	○
Aleve	○	○	○	○
Empty	○	○	○	○
Unknown	○	○	○	○

Initials

the plastic wells to ensure that you add the correct drug to the correct well.

2. You will add each of the drugs to their prescribed well. Aspirin will be added in the first row under Marquis, under Tannic Acid, under Ferric Chloride, and under Nitric Acid. There should be four wells filled with aspirin from left to right as indicated by the diagram. Follow the same procedure with each drug. To avoid contamination, use your cut-out slotted 5 × 8 card. Place the card over the plastic mini wells so that the cut-out row is correctly positioned for the drug that you will be adding. This way, the other rows are covered and will not become contaminated.

The drugs to be tested (Aspirin, Tylenol, Motrin, Aleve, and the unknown) will be located at a designated lab station. Each station will have a small vial of white powder and a toothpick.

3. Using the flat end of the toothpick, place a pinhead-sized amount of aspirin powder into the four wells as indicated by the diagrams. Wipe off the 5 × 8 slotted card with a clean paper towel after adding each row of drugs.

4. Repeat step 3 with each of the other three remaining powders. Place powders in mini wells as shown in the diagrams.

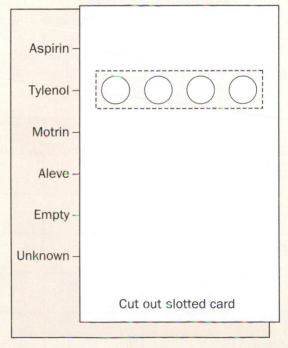

	Marquis solution	Tannic acid	Ferric chloride	Nitric acid
Aspirin				
Tylenol	○	○	○	○
Motrin				
Aleve				
Empty				
Unknown				

Cut out slotted card

5. Using a toothpick, place a pinhead-sized amount of the unknown powder in row 6.

(If the lab needs to be completed on a different lab period, cover your mini wells with the plastic lid. You can add the chemical reagents during your next lab period.)

Addition of Chemical Reactions

1. Into column M, add two drops of Marquis solution to each of the five different drugs. Use the slotted card to shield the other rows to avoid contamination. You will need to rotate the slotted card in a vertical position to align with the columns.

2. Into column T, add two drops of tannic acid solution to each of the five different samples. Use the slotted card to shield adjacent wells from contamination.

3. Into column F, add two drops of ferric chloride solution to each of the five different samples. Use the slotted card to shield adjacent wells from contamination.

4. Into column N, add two drops of nitric acid solution to each of the five different samples. Use the slotted card to shield adjacent wells from contamination.

5. Gently agitate the mini tray, being very careful not to spill any of the mixtures.

6. Observe any changes that occur in the mini wells after adding the reagents. Record the changes in the Data Table 1. In particular, note

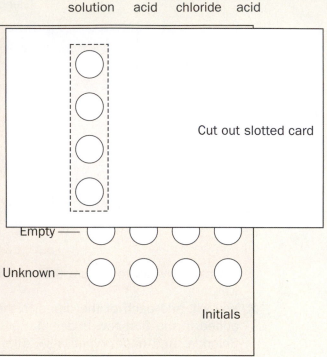

Data Table 1: Drug Testing Results

Row	Drug Tested	Marquis (Column M)	Tannic Acid (Column T)	Ferric Chloride (Column F)	Nitric Acid (Column N)
Row 1	Aspirin				
Row 2	Tylenol				
Row 3	Motrin				
Row 4	Aleve				
Row 5	Unknown				

changes in color, bubble formation, or precipitation. Use NR if no reaction occurred.

7. Sketch the appearance of your results on Data Table 2. Use colored pencils to indicate any color changes.

8. Examine your unknown drug. Compare the reactions of the unknown drug with the four known drugs. Based on these tests, can you identify the unknown drug?

Questions:

1. Refer to the opening scenario about the young man found dead in front of his television with 15 white pills next to him. After completing these preliminary tests, how could you determine if the victim took Aspirin, Tylenol, Motrin, or Aleve?

2. Justify your answer using supporting data from your experiment.

3. How would it be possible to determine if the victim took an overdose of pills? What procedures would be done at the autopsy to determine how much of these drugs the victim took?

4. Why was it important to wear both goggles and gloves when doing this experiment?

Data Table 2: Sketch Your Results
Use colored pencils to indicate color changes.
Write NR to indicate no reaction.

	Marquis solution	Tannic acid	Ferric chloride	Nitric acid
	M	T	F	N
Aspirin	○	○	○	○
Tylenol	○	○	○	○
Motrin	○	○	○	○
Aleve	○	○	○	○
Empty	○	○	○	○
Unknown	○	○	○	○

Initials

Further Study

1. Investigate how forensic scientists test for the presence of the following drugs:
   a. Cocaine
   b. Heroin
   c. Amphetamines
   d. Barbiturates, rohypnol, PCP, glue sniffing

2. If a person is found unconscious as a result of an overdose of pills, he or she may be taken to a hospital to have the stomach pumped.
   a. What is this procedure?
   b. Is there any danger in having this procedure done?
   c. Would a stomach pump be of value if someone had injected the drug into his or her system? Explain your answer.

CHAPTER 10

Handwriting Analysis, Forgery, and Counterfeiting

MASTER FORGERS

Frank W. Abagnale, a reformed master forger, describes in his book, *The Art of the Steal,* how a visitor from Argentina was issued a parking ticket on a rental car in Florida. Although the fine was $20, he placed $22 in an envelope and mailed it to the Miami city clerk. On receipt of the money, the clerk issued a $2 refund. On receiving the check, the man scanned it into his computer, changed the amount to $1.45 million, and deposited the check into his account in a bank in Argentina. The check was cashed, and the money was transferred. He was never arrested, and the money was not recovered.

According to Abagnale, the Argentinean example is not uncommon. Stolen money is often not recovered, and thieves are not caught. Abagnale tells of his own life of forgery and fraudulence in the book *Catch Me if You Can.* He began his life of crime as a teenager, when he changed a number on his driver's license to make himself appear 10

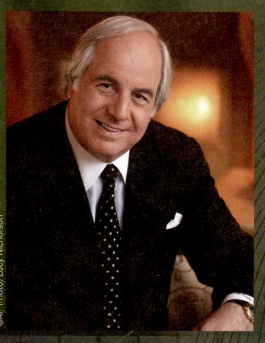

Frank Abagnale, once a wanted forger, is now a leading consultant in the area of document forgery and fraudulence.

©AP Photo/Lucy Nicholson

years older. After acquiring a small amount of money, he opened a bank account. He came up with the idea to print his account number in magnetic ink on deposit slips and return them to the bank counter. By the time the bank discovered his fraudulent scheme, he had made over $40,000, and he had already changed his identity. Working with eight different identities, he passed more than $2.5 million in fraudulent checks in 26 countries and throughout the United States.

Frank Abagnale is now a leading consultant in forgery, fraudulence, and secure documents. For more than 25 years, he has consulted with many financial institutions, corporations, and government agencies, such as the Federal Bureau of Investigation (FBI). Today, he teaches and lectures on how to detect forgery, avoid consumer fraud, and prevent crime. Abagnale says that the best way to deal with fraud is to prevent it from happening in the first place.

OBJECTIVES

By the end of this chapter, you will be able to

✔ Describe 12 types of handwriting exemplars that can be analyzed in a document.

✔ Demonstrate an example of each of the 12 exemplars of handwriting traits.

✔ Identify the major goals of a forensic handwriting analysis.

✔ Describe some of the technology used in handwriting analysis.

✔ Distinguish between the terms *forgery* and *fraudulence*.

✔ Identify several ways in which businesses prevent check forgery.

✔ Describe four features of paper currency that are used to detect counterfeit bills.

TOPICAL SCIENCES KEY

VOCABULARY

counterfeiting the production of an imitation of currency, works of art, documents, and name-brand look-alikes for the purpose of deception

document analysis the examination of questioned documents with known material for a variety of analyses, such as authenticity, alterations, erasures, and obliterations

document expert a person who scientifically analyzes handwriting

exemplar a standard document of known origin and authorship used in handwriting analysis

forgery the making, adapting, or falsifying of documents or other objects with the intention of deceiving someone

fraudulence when a financial gain accompanies a forgery

questioned document any signature, handwriting, typewriting, or other written mark whose source or authenticity is in dispute or uncertain

INTRODUCTION

Document analysis, a very broad area in the field of forensics, is the examination and comparison of questioned documents with known material. Experts establish the authenticity of the documents and detect any changes, erasures, or obliterations that may have occurred. A **questioned document** is any signature, handwriting, typewriting, or other written mark whose source or authenticity is in dispute or uncertain. Checks, certificates, wills, passports, licenses, money, letters, contracts, suicide notes, receipts, and even lottery tickets are some of the questioned documents of interest. Experts in this field may examine handwriting, typewriting, commercial printing, paper material, and the type of ink in these documents. Figure 10-1 shows an example of a historical document.

A **document expert** is a specially trained person who scientifically analyzes handwriting and other features in a document. For example, a document expert may be called into a crime-scene area or to the lab to examine the handwriting of threatening notes, ransom notes, or suicide notes. Investigators analyze and compare various traits, such as the appearance of letters, of suspicious documents with known samples to help identify the author of the document. Investigators might also be asked to detect changes that may have occurred in an original document.

A document expert is different from a graphologist, who studies the personality of the writer based on handwriting samples. The study of graphology is not necessarily accepted as part of forensic science, but it can be used as a possible indicator of the writer's personality type. The scientific analysis of handwriting is the focus of this chapter.

Figure 10-1. *Historical documents are often targets for forgers.*

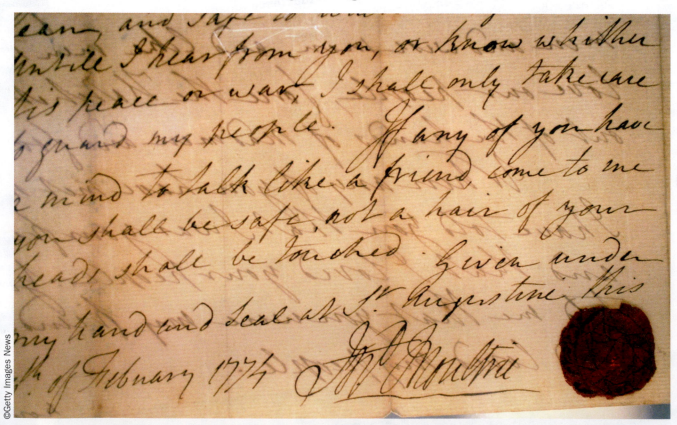

©Getty Images News

HISTORY OF FORENSIC HANDWRITING ANALYSIS

Like fingerprints, every person's handwriting is unique and personalized. Because handwriting is difficult to disguise and forge, handwriting analysis is a good tool for including or excluding persons when determining a match between known material, known as an **exemplar**, and a questioned document.

In the 1930s, handwriting analysis played an important forensic role during the trial of Bruno Richard Hauptmann for the kidnapping and murder of the son of world famous aviator Charles Lindbergh. Handwriting analysis of the many ransom notes, along with known handwriting samples and other evidence, led to Hauptmann's conviction and execution (Figure 10-2). Today, Hauptmann's involvement in the crime has come into question because of the manner in which samples were collected and how the evidence was handled.

Figure 10-2. *Comparative handwriting samples from Hauptmann used in the Lindbergh kidnapping case.*

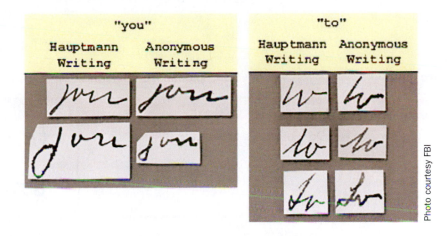

Photo courtesy FBI

The courts have not always accepted handwriting analysis as a creditable form of evidence. This changed in 1999, when the U.S. Court of Appeals determined that handwriting analysis qualified as a form of expert testimony. Handwriting evidence is admissible in court, provided that scientifically accepted guidelines are followed. Scientific analysis of handwriting is now an important tool for forensic document examiners. Scotland Yard, the F.B.I., and the Secret Service use handwriting analysis in solving important cases.

INTRODUCTION TO HANDWRITING

Everyone's handwriting exhibits natural variation depending on several factors. The use of different types of writing instruments, such as a pen, pencil, marker, or crayon, can affect our handwriting. Our mood, our age, and how hurried we are all contribute to the differences we notice in our own handwriting.

Despite these minor variations in handwriting, each person has a unique handwriting style. Characteristics such as the slant and curl of the letters, the height of the letters, or even how the page is filled with text can distinguish our identity. This is because the brain is doing the writing! Adults show only slight variation in handwriting, because as children we learn to write through basically the same method. However, once a person starts writing subconsciously, with characters formed as a result of habit, unique handwriting is formed.

CHARACTERISTICS OF HANDWRITING

A person's handwriting exhibits unique characteristics that make it distinguishable from other samples. Handwriting experts examine 12 major categories of exemplars. These 12 characteristics are functions of letter form, line form, and formatting.

Letter form includes the shape of letters, curve of letters, the angle or slant of letters, the proportional size of letters, and the use and appearance of connecting lines between letters. It also includes whether letters are shown correctly, such as a dotted "i" and a crossed "t."

Line form includes the smoothness of letters and the darkness of the lines on the upward compared to the downward stroke. Line form is influenced by the speed of writing and the pressure exerted while writing. The choice of writing instrument can also influence line form.

Formatting includes the spacing between letters, the spacing between words and lines, the placement of words on a line, and the margins a writer leaves empty on a page. Some characteristics studied by handwriting experts are shown in Figure 10-3.

Figure 10-3. Characteristics of handwriting. (continued on the next two pages)

Specific Trait	Description	Example
Line quality	Do the letters flow or are they erratic or shaky?	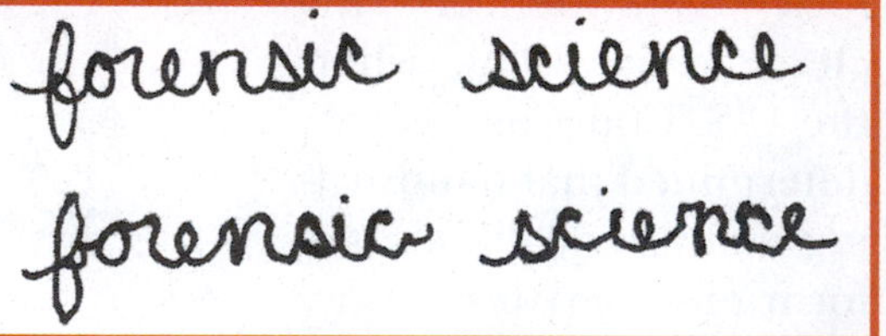 *forensic science* *forensic science*
Spacing	Are letters equally spaced or crowded?	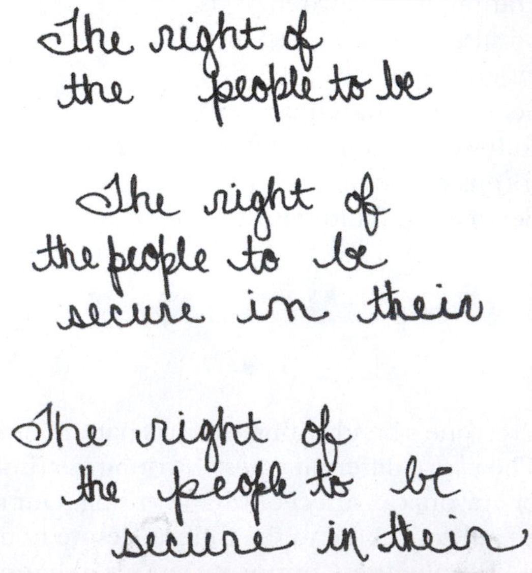 *The right of the people to be* *The right of the people to be secure in their* *The right of the people to be secure in their*
Size consistency	Is the ratio of height to width consistent?	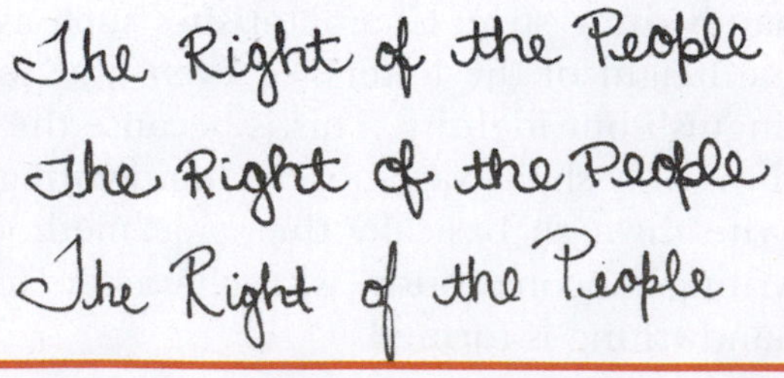 *The Right of the People* *The Right of the People* *The Right of the People*

Specific Trait	Description	Example
Continuous	Is the writing continuous or does the writer lift the pen?	*forensic science* *forensic science*
Connecting letters	Are capitals and lower-case letters connected and continuous?	*The Right of the* *The Right of the*
Letters complete	Are letters completely formed? Or, is a part of the letter missing?	*the right of the people* *the right of the people*
Cursive and printed letters	Are there printed letters, cursive letters, or both?	*Forensic Science* *Forensic Science* *Forensic Science*
Pen pressure	Is pressure equal when applied to upward and downward strokes?	*forensic science* *forensic science* *forensic science*
Slant	Left, right, or variable?	*forensic science* *forensic science* *forensic science*
Line habits	Is the text on the line, above the line, or below the line?	*forensic science* *forensic science* *forensic science*

Specific Trait	Description	Example
Fancy curls or loops	Are there fancy curls?	*Forensic Science*
Placement of crosses on t's and dots on i's	Correct or misplaced? Are t's crossed, crossed in the middle, toward top, or toward the bottom? Are i's dotted, dotted toward the right, left, or centered?	*right right right right*

HANDWRITING EXAMINATION

The goal of forensic handwriting analysis is to answer questions about a suspicious document and determine authorship using a variety of scientific methods. Methods are based on the principle of identification in that "two writings are the product of one person if the similarities . . . are . . . [unique] and there are no fundamental unexplainable differences." Document experts often compare handwriting characteristics of a questionable document to those of a known sample or *exemplar* to try to determine if the same person wrote the document. These analyses can also help detect forgeries. **Forgeries** are documents made, adapted, or falsified with the intention of deceiving someone.

ANALYZING A HANDWRITING SAMPLE

There are three basic steps in the process of analyzing a handwriting sample. First, the questioned document and the standards (exemplars) are examined and detectable characteristics are recorded. Obtaining a standard may require a suspected author to write a sample for the investigators under supervision. If possible, a handwriting sample should be obtained without first informing someone of the intention of comparison, such as asking for a written statement by the involved parties. The best exemplars tend to be letters, diaries, greeting cards, or personal notes. The obtained exemplar should also contain several of the words or letter combinations found in the questioned document. Next, the characteristics of the questioned item are compared with the known standard. For those samples that appear to be similar to the questionable document, there must be a thorough analysis addressing all of the handwriting characteristics in each document. Finally, experts determine which characteristics are valuable for drawing a conclusion about the authenticity and authorship of the questioned document (Figure 10-4).

If there are obvious differences between a standard and a questioned document, then it is likely that the documents have different authors. Those samples can be visually eliminated without even having to assess the list of handwriting characteristics. However, similarities do not necessarily guarantee common authorship, because it is possible that unique characteristics of a person's handwriting may occur in another's handwriting. Highly trained document experts must take into account a great number of factors and a statistically significant repetition of similarities in

Figure 10-4. *A document expert scientifically analyzes the handwriting in a document.*

©Mikael Karisson/Alamy

their analyses. Professionals also have ways of determining whether a person has tried to disguise his or her handwriting or to copy someone else's handwriting, known as a conscious writing effort. Many things can be done to minimize this conscious writing effort, such as the following: (1) a suspect should not be shown the questioned document; (2) a suspect should not be given any instructions about punctuation or spelling; and (3) the pen and paper should be similar to that of the questioned document.

TECHNOLOGY USED IN HANDWRITING ANALYSIS

Initial comparisons of documents are done with the naked eye, a handheld lens, or a microscope. However, even more advanced technology available today can assist the examiner with more technical aspects of the writing and document. Specialized equipment can reveal minor details about how a document was changed. For example, examination using an infrared spectroscope can determine if more than one kind of ink was used on the document. This is because of the way that different inks may absorb or reflect different wavelengths of light such as infrared.

Biometric Signature Pads

 The biometric pad, a new research tool, has been designed for identity authentification. The computerized pad recognizes your signature based on the speed, pressure, and rhythm of signing your name (Figure 10-5). Forgeries can be recognized by slight differences that are detected by the pad.

Computerized Analysis

Computerized analysis of handwriting samples has the advantage of being faster and more objective than analysis by an individual. For example, if the pen pressure is being reviewed, an examiner looking at the sample uses his or her subjective opinion. However, if the handwriting is first scanned into a computer, the pen pressure can be objectively analyzed by the shading in the pixels.

The Forensic Information System for Handwriting (FISH) is a computerized handwriting database used and maintained by the Secret Service (Figure 10-6). Investigators scan in handwritten documents for a comparative analysis. Once the sample is scanned, it can be compared to other existing handwriting in the database. This system has verified that no two writers pen their words exactly the same, nor do they have the same combination of handwriting characteristics.

HANDWRITING EVIDENCE IN THE COURTROOM

After handwriting samples are scientifically analyzed, the expert handwriting witness prepares a written report of the analysis to present to a jury. Both the defense and prosecuting attorneys ask the handwriting expert questions about the analyses. The expert witness demonstrates how document comparisons were made and how they were used to indicate the suspect's guilt or innocence. The expert witness validates comparisons by showing the jury

Figure 10-5. *Advanced technology today, such as biometric signature pads, allows for more accurate analyses of handwriting samples, even at the grocery checkout line.*

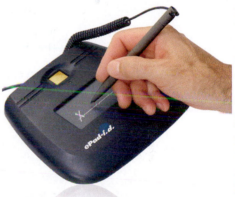

ePad-i.d., Handwritten electronic capture device by Interlink Electronics, Inc.

Figure 10-6. *The Forensic Information System for Handwriting (FISH) is a database of handwriting features that is categorical and quantifiable.*

©Mikael Karlsson/Alamy

examples of similarities or dissimilarities that led to the final conclusion. In court, the expert must be able to defend his or her findings, because the defense will likely hire their own document examiner to refute the prosecution's expert witness.

SHORTCOMINGS IN HANDWRITING ANALYSIS

Although an experienced document expert can detect many cases of forgery, some may be missed. One limitation is that the quality of the standards obtained often determines the quality of a comparison analysis, and good standards may be difficult to obtain. For example, analysis errors have occurred in history when the standard documents that experts used in their comparisons turned out to be forgeries as well. Another limitation is the effects of mood, age, drugs, fatigue, and illness on a person's handwriting.

Because document analysis has become well accepted in forensics alongside other evidentiary types, programs have evolved to certify the training. The American Board of Forensic Document Examiners is one establishment that offers such a training program. Although it is still important that handwriting evidence be used in combination with other sources of evidence, handwriting analysis is considered a reproducible and peer-reviewed scientific process.

FORGERY

As discussed previously, forgery is the process used by criminals to make, alter, or falsify a person's signature or another aspect of a document with the intent to deceive another. Forged documents might include checks, employment records, legal agreements, licenses, and wills. When a material gain, such as money, accompanies a forgery, it is called **fraudulence**. Generally, the primary purpose of forging something is to profit from the fake or alteration. For example, Martin Coneely was a fraudulent forger in the United States in the 1900s. In 1937, after selling a forged Abraham Lincoln letter (Figure 10-7), Coneely was arrested and spent three years in prison. Ironically, his forgeries are collector's items today.

Figure 10-7. *Martin Coneely forged Lincoln's writing and signature, but was caught and sent to prison.*

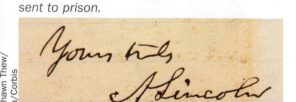

©Shawn Thew/ epa/Corbis

CHECK FORGERY

Americans write more than 70 billion checks a year. Approximately $27 million in illegitimate checks are cashed each day. Criminals can alter or acquire checks in many ways, including:

• Ordering someone else's checks from a deposit slip

• Directly altering a check

• Intercepting someone's check, altering it, and cashing it

• Creating forged checks from scratch

PREVENTING CHECK FORGERY

Reformed master forger Frank Abagnale once said that the best way to deal with fraud is to prevent it from happening in the first place. How do companies protect themselves against forgeries? Several techniques are used to protect businesses, banks, and the public from forged and altered checks, as shown in Figure 10-8. However, these are all aspects of the paper, and they require someone to be knowledgeable about these security features and willing to look for them. In their attempt to prevent check fraud, many banks hope to eventually eliminate checks alltogether. In fact, many banks and credit unions encourage the use of their debit and check cards for this very reason.

Figure 10-8. *Methods used to prevent check forgery.*

Print checks on chemically sensitive paper
Use a large font size that requires more ink and makes alterations more difficult
Use high-resolution borders on the checks that are difficult to copy
Print checks in multiple color patterns
Embed fibers in checks that glow under different types of lights
Use chemical-wash detection systems that change color when a check is altered

LITERARY FORGERY

Literary forgery refers to forgery of a piece of writing, such as an historic letter or manuscript. Letters written by famous people are often valuable, especially if the writer was an important world figure, developed a famous theory, or was a notable writer. For example, a letter written by Adolf Hitler, Albert Einstein, or Charles Darwin would be treasured because it might provide insight into the thinking of the writer.

The best literary forgers try to duplicate the original document, so the materials used are similar to those used in the original document. They do this by collecting old paper or old books, from which they can cut out properly aged paper for their forgeries. Because the process of papermaking has changed, it is essential for forgers to use aged paper to pass the microscopic examination tests. Inks have also changed; so intelligent forgers

must mix their own inks from material that would have been used at the time. Watermarks impressed in the paper when it was made also help to age a piece of paper as shown in Figure 10-9. Handwriting tools and styles of penmanship popular at the time of the printing are also considered.

Documents are sometimes chemically treated to make them look older. Chemicals may be added to the paper to age both

Figure 10-9. *Forgers have used watermarks in the past to help age a document.*

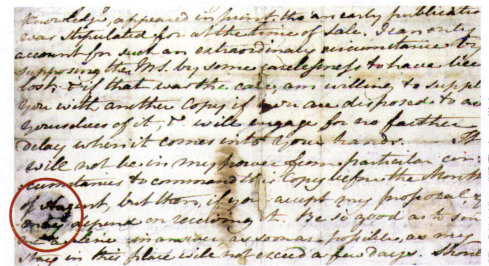

Figure 10-10. An injured Mark Hofmann is arrested for murder.

©AP Photo/The Desert News

the paper and the ink. In the early 1980s, Mark Hofmann, a document dealer and master forger, created several hundred forged documents using this method. Besides forging documents, Hofmann also forged coins and banknotes. One of Hofmann's most significant forgeries was his creation of 116 pages of a supposedly lost Mormon document. He sold this document for a fortune to a Mormon collector. Hofmann also forged works attributed to Emily Dickinson, Abraham Lincoln, and Mark Twain. In 1985 Hofman devised a plan to forge another collection of Mormon documents. Unable to produce the forgeries in time, he used a bomb to buy time and escape detection. His bombs killed an innocent Mormon business leader, Steven Christensen, and Kathy Sheets, wife of Christensen's business partner, Gary Sheets. A third bomb exploded unexpectedly in Hofmann's car, severely injuring him and attracting police attention. Hofmann was tried and convicted of forgery and murder and is currently serving his life sentence. Figure 10-10 shows Hofmann after his arrest, holding his injured hands.

COUNTERFEITING

When false documents or other items are copied for the purpose of deception, it is called **counterfeiting**. Travelers' checks, certain bonds, and currency are among the most often counterfeited items. Other examples of counterfeited items include coins, food stamps, postage stamps, and paper money. Counterfeiting of money is one of the oldest crimes. Under U.S. law, counterfeiting is a federal felony punishable with up to 15 years in prison. The U.S. Secret Service is the federal agency in charge of investigating counterfeit U.S. currency.

COUNTERFEIT CURRENCY

Did You Know?

U.S. law enforcement agencies forecast that companies lose approximately $400 billion to $450 billion annually to counterfeiters.

In the past, with access to a scanner and color printer, it was not very difficult to create counterfeit currency. Scanning could pick up the intricate lines and details found on currency. The Secret Service, with the aid of technology, has added features to paper currency that, when scanned, prevent the currency from being copied. If the currency was successfully scanned, a counterfeiter would still encounter difficulty printing it. The most sophisticated printers cannot reproduce this microscopic detail, because of the built-in security features of money (Figures 10-11 and 10-12). Counterfeit

Figure 10-11. *Different parts of paper money contain tiny, intricate lines and details that cannot scan well.*

©Visions of America, LLC/Alamy

Figure 10-12. *As you can see here, the tiny, intricate lines and details on paper money do not always print well in counterfeit bills.*

©Visions of America, LLC/Alamy

In 2004, a woman tried to buy more than $1,000 worth of items at Wal-Mart using a fake $1 million bill. The U.S. Treasury does not make a $1 million bill, so she was sent to jail.

money also feels different because real money is printed on special paper. In fact, the number-one way that people suspect fakes and scrutinize money is because it doesn't feel right. The paper itself is therefore the number-one security feature.

The government continues to change the design of paper money to make currency more difficult to copy and to prevent counterfeiting. The new series of currency changes began with a revision to the $20 bill on October 9, 2003, followed by the $50 bill on September 28, 2004, and the $10 bill on March 2, 2006. More recently, the U.S. government announced that a redesigned $5 note should be released in early 2008. In Activity 10-3 you will make observations about the newly designed bills.

DETECTING COUNTERFEIT CURRENCY

It is relatively easy to detect counterfeit currency. Counterfeit-detecting pens are inexpensive special pens and markers containing the element iodine. When they come in contact with a counterfeit bill, the paper marked with the pen will change to a bluish-black color. The color change is caused by a chemical reaction involving starch, a compound found in regular printer paper. By contrast, real currency does not contain starch. Many currencies are also printed on paper containing a fiber. When real money is marked with the counterfeit-detecting pen, the pen will leave a pale yellow color on the bill, which fades within a very short time. Figure 10-13 shows features found in real currency.

Pen manufacturers claim the counterfeit-detecting pen is 98% effective. However, the U.S. government does not concede this level of effectiveness and uses additional criteria for judging whether currency is counterfeit. These criteria are important, because some counterfeiters actually bleach small bills to provide the correct paper for use. For example, a counterfeiter might bleach a $1 bill for reprinting into a $50 or $100 denomination. This bill will pass the counterfeit-pen test, but not some of the other safety measures found on real currency. Furthermore, there is currently a global movement to change to polymer money, a type of plastic money. Plastic money is much more difficult to counterfeit and less expensive to print.

Figure 10-13. *Features found in real currency: making counterfeiting money difficult.*

Number	Some Features Found in Real Currency
1	Portrait stands out from the background and appears raised off the paper.
2	There is minute microprinting on the security threads, as well as around the portrait.
3	Serial number is evenly spaced and the same color as the Treasury seal.
4	Check Letter and Quadrant Number
5,6	Federal Reserve seal (5) no sharp points, and Treasury seal (6) with clear, sharp sawtooth points
7	Clear red and blue fibers are woven throughout the bill. Security thread is evident, consisting of a thin, embedded vertical line or strip with the denomination of the bill written in it.
8	Federal Reserve Number and Letter
9	Series
10	Check Letter and Face Plate Number
11	Watermark appears on the right side of the portrait of the bill in the light.
12	When a new series bill is tilted, the number in the lower right-hand corner makes a color shift from copper to green resulting from color-shifting ink.
13	Clear, distinct background details and lines
14	Clear, distinct border edge

SUMMARY

- Fraudulence is attempting to get financial or other gain from forgery.

- Handwriting analysis is the examination of questioned documents compared with exemplars by document experts to establish the authenticity and/or authorship of the documents.

- Document experts use scientific tools and protocols to compare handwriting characteristics of a questionable document to those of an exemplar to help identify authors and detect any alterations, erasures, and obliterations.

- Certain aspects of a person's handwriting style, such as letter form, line form, and formatting, can be analyzed to ascertain authenticity or authorship.

- Handwriting analysis has become an important tool, especially for forensic examiners. Handwriting experts help financial, legal, and governmental institutions, as well as the general public, detect and prevent forgery, counterfeiting, and other fraudulent crimes.

- Technological advances, such as the biometric signature pad and the use of the infrared spectroscope, have greatly enhanced the detection of forged documents.

- Countries continue to refine methods to protect their currency from counterfeiters.

CASE STUDIES

John Magnuson (1922)

A package mailed to the rural home of James Chapman exploded as it was unwrapped. James Chapman's wife actually opened the package, thinking it was a Christmas present. She was killed, and James was injured. John Magnuson, a neighbor, was a suspect because he had recently quarreled with Chapman over property drainage rights. John Tyrell was called in to analyze the handwriting on the package. He concluded that Magnuson's handwriting matched the handwriting on the package. In addition, many of the misspellings indicated a reliance on phonetic spelling and a person of Swedish ancestry. John Magnuson was the only person of Swedish descent (ancestry) in the area and lived less than four miles from Chapman's home. The pen point and ink mixture used on the bomb's label also matched supplies found at Magnuson's house. Magnuson was sentenced to life imprisonment.

The Hitler Diaries (1981)

In February 1981, three diaries supposedly written by Adolf Hitler were discovered. Document experts authenticated the documents by comparing them with forged samples. A bidding war followed, with the price of some of the manuscripts reaching $3.75 million. Eventually, the paper on which the documents were written exposed the hoax. A paper whitener found in many of the pages of the documents had not been developed until nine years after the war ended and Hitler committed suicide. The inks used were also from the postwar era. It was determined that the documents had been written less than a year before their discovery. Konrad Kujau, the West German memorabilia dealer who had written and forged the diaries, was located and imprisoned for four years. The hoax was said to have cost more than $16 million in lost revenues to those who had purchased the alleged diaries.

Digging Deeper
with Forensic Science e-Collection

Fraudulent secret diaries of Hitler's that spanned the years 1932 through 1945 were discovered in 1983. These fake diaries set the standard for literary hoaxes. The diaries were at best an amateur job by forger Konrad Kujau, when it comes to forgery standards. So, how did they acquire such worldwide fame? Go to the Gale Forensic Sciences eCollection on school.cengage.com/forensicscience and research the Hitler Diaries. Make your own investigation by reading the primary sources available on the web site. Write a summary of the case that covers (1) the motives of the forger, (2) the involvement of the German magazine *Stern*, (3) the errors that were uncovered in the analysis of the diaries, and (4) the information that was unveiled at the trial.

 Think Critically You work at a bank. One of your customers has misplaced his or her checkbook. Write a letter to the customer explaining the safeguards that will protect him or her.

Lloyd Cunningham, Document Expert

Lloyd Cunningham is the world's leading handwriting expert of San Francisco's famous fugitive killer known as the Zodiac. In the 1960s, the Zodiac, a serial killer who was never identified, mocked the police with handwritten notes telling of his crimes. For more than 25 years, Cunningham has evaluated numerous documents potential suspects that are submitted each year by police officers, news reporters, and detectives. Cunningham carefully analyzes the handwriting of each submitted sample and compares it to Zodiac's original documents with hopes of finding clues and answers.

Lloyd Cunningham first became interested in the Zodiac case as a San Francisco police officer. In 1969, he was among the many police officers who came to Presidio Park following the shooting of cab driver Paul Stine, the last of the Zodiac's verified victims. Cunningham eventually trained with the U.S. Secret Service. In 1980, he became the U.S. Secret Service's first forensic document examiner. He began investigating Zodiac soon after he finished his forensic training. Since then, Cunningham has analyzed hundreds of famous documents, including the ransom letter in the JonBenet Ramsey case. He retired as a police officer in 1991, but has continued to work on the Zodiac case as a private consultant.

Over the years, Cunningham has memorized Zodiac's handwriting, including his unique letter formations and style of formatting. The mystery killer apparently crossed his "t" low on the vertical line and had large spaces between his lines. So, how does Lloyd Cunningham determine if a newly submitted Zodiac sample is just another hoax? Cunningham says, "There's a rhythm in writing;

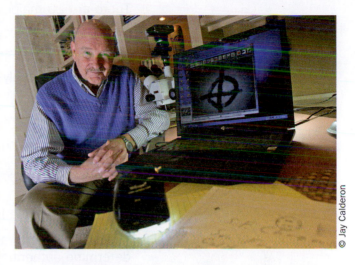

© Jay Calderon

Lloyd Cunningham has become the world's leading handwriting expert of the Zodiac killer.

when people jot notes or sign documents, they write quickly and confidently. But if someone tries to copy or disguise their handwriting, it's no longer spontaneous, and an expert can see signs of the effort in the script." Cunningham explains about his lack of frustration in his continued efforts to investigate new samples because "Who knows? Maybe one of them is right."

There is not a specialized degree in forensic document examination. Although one is not required, many investigators start out with a degree in the sciences and the proper training in scientific analysis. Once a university degree is attained, skills in document examination can be acquired through job experience and certified training programs that are often found in government crime laboratories.

Learn More About It

To learn more about the work of a document expert or handwriting analysts, go to school.cengage.com/forensicscience.

CHAPTER 10 REVIEW

True or False

1. There are 10 major categories of handwriting characteristics.
2. To prevent forgery, some checks have embedded fibers that glow under special lights.
3. In handwriting analysis, a person's handwriting is compared to several exemplars.
4. A person's handwriting is so consistent that nothing alters it.
5. Some forgers now use chemicals to "age" paper.
6. A biometric pad measures the speed, rhythm, and pressure of your handwriting.
7. The Secret Service is charged with the security of U.S. currency.
8. Document experts analyze the paper and ink as well as the writing to determine authenticity.
9. Forgers rarely spend time trying to forge documents by famous people.
10. U.S. paper money has special thread embedded in it as a guard against counterfeiting.

Short Answer

11. Describe three different characteristics of handwriting that experts analyze during a forensic investigation.

_____

_____

12. Define an *exemplar*.

_____

_____

13. Summarize the three basic steps in handwriting analysis.

_____

_____

14. Describe some of the technologies used by document experts to analyze handwriting.

_____

_____

15. Distinguish between check forgery and literary forgery.

_____

_____

16. What are some methods used by criminals to make paper and ink appear old when forging a historic document?

_____

_____

17. List the features of money bills that are used to help distinguish genuine money from counterfeit money.

_____

_____

18. What type of change occurs when a counterfeit pen's ink comes into contact with counterfeit money, and why does this reaction not occur when the counterfeit pen is used on genuine money printed in the United States?

_____

_____

Connections

19. We now have handwriting databases of letters and handwriting samples. How can they be analyzed using mathematical methods?

_____

_____

Bibliography

Books and Journals

Abagnale, Frank W. *The Art of the Steal: How to Protect Yourself and Your Business from Fraud, America's #1 Crime.* New York: Broadway, 2002.

Ganzel, Dewey. *Fortune and Men's Eyes: The Career of John Payne Collier.* Oxford, UK: Oxford University Press, 1982.

Grebanier, Bernard. *The Great Shakespeare Forgery.* Portsmouth, NH: Heinemann, 1966.

Haring, J. Vreeland. *Hand of Hauptmann: The Handwriting Expert Tells the Story of the Lindbergh Case.* Montclair, NJ: Patterson Smith (reprint), 1937.

Magnuson, Ed. "Hitler's Forged Diaries," *Time.* May 16, 1983, 36–47.

Web sites

Gale Forensic Science eCollection, school.cengage.com/forensicscience.

http://www.usatoday.com/educate/newmoney/index.htm

http://www.moneyfactory.gov/newmoney/index.cfm

http://www.lib.udel.edu/ud/spec/exhibits/forgery/wise.htm

http://www.crimelibrary.com/criminal_mind/scams/shakespeare/6.html

http://en.wikipedia.org/wiki/Thomas_Chatterton

http://www.secretservice.gov/money_detect.shtml

http://www.myhandwriting.com/celebs/ransom1.html

http://www.csad.ox.ac.uk/CSAD/newsletters/newsletter10/newsletter10c.html

http://www.fbi.gov/libref/historic/famcases/lindber/lindbernes.htm

http://www.celebritymorgue.com/lindbergh-baby/

http://www.myhandwriting.com/celebs/ransom1.html

http://www.crimelibrary.com/forensics/literary/3.htm

http://www.crimelibrary.com/lindbergh/lindcrime.htm

http://www.forgeryfinder.com/Standr.htm

http://www.courttv.com/talk/chat_transcripts/2001/1025baggett.html

http://www.myhandwriting.com/reports

http://www.handwritingsherlock.com

ACTIVITY 10-1
HANDWRITING ANALYSIS

Objectives:

By the end of this activity, you will be able to:

1. Describe the 12 different exemplars used in handwriting analysis.
2. Provide an example of the 12 different characteristics used in handwriting analysis.
3. Identify examples of the 12 different characteristics found in handwriting samples.
4. Analyze your own handwriting sample using the 12 exemplars.
5. Analyze the handwriting sample of a classmate's handwriting using the 12 different characteristics.

Time Required to Complete Activity: 40 minutes

Materials:

(per student)
pens or pencils and lined paper
colored pencils or highlighters
ruler (mm) or calipers
two handwriting samples of the Fourth Amendment provided by your partner

Safety Precautions:

None

Procedure 1: Analysis of Your Own Handwriting

1. Copy the Fourth Amendment from the overhead projector onto a lined sheet of notebook paper. Prepare two copies on separate sheets of paper. (Note this sample should be taken prior to the exercise. Students should not be aware that this sample will be used for handwriting analysis.)
2. Review the descriptions and examples of the 12 exemplars of handwriting traits.
3. Use Figure 10-3 to perform a handwriting analysis of your own handwriting (Fourth Amendment) by completing Data Table 1.
   a. Begin by examining the handwriting for line quality. Answer the question, yes or no, on the Data Table. Include a description to qualify your answer.
   b. Use highlighters or colored pencils to circle letters, words, or lines that demonstrate unusual characteristics or traits.
   c. For some of the exemplars, it is important that you use a ruler or a caliper to measure the letters or spacing. For example, in exemplar 2, you will need to measure the spacing between words. Note if it is consistent. Note the size of the distance between words. Include your measurements under the description heading.
   d. You will need to complete the entire Data Table 1 as you analyze the handwriting using the different exemplars.

Data Table 1: Analysis of Your Own Handwriting

Characteristic #	Yes	No	Comments (and measurements in mm) if required
1. Is line quality smooth?			
2. Are words and margins evenly spaced?			Margins: Words:
3. Is the ratio of small letters to capital letters consistent? What is the ratio?			
4. Is the writing continuous?			
5. Are letters connected between capitals and lowercase letters?			
6. Are letter formations complete?			(Be specific, which letters?)
7. Is all of the writing cursive?			(Be specific, which words?)
8. Is the pen pressure the same throughout?			
9. Do all letters slant to the right?			
10. Are all letters written on the line?			
11. Are there fancy curls or loops?			(which letters?)
12. Are all i's and t's dotted and crossed? (top, middle, or not)			i's t's

Procedure 2: Analysis of a Classmate's Handwriting

1. After completing the analysis of your own handwriting, exchange a handwriting sample with a classmate. (Be sure to give them a clean copy of your handwriting, not the copy that you just marked.)
2. Analyze a classmate's handwriting by completing Data Table 2. Be sure to use highlighters or colored pencils to mark any unusual traits. Include measurements where necessary.
3. After completing the analysis, answer the questions on the next page.

Data Table 2: Analysis of Your Partner's Handwriting

Characteristic #	Yes	No	Comments (and measurements in mm) if required
1. Is line quality smooth?			
2. Are words and margins evenly spaced?			Margins: Words:
3. Is the ratio of small letters to capital letters consistent? What is the ratio?			
4. Is the writing continuous?			
5. Are letters connected between capitals and lowercase letters?			
6. Are letter formations complete?			(Be specific, which letters?)
7. Is all of the writing cursive?			(Be specific, which words?)
8. Is the pen pressure the same throughout?			
9. Do all letters slant to the right?			
10. Are all letters written on the line?			
11. Are there fancy curls or loops?			(which letters?)
12. Are all i's and t's dotted and crossed? (top, middle, or not)			i's t's

Questions:

1. Were the handwritings samples of you and your partner similar, or could you easily tell that the two samples of handwriting were not from the same person by simply glancing at them? Explain your answer.
2. Review your Data Tables of the two handwriting analyses. Did the two handwriting samples have any characteristics that were the same? Explain your answer.
3. Review your Data Tables and state what characteristics of your own handwriting were very different from your partner's handwriting.
4. Reviewing your Data Tables, did any handwriting traits found in either of the two handwriting samples seem to be unique and could easily be used to help identify any other handwriting samples written by either you or your classmate? If so, describe the trait below:
5. Unique letter combinations is another characteristic that could be added to this list of the 12 exemplars used for handwriting analysis. For example, many people may have very distinctive ways of writing double Ls, such as in the word galloping. Other people may have a unique way of writing the letters "th," such as in the words *the, them*, or *their*. Describe a different example of an exemplar that you would like to see added to the 12 characteristics of handwriting used in handwriting analyses.
6. Why was it important to have your handwriting samples prepared in advance of this activity?

ACTIVITY 10-2
ANALYSIS OF RANSOM NOTE AND EXPERT TESTIMONY

Scenario:

Someone abducted a 10-year-old child from a well-to-do, private, residential school. His wealthy and famous parents received a ransom note requesting a large sum of money in exchange for the safe return of their son. They immediately contacted the police and gave them the ransom note.

The police got a lead implicating six members of a group, and all six young men were taken to the police station for questioning. The six young men were separated from each other so that they could not collaborate on their story. The first thing the police asked them to do was to write down their whereabouts for the past 48 hours. The police actually wanted a handwriting sample from each of the men. It was important not to tell them that a handwriting sample was being collected, because then the men might not write normally or spontaneously.

When the police obtained all six handwriting samples and the ransom note, they called in a renowned handwriting expert (you!) to analyze the ransom notes and the six suspect notes. Did any of the handwriting samples from the six men match the handwriting sample found in the ransom note?

Objectives:

By the end of this activity, you will be able to:
1. Act as the expert handwriting witness requested to testify that at least one of the suspect's handwriting samples matches the handwriting sample found in the ransom note.
2. Present your findings in a written report that will be submitted to the jury before your testimony at the trial.

Time Required to Complete Activity:

Two 40-minute periods are required to analyze the ransom note and the six suspect handwriting samples. Additional time is required to prepare the written report.

Materials:

(2 students/group)
6 different handwriting samples
1 ransom note in the handwriting of one of the above writers
1 six-inch mm ruler or calipers
several colored pencils or highlighters

Safety Precautions:

None

Procedure:

1. Study the ransom note provided by your instructor. Perform an analysis of the handwriting sample using the 12 exemplars (see Figure 10-3). Record your findings in Data Table 1.
2. Analyze the six suspects' handwriting samples.
   a. If possible, visually eliminate some of the samples without using the list itemizing the characteristics. If you can eliminate any handwriting samples without performing the 12-exemplar handwriting analysis, you will need to write a statement showing evidence that the handwriting samples are obviously very different. Record this information on a separate sheet of paper using the Visual Elimination Format provided. If you eliminate the suspect by this visual inspection, you will not need to complete the 12-exemplar section of the Data Table.
   b. For those samples that appear to be similar to the ransom note, perform a handwriting analysis using the 12 different handwriting characteristics. Record your results in seperate Data Tables for each suspect's handwriting sample.
3. After analyzing both the ransom note and the six suspects' handwriting samples, determine which of the suspect's handwriting matches the handwriting of the ransom note. You will need to prepare your findings in a written report to be submitted to the jury. You will also need to testify at the hearing.

Written Lab Report

1. The purpose of the written report is to convince the members of a jury that you are an expert in the area of handwriting analysis.
2. You are able to analyze handwriting samples and show the jury evidence of how the samples are similar or dissimilar. Through your investigation, you have been able to conclude that one or more of the handwriting samples matches the handwriting of the ransom note.
3. Keep in mind that most juries have no knowledge of handwriting analysis. They may be highly educated, or they may have very little formal education. Therefore, any terms you use must be clearly defined.
4. Your report should be typed and spell-checked.
5. Print out a rough draft. Ask a partner to proofread your rough draft and help you edit your first copy. Your editor needs to sign the bottom of your rough draft after editing.
6. You should submit both the edited rough draft and your final copy.

The format for your written report to the jury is outlined as follows:

I. Introduction (10 points)
   a. State the purpose of your report.
   b. No factual or detailed information should be in the introduction.
   c. State how you analyzed the handwriting.
      · State how many different characteristics you used.
      · State if it was possible to prove someone wrote the note, or merely that the handwriting was similar.

II. Several Body Paragraphs (at least six) (60 points)
   a. One main idea or exemplar in each paragraph (at least six)
   b. For each trait, you need to:
      · Describe the trait.

- Explain how the ransom note writer's handwriting matched or didn't match the trait you are describing.
- Remember that you need to explain these terms to the jury and convince them that your comparisons are correct.

Example: Exemplar 2
- Did you look at the spacing?
- Did you measure the spacing with a ruler?
- If so, what was the spacing?
- What is the ratio of lowercase letters to uppercase (capital) letters?
- Is the ratio consistent?
- Is the ratio the same in both the ransom note and the suspect's handwriting sample?

III. Conclusion (10 points)
    a. Summarize your findings.
    b. Do not repeat detailed information.
    c. How reliable is your conclusion?
    d. Is handwriting evidence enough to convict someone?
    e. Is this an important piece of evidence?

Data Table 1: Ransom Note Analysis

Characteristic #	Yes	No	Comments (and measurements in mm) if required
1. Is line quality smooth?			
2. Are words and margins evenly spaced?			Margins: Words:
3. Is the ratio of small letters to capital letters consistent?			
4. Is the writing continuous?			
5. Are letters connected between capitals and lowercase letters?			
6. Are letter formations complete?			(Be specific, which letters?)
7. Is all of the writing cursive?			(Be specific, which words?)
8. Is the pen pressure the same throughout?			
9. Do all letters slant to the right?			
10. Are all letters written on the line?			
11. Are there fancy curls or loops?			(Which letters?)
12. Are all i's and t's dotted and crossed?			i's t's

Characteristic #	Yes	No	Comments (and measurements in mm) if required
1. Is line quality smooth?			
2. Are words and margins evenly spaced?			Margins: Words:
3. Is the ratio of small letters to capital letters consistent?			
4. Is the writing continuous?			
5. Are letters connected between capitals and lowercase letters?			
6. Are letter formations complete?			(Be specific, which letters?)
7. Is all of the writing cursive?			(Be specific, which words?)
8. Is the pen pressure the same throughout?			
9. Do all letters slant to the right?			
10. Are all letters written on the line?			
11. Are there fancy curls or loops?			(Which letters?)
12. Are all i's and t's dotted and crossed?			i's t's

Visual Elimination Format:

If any of the suspects' handwriting can be quickly eliminated without performing a 12-character analysis, you will need to justify your elimination with a brief statement explaining why the handwriting is obviously not the same as the handwriting found in the ransom note. Use the following format:

Suspect # _____

Reasons for quickly eliminating this suspect:

1._____

2._____

3._____

4._____

5._____

ACTIVITY 10-3
EXAMINATION OF U.S. CURRENCY: IS IT REAL OR A FORGERY?

Scenario:

Camille handed the cashier her $50 bill. The cashier held it up against the light and looked at it. Perplexed, Camille asked the cashier why he held the $50 bill up to the light. He told her that cashiers were required to examine all $50 bills to be sure that they were legitimate and not counterfeit bills. Camille couldn't imagine how holding the bill up against the light could help him determine if it was a genuine-issue bill or a counterfeit bill. What was he looking for?

Objectives:

By the end of this activity, you will be able to:
1. Identify who is on the front of $1, $5, $10, and $20 bills.
2. Describe what images appear on the back of the bills.
3. Describe the seals, signatures, and images that appear on American currency bills.
4. Given a counterfeit-detecting pen, determine if a bill is genuine or a forgery.
5. Given U.S. paper currency, describe methods used to determine if the currency is counterfeit or legitimate.
6. Explain why it is difficult to counterfeit U.S. currency.

Time Required to Complete Activity:

Part A: Pre-test (5 minutes)
Part B: $1 examination (30 minutes)
Part C: Hidden feature exploration (30 minutes)
Part D: $10 bill analysis (30 minutes)
Part E: Internet tutorial (30 minutes)

Materials:

(students should work in pairs)
stereo or compound microscopes or hand lens
counterfeit-detecting pen (to share with other groups)
an assortment of various denominations of U.S. currency ($1, $10, $20, $50) to share
computers (optional)
digital camera (optional)

Safety Precautions:

None

Procedure:

1. Complete the pre-test questions in Part A
2. Complete Part B: $1 bill examination
3. Complete Part C: Hidden feature exploration
4. Complete Part D: $10 bill analysis
5. Complete Part E: Internet tutorial

If computers are available, examine the following web site:
http://moneyfactory.gov/newmoney/main.cfm/learning/download

Part A

Take the pre-test before starting the lab to determine how much you know about our paper currency. For this part of the lab, you should not be looking at any money but answering the questions from memory. *Record your answers on Data Table 1.*

Procedure:

Take the pre-test before starting the lab.

Pre-test:

1. Whose face appears on the front of a $1 bill?
2. Whose face appears on the front of a $5 bill?
3. Whose face appears on the front of a $20 bill?
4. What building is pictured on the back of a $5 bill?
5. What building is pictured on the back of a $10 bill?
6. What building is pictured on the back of a $20 bill?
7. What pictures appear on the back of a $1 bill?
8. On the front of $1, $5, and $10 bills, what words are written?
9. On the back of the $1, $5, and $10 bills, what words are written?
10. Is the date the bill was issued printed on the front or back of the bill?
11. What seals appear on the front of a bill?

True/False:

12. The Secretary of the Treasury and the U.S. Treasurer are the same.
13. The serial number is printed in two places on the front of a bill.
14. Newer bills contain more colors than the older bills.
15. There is only one signature located on the front of a bill.
16. There is a picture of a building located on the back of $1, $5, $10, and $20 bills.
17. The White House appears on the back of the $20 bill.
18. Because of the separation of church and state, no mention of a higher being or deity can be printed on the bills.
19. There are "hidden images" on the front side of a bill that can only be seen if you hold the bill up to the light.
20. On the back of $10 and $20 bills, small yellow numbers indicating their denominations is stamped in the area surrounding the picture.

Data Table 1: Pre-test

Question	Answer
1	
2	
3	
4	
5	
6	
7	
8	
9	
10	
11	
12	
13	
14	
15	
16	
17	
18	
19	
20	

Part B: Observation of $1 Bills

After reviewing your answers to the pre-test, you will be given some time to study a $1 bill. To help guide you in your observations, answer the following questions and place your answer in Data Table 2. You will need to look at the bill using a hand lens or a stereomicroscope.

Front of the $1 bill

1. Whose picture is on the front of the $1 bill?
2. What is written across the very top of the front of the $1 bill?
3. What is printed on the very bottom of the front of the $1 bill?
4. What seal appears on the front, left-hand side?
5. What seal appears on the front, right-hand side?
6. Find the date on the $1 bill. Record its date in Data Table 1.
7. Who was the Secretary of Treasury at the time this bill was issued?
8. Who was the U.S. Treasurer at the time the bill was issued?
9. Record the serial number for this bill.
10. How many places on the check is the serial number printed?

Back of the $1 bill

11. What words are printed on the top line?
12. What words are printed on the bottom line?
13. What image appears on the back on the left side?
14. What image appears on the right side?
15. What reference to God appears on the back of the bill?

Data Table 2: $1 Bill Examination

Question	Answer
Front	
1	
2	
3	
4	
5	
6	
7	
8	
9	
10	
Back	
11	
12	
13	
14	
15	

Part C: How Many Hidden Images Can You Find?

Courtesy of www.moneyfactory.gov

For this part of the lab, you will need new $10 bills.

1. Form groups of four students each.
2. Each group should have a $10 bill, hand lens, and stereomicroscope (optional).
3. Each team is to try to identify as many hidden images on the front and back of the $10 bill. These will include images that can only be noticed by:
   a. Holding the bill up against the light
   b. Viewing the bill with a hand lens or stereomicroscope
   c. Looking for numbers or words that cannot be seen without the aid of some type of magnification
4. Cooperative Learning Teams: Each team should assign one person to each of the following jobs:
   a. *Recorder*: Person who will write down each of the discoveries as they are noted. Record the notes in Data Table 3.
   b. *Presenter*: Person who has the job of reporting to the class what the team has discovered
   c. *Light specialist*: Person who will hold the bill up against the light source and find hidden images

d. *Magnification specialist*: Person who will use a hand lens (or stereomicroscope) to view hidden images that are visible only with increased magnification
5. Your team will be allowed a limited amount of time to discover the hidden images.
6. The team with the greatest number of discoveries reports to the entire class first. Any of the other teams will report any discoveries that were not already mentioned by the first team.

Data Table 3: Hidden Images on the $10 Bill

Location	Images	Numbers
Front of the Bill		
Back of the Bill		

Part D: Analysis of a $10 Bill

Using a hand lens or dissecting microscope, check your bill for the following:
1. The portrait appears flat on the genuine bills, but appears raised on counterfeit bills.
2. For newer $10 and $20 bills, the oval around the portrait is gone.
3. The background details of the portrait are clear and distinct on genuine bills.
4. The border edge of the genuine bill is clear and distinct.
5. Note the hidden numbers and words embedded in fine print.

6. On genuine bills, the Treasury seals have clear, sharp, sawtooth points.

Hidden numbers and letters

Written above Hamilton's name

7. On genuine bills, the serial number is evenly spaced and the same color as the Treasury seal.

8. Genuine paper currency has red and blue fibers woven throughout the bill. You may not be able to see these red and blue fibers without a hand lens or a stereo-microscope.

Serial number

9. Counterfeit currency uses red and blue inks that are often blurred. This inking may be detected with a hand lens.

10. Examine a bill looking for the following:

a. *Security thread.* Hold the bill up to the light, and a thin line appears with the denomination of the bill written in it. The position of the thread varies from denomination to denomination but always runs from top to bottom.

b. *Color-shifting ink.* When the bill is tilted, the color of the left-corner 10 shifts from copper to green.

c. *Watermark.* Appears on the right side of the face of the bill if it is held up to a light. The image also appears on the left side of the bill if viewed from the back of the bill.

d. *Color.* The background color on both sides of the bill is enhanced.

e. *Symbols of freedom.* A large, red image of the Statue of Liberty's flame is printed to the left of Hamilton, and a smaller, red metallic

Security thread

Color-shifting ink

Watermark

Symbols of freedom

Enhanced portrait

image is found to the right ($10). Other seals are affixed to other denominations in the same position.

f. *Enhanced portrait*. The oval border around the portrait has been removed, and the shoulder extends to the border of the bill. The portrait appears to be in front of the bill.

g. *Multiple 10s, 20s, 50s, etc*. Small yellow 10s, 20s, and 50s are printed on the front, back, or both sides of the bill designating its denomination.

11. Using a counterfeit-detecting pen, mark the edge of the bill and examine the color. A genuine bill will be pale yellow to tan, whereas a counterfeit bill will turn brown.

The old $10 note, front

The series 2001 $10 note, front

Part E: Internet Tutorial

Go to the following web site and click on Interactive Notes.
http://moneyfactory.gov/newmoney/ main.cfm/learning/download

Questions:

1. Counterfeiters sometimes collect dollar bills and bleach them to remove the ink. Using a printer, they will print images of a higher-denomination bill on the bleached paper. What is the advantage of bleaching the dollar bill over just printing the higher-denomination bill onto clean paper?

2. Why has it been necessary to make so many changes to our paper currency in the past 30 years as compared to the last 100 years?

3. Of all the safeguards added to our higher-denomination currency, which do you consider the most important and why?

4. Counterfeiters try to pass off their counterfeit money at public events where many people gather. The Olympics held in Salt Lake City, Utah, employed many volunteers. These people are not necessarily trained in checking larger bills to see if they were genuine or counterfeit. Provide a list of four items to quickly and easily check the authenticity of a $10 bill.

CHAPTER 11

Death: Meaning, Manner, Mechanism, Cause, and Time

MYSTERIOUS DEATH AT THE FAIR

The Washington County Fairgrounds in upstate New York was the site of the 1999 annual county fair. Well water, the source of drinking water for the event, became contaminated by runoff from a nearby cattle barn after a recent storm. The cattle and their manure carried a type of the bacteria called *Escherichia coli*. *E. coli* is a natural, and necessary, inhabitant of our digestive systems, but one strain carried by cattle produces a powerful toxin. The cattle that carry this strain of *E. coli* are unharmed, but humans can become very sick and die from an infection. Two of the 127 confirmed cases of *E. coli* poisoning from the fair died from the infection. As is often the case with *E. coli* poisoning, the deaths were from among the youngest and eldest of the group infected, a 79-year-old man and a 3-year-old girl. Manner of death—accidental; cause of death—food poisoning/water contamination; mechanism of death—kidney failure.

Escherichia coli is a leading cause of foodborne illness. Scientists estimate there are 73,000 cases of infection and 61 deaths in the United States each year. In addition to eating undercooked meat, people can become infected in a variety of ways. People have

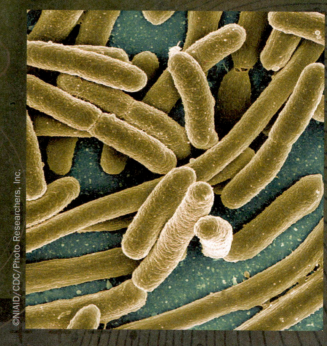

©NIAID/CDC/Photo Researchers, Inc.

E. coli bacteria.

become ill from eating contaminated bean sprouts or fresh leafy vegetables, or through person-to-person contact in families and child care centers. Infection also can result from drinking raw milk or swimming in or drinking sewage-contaminated water.

OBJECTIVES

By the end of this chapter you will be able to

✔ Discuss the definition of death.

✔ Distinguish between four manners of death: natural, accidental, suicidal, and homicidal.

✔ Distinguish between cause, manner, and mechanisms of death.

✔ Explain how the development of rigor, algor, and livor mortis occurs following death.

✔ Use evidence of rigor, algor, and livor mortis to calculate the approximate time of death.

✔ Describe the stages of decomposition of a corpse.

✔ Use evidence from the autopsy's report on stomach contents to estimate time of death.

✔ Explain how time of death can be estimated using insect evidence.

✔ Provide an example of the succession of different types of insects that are found on a body as it decomposes.

✔ Given insect evidence, livor, rigor, and algor mortis data, be able to estimate time of death.

✔ Describe how various environmental factors may influence the estimated time of death.

TOPICAL SCIENCES KEY

BIOLOGY · EARTH SCIENCES · CHEMISTRY · PHYSICS · PSYCHOLOGY · MATHEMATICS

VOCABULARY

algor mortis the cooling of the body after death

autolysis the spontaneous breakdown of cells as they self-digest

cause of death the immediate reason for a person's death (such as heart attack, kidney failure)

death the cessation, or end, of life

decomposition the process of rotting and breaking down

forensic entomology the study of insects as they pertain to legal issues

instar one of the three larval stages of insect development

larva (plural **larvae**) immature form of an animal that undergoes metamorphosis (for example, a maggot)

livor mortis the pooling of the blood in tissues after death resulting in a reddish color to the skin

manner of death one of four means by which someone dies (i.e., natural, accidental, suicidal, or homicidal)

mechanism of death the specific body failure that leads to death

pupa (plural **pupae**) the stage in an insect's life cycle when the larva forms a capsule around itself and changes into its adult form

rigor mortis the stiffening of the skeletal muscles after death

INTRODUCTION

In the 17th century, before the stethoscope was invented, anyone in a coma or with a weak heartbeat was presumed dead and was buried. The fear of being buried alive led to the fad of placing a bell in the coffin. If someone was buried by mistake and awoke, he or she could ring the bell to get someone's attention. This is how the phrase "saved by the bell" might have originated.

Today, people no longer fear being buried alive. It is, however, sometimes difficult to tell if a person is dead or not. The outward signs of death, such as being cold to the touch and comatose, can be present even though a person is still alive. One definition of **death** is the cessation, or end, of life. To be more precise, death is sometimes defined as the "irreversible cessation of circulation of blood." In other words, the heart stops beating and cannot be restarted. Death might also be defined as the cessation of all brain activity. Even this definition is not perfect. Experts cannot agree on a single definition for death. Is a person with a heartbeat alive even if there is no brain activity? This is not an easy question to answer.

It remains difficult to precisely pinpoint the moment that someone dies. For one thing, death is a process rather than an instant event. The moment of death is usually considered the point of no return. According to physiologists, when the heart stops beating, the cells of the body begin to die because they no longer receive a fresh supply of oxygen. As oxygen levels drop, the basic processes of the body fail to occur. Nerves, muscles, organs, and the brain stop working. This is the first stage of death—stoppage.

When a cell dies, it breaks down. Once enough cells begin to break down, life cannot be restarted. Cell breakdown is called **autolysis** (Figure 11-1). When the cell membrane dissolves, enzymes and other cell contents spill out and digest surrounding tissues.

In cases of suspicious or unnatural deaths, a medical doctor called a forensic pathologist conducts an examination on the deceased. This examination is called an *autopsy*. The autopsy is conducted to determine the manner, cause, and mechanism of death, described in the following sections.

Did You Know?

To avoid burying people before they were dead, "waiting mortuaries" were established in the 17th century. Those people thought to be dead were placed on cots and observed until the body began to rot. Only then was the person declared dead.

Figure 11-1. *Autolysis occurs when cells break down.*

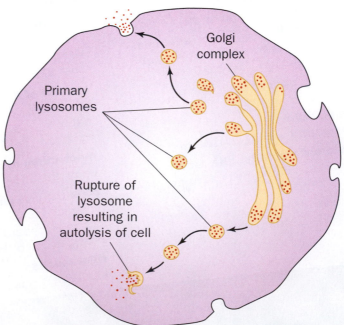

Golgi complex

Primary lysosomes

Rupture of lysosome resulting in autolysis of cell

THE MANNER OF DEATH

There are four ways a person can die, referred to in official terms as the **manner of death**: natural death, accidental death, suicidal death, and homicidal death. A fifth manner of death, undetermined, perhaps should be added because it is occasionally the official cause recorded on a death certificate. Natural death is caused by interruption and failure of body functions resulting from age or disease. This is the most common manner of death. Accidental death is caused by unplanned events, such as a car accident or falling from a ladder. Suicide occurs when a person purposefully kills oneself, whether by hanging, drug overdose, gunshot, or some other method. A homicide is the death of one person caused by another person.

Sometimes it is difficult to determine if the manner of death was a suicide or an accidental death. Did the person deliberately take an overdose of pills, or was it an accident? Did a person mean to shoot himself, or was it a mistake? In some cases, the coroner cannot make this determination and marks the manner of death as unknown on the death certificate.

Consider the following two examples. How would you categorize the manner of death?

- A man with a heart condition is attacked and dies from a heart attack during the assault. Is the manner of death accident or homicide?

- An elderly woman dies after being kept from receiving proper health care by her son. Is the manner of death natural or homicide?

In both cases, homicide would be the manner of death. Proving in court that the manner of death was a homicide, however, may be difficult.

CAUSE AND MECHANISM OF DEATH

Figure 11-2. *Cause of death describes the event that led to a person's death (e.g., hanging).*

©Comstock Images/Jupiter Images

The reason someone dies is called the **cause of death** (Figure 11-2). Disease, physical injury, stroke, and heart attack can all cause death. Examples of causes of death by homicide include bludgeoning, shooting, burning, drowning, strangulation, hanging, and suffocation. Have you ever heard the term "proximate cause of death"? It refers to an underlying cause of death, as opposed to the final cause. If someone is exposed to large amounts of radiation and then develops cancer, the proximate cause of death is the radiation exposure.

Mechanism of death describes the specific change in the body that brought about the cessation of life. For example, if the cause of death is shooting, the mechanism of death might be loss of blood, exsanguination, or it might be the cessation of brain function. If the cause of death is a heart attack, the mechanism of death is the heart stopping to beat or pulmonary arrest.

Figure 11-3. *The official death certificate lists the cause and sometimes the mechanism of death.*

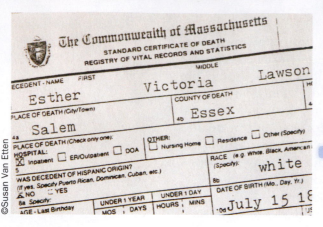

©Susan Van Etten

A forensic pathologist's report may indicate the cause and mechanisms of death in a single statement (as do some death certificates, Figure 11-3). For example, someone killed in a car accident may be said to have died from "massive trauma to the body leading to pulmonary arrest." Trauma to the body is the cause of death; respiratory arrest is the mechanism of death.

TIME OF DEATH

During an autopsy, the forensic examiner wants to determine when the person died. By establishing the time of death, a suspect may be proved innocent simply because he or she was not in the same place as the victim at the time of death. On the other hand, the suspect may remain a person of interest if he or she was in the same area at the time a person died. Many factors are used to approximate the time of death. These factors are discussed in more detail in the following sections.

LIVOR MORTIS

Livor mortis means, roughly, death color. As a body begins to decompose, blood seeps down through the tissues and settles into the lower parts of the body. The red blood cells begin to break down, spilling their contents. Hemoglobin, the substance in red blood cells that carries oxygen and gives blood its red color, turns purple when it spills out of the cells. This purplish color is visible on parts of the skin wherever the blood pools. The process of livor mortis takes time.

Pooling of blood in the body, known as *lividity*, provides a clue as to how long the person has been dead. Lividity first begins about two hours after death. The discoloration becomes permanent after eight hours. If death occurred between two and eight hours, lividity will be present, but if the skin is pressed, the color will disappear. After eight hours, if the skin is pressed, the lividity will remain. The ambient temperature at which a person dies impacts the time it takes for lividity in set in. If the corpse is left outside on a hot, summer day, livor mortis takes place at a faster rate. If the body is left in a cool room, livor mortis is slower. This is why it is so important to record the environmental conditions surrounding a dead body. The extent of livor mortis is also affected by anything impeding the flow of blood, such as tight wristwatches or belts.

Besides providing an approximate time of death, livor mortis can provide other important clues. Because gravity pulls the blood toward the ground, lividity can reveal the position of a corpse during the first eight hours (Figure 11-4). If the corpse were face down in a flat position, blood would pool along the face, chest, abdomen, and portions of the arms and legs close to the floor. If the corpse were positioned on its back, blood would pool along the back, the buttocks, head, and the parts of the arms and legs close to the floor. If the corpse were wedged in a standing position, the blood would collect in the lower legs and feet and the lower arms and hands.

Lividity also can reveal if a body has been moved. For example, if the person

Figure 11-4. *The location of livor mortis can reveal the position of the body during the first eight hours after death.*

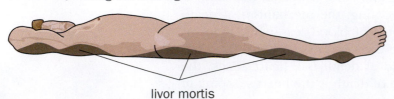

livor mortis

died sitting in a chair, lividity would appear on the back of the thighs, buttocks, and the bottom of the feet. If the corpse were then moved so that the body was lying face down on the floor, lividity would also be found on the face, chest, abdomen, and front surface of the legs. Dual lividity could occur if the body was kept in one position two hours after death and then moved to a second position before the lividity became permanent. This is not uncommon if a murder victim is killed in one place and then transported somewhere else.

RIGOR MORTIS

Have you ever seen a dead animal in the road? Were all four of its legs stiff and sticking straight up in the air (Figure 11-5)? If the animal was still there a few days later, you may have noticed that the animal was no longer stiff. **Rigor mortis** means, roughly, death stiffness. It is temporary and can be very useful in determining the time of death.

Rigor mortis starts within two hours after death. The stiffness starts in the head and gradually works its way down to the legs. After 12 hours, the body is at its most rigid state. The stiffness gradually disappears after 36 hours. Sometimes, depending on body weight and temperature in the area, rigor may remain for 48 hours. If a body shows no visible rigor, it has probably been dead less than two hours or more than 48 hours. If a body is very rigid, then the body has been dead for about 12 hours. If the body exhibits

If a body needs to be transported when rigor is at its peak, it might be necessary to break bones to change the position of the body.

rigor only in the face and neck, then rigor has just started, and the time of death is just over two hours. If there is some rigor throughout the body, but a lack of rigor in the face, then the body is likely to be losing rigor, and the death occurred more than 15 hours ago.

The stiffness occurs because the skeletal muscles are unable to relax and remain contracted and hard. In life, the flexing and relaxing of muscles happens as the muscle fibers slide back and forth. Whenever muscles contract, they release calcium. In healthy, live muscles, the calcium molecules are removed from the cells. This requires energy, and for cells to get energy, they need oxygen. After death without circulation, oxygen flow to the cells ceases, and

Figure 11-5. During the first 48 hours of death, the skeletal muscles are stiff—a condition known as rigor mortis.

©AP Photo/Damian Dovarganes

calcium accumulates in the muscle tissue. In the presence of the extra calcium, the muscle fibers remain in the contracted, rigid position (Figure 11-6). Because the muscles control the movement of bones, the joints appear to be rigid as do the muscles. The muscles eventually begin to relax as the cells and muscle fibers begin to dissolve by autolysis.

Many factors affect when rigor mortis sets in and how long it lasts. When trying to estimate the time of death, these factors need to be taken into account:

1. *Ambient temperature.* The cooler the body, the slower the onset of rigor. The warmer the body, the onset of rigor is faster because chemical reactions happen more quickly at higher temperatures.

Figure 11-6. Live muscle fibers slide back and forth; in the first 48 hours of death, the muscle fibers become locked in a flexed position.

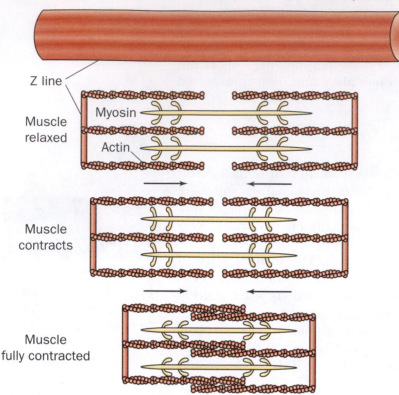

Z line

Muscle relaxed

Myosin

Actin

Muscle contracts

Muscle fully contracted

Figure 11-7. Body weight and physical activity affect the timing of rigor mortis.

©Pixland/Jupiter Images

©Workbook Stock/Jupiter Images

Figure 11-8. Progression of rigor mortis.

2. *A person's weight.* Body fat stores extra oxygen and will slow down rigor mortis. A person with less oxygen stored in the body experiences rigor faster (Figure 11-7).

3. *The type of clothing.* Because clothing helps keep a body warm, the presence of clothes accelerates rigor mortis. A naked body cools faster, which slows down the onset of rigor mortis.

4. *Illness.* If a person dies with a fever, the body temperature will be higher, and rigor mortis will set in faster. If a person experiences hypothermia, the onset of rigor will be slower.

5. *Level of physical activity shortly before death.* If a person was exercising or struggling before death, then rigor will progress faster. This is true for several reasons, including the fact that exercise increases body temperature and decreases oxygen availability to the cells in the body and increases lactic acid levels.

6. *Sun exposure.* A body exposed to direct sunlight will be warmer, and rigor mortis would occur faster.

Because so many variables can affect how fast rigor mortis progresses, a precise time of death cannot be determined, it can only be estimated. However, when rigor mortis is combined with other factors, a more accurate time of death can be established (Figures 11-8 and 11-9).

Time After Death	Event	Appearance	Circumstances
2 to 6 hours	Rigor begins	Body becomes stiff and stiffness moves down body.	Stiffness begins with the eyelids and jaw muscles after about two hours, then center of body stiffens, then arms and legs.
12 hours	Rigor complete	Peak rigor is exhibited.	Entire body is rigid.
15 to 36 hours	Slow loss of rigor	Loss of rigor in small muscles first followed by larger muscles	Rigor lost first in head and neck and last in bigger leg muscles.
36 to 48 hours	Rigor totally disappears	Muscles become relaxed.	Many variables may extend rigor beyond the normal 36 hours.

Figure 11-9. *Factors Affecting Rigor Mortis.*

Factors Affecting Rigor	Event	Effect	Circumstances
Temperature	Cold temperature	Inhibits rigor	Slower onset and slower progression of rigor
	Warm temperature	Accelerates rigor	Faster onset and faster progression of rigor
Activity before death	Aerobic exercise	Accelerates rigor	Lack of oxygen to muscle, the build up of lactic acid, and higher body temperature accelerates rigor
	Sleep	Slows rigor	Muscles fully oxygenated will exhibit rigor more slowly
Body weight	Obese	Slows rigor	Fat stores oxygen
	Thin	Accelerates rigor	Body loses oxygen quickly and body heats faster

ALGOR MORTIS

Algor mortis means, roughly, death heat and describes the temperature loss in a corpse. When a person is alive, the body maintains a constant temperature. To keep our temperature within a normal range, many parts of our body work together, including the circulatory, respiratory, and nervous systems. In death, the body no longer generates heat and begins to cool down.

To take a corpse's temperature, forensic investigators insert a thermometer into the liver. Having a standard location for taking body temperature ensures that investigators can compare their results.

How fast a corpse loses heat has been measured, and investigators can determine how long ago death occurred by its temperature. Approximately one hour after death, the body cools at a rate of 0.78°C (1.4°F) per hour. After the first 12 hours, the body loses about 0.39°C (0.7°F) per hour until the body reaches the same temperature as the surroundings. This is just an estimate and will vary depending on surrounding temperature and conditions. In cooler environments, the body will lose heat faster than in hotter environments. If it is windy, heat loss will occur faster. The surrounding air temperature and other environmental factors are noted when a body is found, because the environment will affect the rate at which the body loses heat. The excess body fat and the presence of clothing will slow down heat loss. Time of death determined by body temperature calculations is always expressed as a range of time because it cannot be calculated exactly; however, a rule of thumb is to expect a heat loss of approximately 1 degree F per hour.

STOMACH AND INTESTINAL CONTENTS

Medical examiners help determine the time of death by studying the corpse's stomach contents. In general, it takes four to six hours for the stomach to empty its contents into the small intestine and another 12 hours for the food to leave the small intestine (Figure 11-10). It takes approximately 24 hours from when a meal was eaten until all undigested food is released from the large intestines. From this, it can be concluded that:

1. If undigested stomach contents are present, then death occurred zero to two hours after the last meal.

Figure 11-10. *The state of food digestion provides a clue to the time of death.*

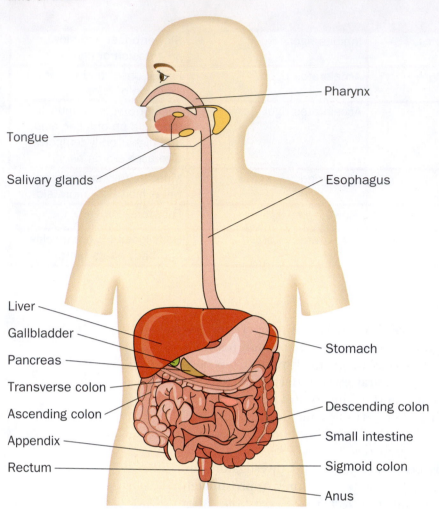

- Pharynx
- Tongue
- Salivary glands
- Esophagus
- Liver
- Gallbladder
- Pancreas
- Transverse colon
- Ascending colon
- Appendix
- Rectum
- Stomach
- Descending colon
- Small intestine
- Sigmoid colon
- Anus

2. If the stomach is empty but food is found in the small intestine, then death occurred at least four to six hours after a meal.

3. If the small intestine is empty and wastes are found in the large intestine, then death probably occurred 12 or more hours after a meal.

CHANGES OF THE EYE FOLLOWING DEATH

In life, the surface of the eye is kept moist by blinking. Following death, the surface of the eye dries out. A thin film is observed within two to three hours if the eyes were open at death and within 24 hours if the eyes were covered at death.

Following death, potassium accumulates inside the vitreous humor. Because decomposition progresses at a predictable rate, the buildup of potassium may be used to estimate the time of death. This method is still being refined and is not yet used as an accurate measure of time of death.

STAGES OF DECOMPOSITION

A corpse decomposes in predictable ways over time that can help examiners judge when death occurred:

1. Within two days after death:

   - Cell autolysis begins following death.

   - Green and purplish staining occurs from blood decomposition.

   - The skin takes on a marbled appearance.

   - The face becomes discolored.

2. After four days:

   - The skin blisters.

   - The abdomen swells with the gas carbon dioxide that is released by bacteria living in the intestines.

3. Within six to ten days:

   - The corpse bloats with carbon dioxide as bacteria continue to feed on tissues. Eventually, the gas causes the chest and abdominal cavities to burst and collapse.

- Fluids begin to leak from the body openings as cell membranes rupture.

- Eyeballs and other tissues liquefy.

- The skin sloughs off.

Figure 11-11 provides more information about the stages of decomposition.

Figure 11-11. *The stages of decomposition provide information about time of death.*

Stage	What Happens During Decomposition
Initial Decay	Corpse appears normal on the outside, but is starting to decompose from the actions of bacteria and autolysis.
Putrefaction	Odor of decaying flesh is present and the corpse appears swollen.
Black Putrefaction	Very strong odor. Parts of the flesh appear black. Gases escape and the corpse collapses.
Butyric Fermentation	Corpse is beginning to dry out. Most of the flesh is gone.
Dry Decay	Corpse is almost dry. Further decay is very slow from lack of moisture.

Ultimately, this leads to the **decomposition,** or rotting, of all tissues and organs. Bacteria and other microorganisms also help decompose a human body, just as they decompose plants and animals in the environment.

The speed of decomposition depends on the person's age, size of the body, and the nature of death. Sick individuals decompose faster than healthy people. The young decompose faster than the elderly. Overweight people with rich deposits of fat and body fluids break down faster than people of normal weight.

Just as environmental conditions affect rigor, they also influence decomposition. Naked bodies decompose faster than clothed bodies. Bodies decompose fastest in the 21–37°C (70–99°F) temperature range. Higher temperatures tend to dry out corpses, preserving them. Lower temperatures tend to prevent bacterial growth and slow down decomposition. Moist environments rich in oxygen speed up decomposition. Bodies decompose most quickly in air and slower in water or if buried.

Digging Deeper
with Forensic Science e-Collection

The science of forensics is always looking to improve its techniques or invent new ones to uncover the truth with greater accuracy. The article "A study of volatile organic compounds evolved from the decaying human body" explores a new technique for gathering information from a corpse. Search the Gale Forensic Sciences eCollection on school.cengage.com/forensicscience for the article. From what you have learned about the nature of tissue decomposition, evaluate this new technique and give a one-page argument about how this technique will be useful or not. Support your answer with information found in this textbook or other reputable sources.

Source: M. Statheropoulos, C. Spiliopoulou, and A. Agapiou, "A study of volatile organic compounds evolved from the decaying human body," *Forensic Science International* 153.2–3 (Oct 29, 2005): pp. 147(9). From *Forensic Science Journals*.

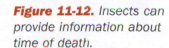

Figure 11-12. *Insects can provide information about time of death.*

INSECTS

Insects can provide detailed information about time of death in several ways. In fact, insects are so useful in crime investigation that there is an entire field dedicated to this study and practice called **forensic entomology** (Figure 11-12). A forensic entomologist at a crime scene observes and records data

Digging Deeper
with Forensic Science e-Collection

Petechial hemorrhages are a common forensic tool that can provide evidence of time of death as well as some clues about the cause of death. Search the Gale Forensic Sciences eCollection on school .cengage.com/forensicscience for the article "Factors and circumstances influencing the development of hemorrhages in livor mortis." Read the article and then research the topic online. In 500 words or less, describe and characterize petechial hemorrhages: what they are, what they look like, what causes them, how forensic scientists use them, and how reliable they are as a tool for determining time of death.

Source: Britta Bockholdt, H. Maxeiner, and W. Hegenbarth, "Factors and circumstances influencing the development of hemorrhages in livor mortis," *Forensic Science International* 149.2–3 (May 10, 2005): pp. 133(5). From *Forensic Science Journals*.

Did You Know?

To establish how different environmental conditions affect the appearance and rate of development of insects, University of Tennessee professor William Bass established the Body Farm in 1980. This has helped forensic experts interpret the evidence about the time of death much more accurately.

about the environmental conditions, including temperature, moisture, and wind, around the body as well as below it. The forensic entomologist collects insect evidence from on, above, and below the victim's body as well as insect evidence from the immediate area around the victim.

Within minutes of a death, certain insects arrive to lay their eggs on the warm body, attracted by the smell of the first stages of decomposition. The eggs will hatch and feed on the nutritious decomposing tissues. Blowflies are a common example. Blowflies are attracted to two gases of decomposition that have only recently been discovered by scientists, called putrescine and cadaverene. As a corpse progresses through the stages of decomposition, other kinds of insects will arrive. Tiny wasps come to lay their eggs on maggots already present on the body. Wasp **larvae** live as parasites inside the maggots, feeding on their flesh. The cheese skippers arrive once putrefaction is underway; they are attracted by the seepage of body fluids (Figure 11-13). The last groups of insects to arrive are those that favor drier conditions, such as the mites and beetles that feed on dry tissues and hair as shown in Figures 11-14, 11-15, and 11-16.

Blowflies are one of the first insects to arrive at a dead body and are very useful in determining the time of death. Like other insects, blowflies exhibit different stages as they develop from egg, larva stages (also known as **instars**), pupa, to adult. Refer to the blowfly life cycle table for more information (Figure 11-17).

Figure 11-13. *Cheese skipper.*

Figure 11-14. *Rove beetles feeding on fly larva.*

©A. Skonieczny/Peter Arnold, Inc.

©Nature's Images/Photo Researchers, Inc.

Because scientists know how long it takes for the various stages of development at given temperatures, forensic entomologists can determine when the blowflies arrive by studying the insects on the corpse. More importantly, it is quite easy to identify the stage of blowfly development by noting the change in size, color, mobility, presence or absence of a crop, and number of spiracle slits (Figure 11-18). The stages of blowfly larva can be determined by the number of spiracle slits at their posterior end.

Figure 11-15. *Adult dermestid beetle.*

©Cengage Learning

Figure 11-16. *Dermestid beetle larva.*

©Cengage Learning

Figure 11-17. *Blowfly Life Cycle (times are approximations).*

Stage	Size (mm)	Color	When first appears	Duration in phase	Characteristics	Sketch (not to scale)
Egg	2	white	Soon after death	8 hours	Found in moist, warm areas of body Mouth, eyes, ears, anus	
Larva 1 (instar 1)	5	white	1.8 days	20 hours	Black mouth hooks visible (anterior) Thin body One spiracle slit near anus	
Larva 2 (instar 2)	10	white	2.5 days	15–20 hours	Black mouth hooks (anterior) Dark crop seen on anterior dorsal side Actively feeding Two spiracle slits near anus	
Larva 3 (instar 3)	17	white	4–5 days	36–56 hours	Black mouth hooks Crop not visible, covered by fat deposits Fat body Three spiracle slits near anus	
Pre-Pupa	9		8–12 days	86–180 hours	Larva migrates away from body to a dry area	
Early and late Pupa	9	Light brown Changes to dark brown	18–24 days	6–12 days	Immobile, does not feed Changes to dark brown with age Filled air "balloon" to help split open pupa case prior to adult emerging	
Adult	Varies	Black or green	21–24 days	Several weeks	Incapable of flight for first few hours	

Figure 11-18. *Spiracle slits for larva stages Larva 1, Larva 2, Larva 3.*

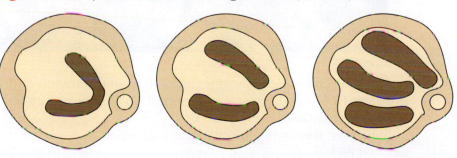

First stage larva   Second stage larva   Third stage larva

Maggots can help detect if a dead person had poison or drugs in the body at the time of death. When the maggots feed on tissues, any toxins or drugs they eat will be built into their own bodies. A "maggot milkshake" is prepared by a forensic entomologist, which is then analyzed for the presence of poisons or drugs.

First-stage larva has one V-shaped slit (1.8 days old); second-stage larva has two slits (2.5 days old); third-stage larva has three slits (4 to 5 days old). For example:

1. If a corpse contains blowfly eggs, then the approximate time of death would be 24 hours or less.

2. If a corpse contains third-stage larvae, then the time of death is approximately four to five days.

3. If a corpse contains pupae, then the time of death would be approximately 18 to 24 days.

Many factors affect insect development, including temperature, moisture, wind, time of day, season, exposure to the elements, and variations among individual insects. Because life cycles are affected by fluctuations in the daily environmental conditions, insects cannot provide an exact time of death, only a close estimate. Insects collected at the crime scene and then raised in the laboratory under the same environmental conditions as those found at the crime scene. This process can provide a more accurate estimate of time of death (Figure 11-19). The process is called Accumulated Degree Hours (ADH). Here is how this is done:

1. Immediately preserve some insects from the crime scene so you know exactly how old they are at the time of discovery of the body.

2. At the crime lab, raise some of the insects from the crime scene in the same conditions as those found at the crime scene.

3. Record the length of time for development under the specific conditions found at the crime scene.

4. Compare the insects raised at the crime lab to those found at the crime scene.

Figure 11-19. *Insects from the crime scene are raised in a laboratory to determine the time of their life cycle.*

©Perennou Nuridansy/Photo Researchers, Inc.

Forensic entomology includes more than an estimation of time of death. If insects from another region are found on a corpse, then it suggests that the corpse may have been moved.

Whatever the cause, mechanism, and time of death, reconstructing a detailed picture surrounding a fatality is critical to any forensic investigation.

SUMMARY

- There are several definitions of death, or the end of life, including the cessation of the heartbeat and the cessation of brain function. Upon death, cells break down and release their contents, resulting in decomposition.

- The manner of death refers to how the death occurs: by suicide, homicide, natural causes, or accident. If one of these four causes cannot be determined, the manner of death may be ruled undetermined.

- The cause of death refers to what led to the death and includes causes such as heart attack, gunshot wound, or cancer.

- The mechanism of death refers to the specific bodily function resulting in death. A heart attack might lead to the heart ceasing to beat, whereas a gunshot wound might lead to loss of blood or the ceasing of brain function that results in death.

- It is often important to determine the time of death in a forensic investigation. There are several means of doing this. Livor mortis, rigor mortis, and algor mortis are changes that happen to a body after it dies. Stomach contents and the condition of the eyes also provide clues. The states of decomposition of a corpse, as well as the insects on the body, provide further evidence of time of death.

CASE STUDIES

The Baby in the Box

In March 1944, the dead body of a newborn child was found wrapped in a blanket and newspaper in a cardboard box. The body had been placed in a pit dug in the forest floor and was covered with leaves. The forensic investigator who examined the body thought it had been abandoned for only a few hours because of the condition of the baby's body. The discovery of more than 20 beetles in the wrappings around the baby demonstrated that the time of death had to be much earlier than this. Recent cold weather had kept the body preserved and had disguised the real time of death.

Beetles on a Ski Mask Lead to the Conviction of a Rapist

A woman was attacked during the summer outside her apartment building in Chicago. The attacker, wearing a ski mask, leapt at her from the shrubs. He escaped. The police began to suspect a man in the building where the woman lived. A search warrant was obtained, and a ski mask was found in the suspect's apartment. The man claimed he had not used the ski mask since the previous winter.

The victim identified the man in a voice lineup, but this evidence was not sufficient for a conviction. On microscopic examination, fragments of a plant that matched a plant at the crime scene were found stuck to the ski mask. Live larvae of a beetle called a weevil were found in the bushes. The species of weevil survives the winter every year as adults, and larvae are found only during the summer months. The weevil larvae were also found on the ski mask. The presence of the weevil larvae on the ski mask proved the suspect was lying. The ski mask had been in contact with the plant at the crime scene that summer. The insect evidence was enough to convince the jury of the suspect's guilt.

When Insect Evidence Fails

In response to a 911 call by a relative, a family of three was found dead in a cabin. Their bodies were decomposing, and maggots were found on their flesh. A shocked relative, Mike Rubenstein, found the bodies. He stated he was the last person at the cabin in mid-November. The insect evidence placed the time of death well after this period, providing Mr. Rubenstein with an alibi, but police were suspicious when he quickly applied for the insurance money. A second look at the bodies by Bill Bass revealed that the stage of decomposition did not match that of the insect evidence. It appeared that the flies did not gain entry to the cabin until weeks after the family was killed. Decomposition evidence placed the time of death in mid-November, at the same time Mike Rubenstein says he was there. Eventually, he was convicted of the triple murder.

 Think Critically Review the Case Studies and the information on insect evidence in the chapter. Then state in your own words how insect evidence impacts a case.

Bibliography

Books and Journals

Blass, Bill. *Death's Acre: Inside the Body Farm* New York: Ballantine Books, 1989.

Dix, Jay. *Death Investigator's Manual*. Public Agency Training Council.

Geberth, Vernon J. *Practical Homicide Investigation*, 3rd ed. Boca Raton, FL: CRC Press, 1996.

Hanzlick, Randy, and Michael A. Graham. *Forensic Pathology in Criminal Cases*, 2nd ed. Lexis Law Publishing, 2000.

Haskell, Neal, and Robert D. Hall. "Forensic (Medicocriminal) Entomology—Applications in Medicolegal Investigations." *Forensic Sciences*. Cyril H. Wecht, General Editor. Matthew Bender & Company, Inc., 2005.

Larkin, Glenn M. "Time of Death." *Forensic Sciences*. Cyril H. Wecht, General Editor. Matthew Bender & Company, Inc., 2005.

Nickell, Joe, and John Fischer. *Crime Science: Methods of Forensic Detection*. Lexington, KY: The University Press of Kentucky, 1999.

Roach, Mary. *Stiff: The Curious Lives of Human Cadavers*. New York: W.W. Norton and Company, 2002.

Sachs, Jeesica. *Time of Death*. New York: New Arrow Books Ltd., 2003.

_____. *Corpse: Nature, Forensics, and the Struggle to Pinpoint the Time of Death*. New York: Perseus Publishing, 2002.

Web sites

Gale Forensic Sciences eCollection, school.cengage.com/forensicscience.

Olkowski, Dorothea, et al. "The Mechanism of Death." Slought Foundation Online Content, September 30, 2004. http://slought.org/content/11254/.

http://www.fair-safety.com/news/washington-county.htm (source)

http://www.brookscole.com/chemistry_d/templates/student_resources/shared_resources/animations/muscles/muscles.html

http://health.howstuffworks.com/muscle2.htm

http://www.phrases.org.uk/meanings/311000.html

http://www.abc.net.au/science/features/death/default.htm

http://drzeusforensicfiles.blogspot.com/2007/01/basic-concepts-cause-manner-and.html

http://www.crimelibrary.com/criminal_mind/forensics/psych_autopsy/index

http://www.arrakis.es/~jacoello/date.pdf

http://www.studyworld.com/basementpapers/papers/stack12_14.html

http://www.rcmp-learning.org/docs/ecdd0030.htm

http://www.rcmp-learning.org/docs/ecdd0030.htm

http://www.forensic-entomology.info/forens_ent/forensic_entomol_pmi.shtml

http://www.crimelibrary.com/criminal_mind/forensics/bill_bass/index.html

http://web.utk.edu/~anthrop/index.htm

William Bass

William Bass was studying psychology. For fun, he enrolled in an elective anthropology course studying the behavior and culture of humans. His professor, a specialist in skeletal remains, was asked to come to the scene of a terrible accident. A collision on the highway resulted in a fire. Three people had died, and one body was burned so badly that identification was difficult and the professor was asked to help. The professor asked Bill for a ride to the crime scene and asked the young psychology student to join him. Bill decided right then and there to switch his studies and his career from psychology to anthropology. He learned all about the human body, skeletal remains, and what they can tell us about the life and death of a person.

William Bass at the Anthropology Research Facility.

©AP Photo/Wade Payne

While at the University of Tennessee, Bill spent many years examining bodies as a forensic expert assisting in solving crimes. In particular he specialized in digging up skeletal remains and learning their secrets, to answer a question of utmost importance in forensic cases, "How long ago did they die?" When Bill started, little information existed to link the physical characteristics of a rotting corpse to a specific time of death. Bill saw a need—and a solution—and in 1971 he approached his university to ask for a small piece of land to do research on decomposition of the human body. His request was granted, and his research on dead bodies has never stopped. Today it is one of the few facilities dedicated to human decomposition. At first, the questions Bill and his team asked were pretty simple, such as, "How long does it take for a limb to fall off of a corpse?" Since then the depth and breadth of the questions being explored has exploded.

At any given time, there are about 40 dead bodies on the three acres of the Anthropology Research Facility. They are rotting away in different circumstances, such as in water, in the shade, in the sun, in shallow burials, and in the trunks of cars. All changes in decomposition are carefully recorded over time. The researchers ask all types of specific questions about the chemical changes in different parts of the body during decomposition, as well as the details of insect growth and development under specific conditions. After the decomposition process is completed, the skeletons are catalogued. The collection of skeletons is, in fact, the largest of its kind in the world and is used to provide a vast array of information about a person from his or her remains. For example, from skeletal comparisons, it is possible to use the length of the thighbone to determine the person's gender, race, and height. The research facility is also used to train FBI personnel. Bodies are buried with evidence planted, and the FBI is sent in to find the body and recover the evidence.

In 1994, Patricia Cornwell, a mystery novel author, wrote a book based on the research facility. She called her book *The Body Farm*. The name has stuck. The bodies on the body farm are not grown, however. Most bodies are donated, either by families of the dead or given in a will. Hundreds of people have given their remains to the cause of improving forensic science.

Bill Bass, the body farmer, is now retired, but he is still involved in his facility and spends a lot of time communicating forensic science to the public. His work and the body farm are featured in documentaries and books, including *Death's Acre*, which Bass wrote.

Learn More About It
To learn more about the work of forensic entomologists, go to
school.cengage.com/forensicscience.

CHAPTER 11 REVIEW

True or False

1. Many factors affect rigor mortis, such as the type of clothing the person was wearing.

2. Experts from different sciences agree that the definition of death is the end of life.

3. Blowflies are one of the first insects to arrive at a dead body and are very useful in determining the time of death.

4. Livor mortis refers to the color of a dead body.

5. The only two manners of death are natural death and homicidal death.

6. Mechanism of death describes what has occurred in the body to cause death.

7. The presence of drugs in a corpse cannot be determined by a chemical analysis of larvae found feeding on the body.

8. In the first 12 hours, a dead body cools about 1 degree F per hour.

9. Within minutes of a death, certain insects are attracted by the smell of the first stages of decomposition.

10. *The Body Farm* is a fictional account of the work of a country coroner.

Short Answer

1. Explain the similarities and differences of the following terms:

a. manner of death and cause of death

_____

_____

_____

b. cause of death and mechanism of death

_____

_____

_____

c. larva and pupa

_____

_____

d. rigor mortis and livor mortis

_____

_____

e. autolysis and decomposition

_____

_____

2. As you drive along a roadside, you and your friend notice a dead deer that apparently was struck by a car. Your friend comments that she has never seen such a fat deer. "Did you see the size of its abdomen? It was huge!" As a student of forensics, how would you explain to your friend why the abdominal region of the dead deer was so large?

_____

_____

3. The forensic examiner tells the detective that he thinks the body was killed in the country before it was later found in an alley in New York City. What type of evidence could be present on the body to lead the forensic examiner to that conclusion?

_____

_____

4. A body is found with rigor mortis present in the face, neck, and upper torso. The young crime-scene investigator claims that the time of death must be at least 15 hours previous to the discovery of the body. The first-responding officer still at the crime scene asks what led her to that conclusion. The crime-scene investigator states that rigor mortis peaks at 12 hours and then gradually fades. Because there was no rigor in the legs, rigor mortis must be disappearing and is now only evident in the face, neck, and upper torso. Do you agree with this time estimation? Provide reasons for your answer.

_____

_____

5. Provide an example of the possible succession of insects that would be found on the body of a dead squirrel. Include in your answer: Which insect is usually the first to arrive on the dead body? Explain your reasoning. Which insects are usually the last to arrive on the dead body? Explain your reasoning.

_____

_____

ACTIVITY 11-1
CALCULATING TIME OF DEATH USING RIGOR MORTIS

Background:

In old detective movies, a dead body was often referred to as a "stiff." The term refers to the onset of rigor mortis that follows soon after death. In this activity, you will estimate the approximate time of death by analyzing the degree of rigor of the deceased body.

Objective:

By the end of this activity, you will be able to:
Estimate the time of death using rigor mortis evidence.

Materials:

paper
pen or pencil
calculator (optional)

Safety Precautions:

None

Procedure:

In pairs, answer the following questions dealing with approximating the time of death based on rigor mortis evidence. Refer to the Rigor Mortis Reference Table in your textbook (Figures 11-8 and 11-9).

Questions:

Part A
Estimate the approximate time of death for the following situations. Explain each of your answers:

1. A body was found with no evidence of rigor.
2. A body was found exhibiting rigor throughout the entire body.
3. A body was found exhibiting rigor in the chest, arms, face, and neck.
4. A body was discovered with rigor present in the legs, but no rigor in the upper torso.
5. A body was discovered with most muscles relaxed, except for the face.
6. A body was discovered in the weight room of a gym. A man had been doing " arm curls" with heavy weights. The only place rigor was present was in his arms.

Part B

Estimate the time of death based on the following information:

7. A frail, elderly woman's body was found in her apartment on a hot summer's evening. Her body exhibited advanced rigor in all places except her face and neck.

8. A body was discovered in the woods. The man had been missing for two days. The average temperature the past 48 hours was 50 degrees Fahrenheit. When the body was discovered, it was at peak rigor.

9. An obese man was discovered in his air-conditioned hotel room sitting in a chair in front of the television. The air conditioner was set for 65 degrees Fahrenheit. When the coroner arrived, the man's body exhibited rigor in his upper body only.

10. While jogging, a young woman was attacked and killed. The perpetrator hid the body in the trunk of a car and fled. When the woman's body was discovered, rigor was noticed in her thighs only.

11. The victim's body is not rigid. How long has she been dead? Explain your answer.

12. The body is completely stiff. How long has he been dead? Explain your answer.

13. The victim was found in a snowbank alongside a road. His body is rigid. How long has he been dead? Explain your answer, remembering the cold temperature.

14. The body of the runner was found in the park one early, hot summer morning. Her body shows rigor in her face, neck, arms, and torso. How long has she been dead? Explain your answer.

ACTIVITY 11-2
CALCULATING TIME OF DEATH USING ALGOR MORTIS

Objective:

By the end of this activity, you will be able to:
Estimate the time of death using algor mortis measurements.

Materials:

paper
pen or pencil
calculator
Rigor Algor Reference Table

Safety Precautions:

None

Procedure:

Working in pairs, answer the following questions using this information:
- For the first 12 hours, the body loses 0.78°C (1.4°F) per hour.
- After the first 12 hours, the body loses about 0.39°C (0.7°F) per hour.

Example 1: What is the temperature loss for someone who has been dead for 12 hours?
Temperature loss = (0.78°C/hour) x 12 hours = 9.36°C

Example 2: Calculate the time of death if a person has been dead for less than 12 hours.
If temperature loss is less than 12 hours (or less than 9.36°C), then you use the rate of 0.78°C per hour to estimate the time of death.
  Temperature of dead body is 32.2°C (90°F).
  Normal body temperature is 37°C. (98.6°F)
  37°C – 32.2°C = 4.8°C lost since death.
How long did it take to lose 4.8°C ?
  0.78 (°C/hour) x (unknown number of hours) = degrees lost
  0.78 (°C/hour) x (unknown number of hours) = 4.8°C lost by body
Solve for the unknown number of hours since death occurred:
  Number of hours = 4.8°C ÷ 0.78 (°C/hour)
  Number of hours = 6.1 hours
Convert 0.1 hours into minutes:
  0.1 hour (60 (min/hour) = 6 minutes
  Hours since death = 6.1 hours or 6 hours and 6 minutes

Example 3: Is the time of death more than 12 hours or less than 12 hours?
Recall that if a body has been dead 12 hours or less, the body loses heat at rate of .78°C per hour. If the body has been dead 12 hours, then .78°C/hour x 12 hours = 9.36°C.

If a body loses 9.36°C, then the person has been dead for 12 hours.

If a body loses more than 9.36°C, then the person has been dead for more than 12 hours.

If they lose less than 9.36°C, then the body has been dead for less than 12 hours.

For each of the following, state if the body had been dead for more than or less than 12 hours based on the number of degrees lost:

1. total loss of 7.9°C
2. total loss of 4.4°C
3. total loss of 11.7°C
4. total loss of 17.2°C
5. total loss of 10.6°C

Example 4: Calculate the time of death if the person was dead for more than 12 hours.

If the body has lost more than 9.36°C, then you know that the victim has been dead for more than 12 hours. Recall that after 12 hours, the body loses heat at a rate of 0.39°C per hour. You need to calculate how many hours beyond the 12 hours that someone died and add it to the 12 hours. Body temperature was given as 22.2°C (72°F).

1. How many total degrees were lost from the time of death until the body was found?

   37°C – 22.2°C = 14.8°C

2. Since 14.8°C is more than 9.36°C, you know that the body was dead longer than 12 hours. How much longer?

   37°C – 22.2°C = total loss of 14.8°C since death
   9.36°C were lost in the first 12 hours
   14.8°C lost since death – 9.36°C lost the first 12 hours =
   5.44°C lost after the first 12 hours

3. Recall that the rate of heat lost after 12 hours is 0.39°C per hour. You need to determine how many hours it took to lose that 5.44°C.

   (0.39°C/ hour) x (unknown # of hours) = degrees lost after 12 hours

   (0.39°C/hours) x (unknown # of hours) = 5.44°C lost after the initial 12 hours

   Solve for unknown number of hours:

   unknown # of hours (x) = 5.44°C ÷ (0.39°C/hour)
   = 13.9 hours total time to lose 13.9°C or approximate time of death

4. First 12 hours there was a loss of 9.36 degrees C 9.36°C
   Next 13.9 hours there was an additional loss of 5.44° C= 5.44°C
   Therefore, the victim has been dead about 25.9 hours. (or approximately 26 hours)

   .8 hours x 60 min/hr = 48 minutes

Questions:

Part A

1. Determine the approximate time of death using evidence from algor mortis. Show your work. Approximately how long has the victim been dead if his body temperature was 33.1°C (91.6°F)?

2. A body found outside in the winter has a temperature of 33.1°C. Has the body been dead a longer or shorter time than in problem 1? Explain your answer.
3. Approximately how long has the victim been dead if his body temperature was 25.9°C (85.2°F)?
4. What is the approximate time of death if the body temperature was 15.6°C (60.8°F)?
5. What is the approximate time of death if the body temperature was 10°C (50°F)?
6. What is the approximate time of death if the body temperature was 29.4°C (84.9°F)?
7. What is the approximate time of death if the body temperature was 24°C (75°F)?

Part B

Describe the impact on time of death for each of the variables listed below. If you based your time of death estimates strictly on temperature loss to be 10 hours earlier, would you reduce your 10-hour estimate or increase your 10-hour estimate if the body had been:

1. Naked
2. Exposed to windy conditions
3. Suffering from an illness prior to death
4. Submerged in a lake

Further Study:

1. Investigate the procedures used by crime-scene investigators to take accurate body temperature readings.
2. What is the significance of determining the time of death? Why is it so important to crime-scene investigators to take the temperature of the deceased body if the person is already known to be dead?

ACTIVITY 11-3
INSECT STUDY

Objective:

By the end of this activity, you will be able to:
Study the behavior and life cycles of insects associated with decomposition.

Introduction:

The blowfly is often the first insect to reach a corpse and lay its eggs. Use Figures 11-17 and 11-18 as references in your study of blowfly development.

Materials:

Data Table: Insect Study
1 lb. cottage cheese or pudding containers
raw calf liver (1 lb.)
sharp knife
warm (or hot) day (but not windy or too hot!)
thermometer
plastic kitchen-sized garbage can with a flip top
plastic garbage bag liner
another cardboard box to cut up into smaller pieces
small cardboard box large enough to hold the plastic cottage cheese
    container and fit inside a garbage can
digital camera (optional)

Safety Precautions:

Wash your hands after handling the flies.

Procedure:

Part A: Setting up the Fly Incubator

1. Line a plastic kitchen-sized flip-top or swinging-lid garbage can with a plastic liner.
2. Cut up a cardboard box and fold the cardboard pieces in half so that the pieces of cardboard don't lie flat.
3. Add the folded cardboard to the bottom of the kitchen garbage bag so that the garbage bag is one-third full. This is important because, during the last stage, the larvae will migrate away from the food to a dry area. If you have ample cardboard pieces available, the larvae will have many areas to hide. (Depending on the size of your garbage can, you might want to add an empty box on top of the cardboard pieces to elevate your fly assembly so that you will have easier access to the fly dish shown on the next page.

Part B: Preparing Your Liver Dish

1. Add raw liver in a plastic cottage cheese container.
2. Cut some slices into the surface of the liver to make gashes within the liver to resemble an open wound.

Part C: Obtaining Flies

1. Leave your open liver container in an area where you want to collect flies. The odor of the liver should attract flies within minutes.
2. Collect flies on a warm day. If it is too cold or too windy, flies will not lay their eggs. Avoid taking fly collections on very windy days or very hot days.
3. Leave the liver container in the open area for at least one hour. Place the dish in an area that will not be disturbed by dogs or cats.
4. Look for very small, white clusters of fly eggs on top of the liver.

Part D: Incubation of Fly Eggs

1. Place the liver container and fly eggs into a small cardboard box. Be sure the cardboard box and the plastic liver dish will fit in the garbage can.
2. Allow the flip-top cover of the garbage can to close. This will still allow other insects to enter the dish while keeping some of the odor inside the garbage can.
3. Keep the garbage can in an area that is not in direct sunlight. Because some odor will be given off, place the garbage can in an area where the odor will not present a problem to others.

Part E: Observations and Data Collection

1. Make observations each day and record on your data table until adult flies have emerged.
2. Take a digital photo of the liver container and any organisms near or on the container. Note: Larvae tend to move away from light, so be ready to quickly take pictures when you view your liver dish.
3. Complete the data table as you make your daily observations. Record the date and time of your daily collections. Record the tempera-

Insect study setup.

Liver inside container   Liver inside container and
in the sun or semi-shade   inside the box

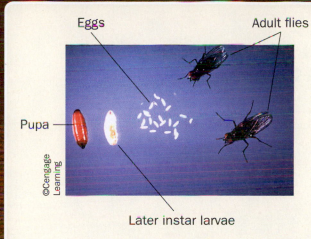

Eggs

Adult flies

Pupa

Later instar larvae

©Cengage Learning

ture inside the fly incubator (garbage can). Make other observations such as the color, size, and behavior of your insects.

Part F: Report

Option 1

1. Create a PowerPoint presentation of the insects collected on the liver. Include your digital photos taken of the different stages of the insects you observed. If you do not have a digital camera, obtain pictures of the insects from the Internet.

2. Include the name of all insects and correctly identify their stage of development. Indicate the preferred food source for each of the insects.

3. Do not place too many words on each frame.

Option 2

Write an autobiography from the viewpoint of the fly as it develops from an egg into adulthood. Include in your diary:

- Physical description of the insect at different stages of development
- Physical description of the insect's environment
- Descriptions of when the insect feeds or does not feed
- Description of the type of food it is eating
- Descriptions of any "travels" or migrations and movements of the insect
- Photos from your digital pictures taken during the study

Your information should be scientifically correct, but feel free to be creative in your insect diary!

Option 3

Prepare a scrapbook from the viewpoint of the insect as it progresses through its different stages. Use the photos from your study. Add notations indicating the:

- Physical descriptions of the insect at the different stages of development
- Physical descriptions of the insect's environment
- Descriptions or comments on the insect's source of food at different stages
- Descriptions of any "travels" or migrations of the insect

Your information should be scientifically correct, but feel free to be creative with this scrapbook!

ACTIVITY 11-4

ESTIMATING TIME OF DEATH USING INSECT, ALGOR, AND LIVOR MORTIS EVIDENCE

Objective:

By the end of this activity, you will be able to:
Estimate the time of death using insect, algor, and livor mortis evidence.

Materials:

paper
pen or pencil
calculator
Rigor Mortis Reference Table: Refer to the tables distributed by your teacher (Figures 11-8 and 11-9)
Insect Reference table: Refer to the table distributed by your teacher (Figure 11-7).

Safety Precautions:

None

Procedure:

Working in pairs, answer the following questions.

Questions

1. A naked, male corpse was found at 8 a.m. on Tuesday, July 9. The air temperature was already 26.7°C (81°F). The body exhibited some stiffness in the face and eyelids and had a body temperature of 34.4°C (93.9°F). Livor mortis was not evident.
   a. Approximately how long ago did the man die?
   b. Justify your answer.
   c. Would clothing on the body have made a difference in determining the actual time of death? Why or why not?
2. At noon, a female corpse was found partially submerged on the shore of a lake. The air temperature was 26.7°C (81°F), and the water temperature was about 15.6°C (61°F). Rigor mortis was not evident, and the body's temperature was 15.6°C. Livor mortis showed a noticeable reddening on the victim's back that did not disappear when pressed. Bacterial activity was not significantly increased, and the lungs were filled with water.
   a. Approximately how long ago did the woman die?
   b. Justify your answer.
3. The body felt cold to the touch. The thermometer gave a reading of 21.1°C (70°F) for the body temperature. No rigor mortis was evident, but livor mortis had set in with blood pooling along the back. There was no noticeable increase in bloating or bacterial activity in the digestive system and no putrefaction. The man had been dead for more than 48 hours. How is that possible?

4. The dead body contained evidence of blowfly infestation. The larvae were collected and reared in a lab in an environment similar to the conditions surrounding the dead body. Adult flies mated and laid eggs. Data was collected, noting the time required to progress from one stage to the other and recorded in the following Data Table.

Stage	Accumulated Time Since Egg Was Laid (Hours)	Accumulated Time Since Egg Was Laid (Days)
Egg	Egg laid minutes after death	0
Larva stage 1	24	1
Larva stage 2	60	2.5
Larva stage 3	96–120	4–5
Pupa	192–288	8–12
Adult	432–576	18–24

a. Record the estimated time since death if the insects recovered from the dead body were in each of the stages below:
egg, larva 1, larva 2, larva 3, pupa, adult.

b. Record the estimated time since death if insects were in the following stages: Some eggs and some larva stage 1, some adults and some pupae, some larva found in stage 2 and stage 3.

5. A dead body of an elderly gentleman was discovered in an abandoned building. Blowfly pupae were found on the body. A missing person's report was filed for an elderly gentleman who had wandered away from home just two days before. The body found was similar in age, height, and weight to the missing person. Could this possibly be the same person as the person described in the missing person's report? Explain your answer.

6. The police received a report about a body found in the woods behind the local shopping center. The forensic investigator collected 5 different types of living insects on the man's body. It's important to stress that investigators found all 5 insects alive on the body at the same time. The insects were sent to the forensic entomology lab, where they were raised under similar conditions to those found around the dead body. The following chart describes the life cycles of each of the five different types of insects found on the dead body. How long has the man been dead? Justify your answer.

Insect	Day									
	1	2	3	4	5	6	7	8	9	10
Blowfly	1	1	1	1	1	1	1	1	1	0
Species A	0	0	0	0	1	1	1	1	1	1
Species B	0	0	0	0	0	0	1	1	1	0
Species C	0	0	0	0	0	0	0	0	1	1
Species D	0	0	0	0	0	1	1	1	1	1

0 = no evidence of fly species; 1 = evidence of egg, larva (maggot) or pupa

ACTIVITY 11-5
TOMMY THE TUB

Background:

Whether you are playing volleyball on a hot beach in August or snowboarding down a mountain on a cold, windy day in January, your body is constantly working to maintain a normal body temperature. Living organisms are equipped with mechanisms that maintain this balance (homeostasis). However, if a person becomes ill or dies, the mechanisms fail.

Objectives:

By the end of this activity, you will be able to:
1. Observe and record the heat loss each hour of a simulated human body, "Tommy the Tub," over a 24-hour period.
2. Compare that heat loss to the projected heat loss of a human corpse.

In this activity, you will create a simulation of a human body and record the heat loss over a 24-hour time period. Because the body is mostly water, you will substitute a tub of approximately the same volume of water as a human body. You are to compare the heat loss of Tommy the Tub to the projected heat loss of 0.78°C (1.4°F) for the first 12 hours and 0.39°C (0.7°F) for the next 12 hours.

Materials:

Activity Sheet for Activity 11-5
66-L plastic tub (with drawing of body, optional)
probe ware interface and two temperature probes or two thermometers
computer or TI-83 calculator (or better)
cart for transporting the tub
graph paper

Safety Precautions:

None

Time Required to Complete Activity:

two consecutive days (30 minutes per day)

Procedure:

1. Fill a tub with approximately 66 liters of hot water, adjusting the temperature to about 37°C (98.6°F).
2. Connect two temperature probes to a computer or TI-83 calculator to record temperature readings over an extended period. One probe should record the ambient air temperature. The second probe should be submerged in the tub to record the "body temperature."
3. Set the probe to measure temperature at one-hour intervals for a 24-hour period.
4. Record tub temperatures on the data table.
5. From the probe ware data or your data table, determine the average air temperature over the 24-hour period.
   Average air temperature = _____°C

6. From the probe ware data or your data table, determine the average loss of tub water temperature for the first 12 hours _____°C, average loss of tub water temperature for hours 13 to 24 _____°C, and average loss of tub temperature for hours 1 to 24 _____°C.

7. Prepare a best-fit graph of Tommy's heat loss over a 24-hour period. Include in your graph:
   a. Title of graph
   b. Appropriate scale for x and y axis
   c. Labeled x and y axis
   d. Units on x and y scale
   e. Draw the best-fit line (This line is approximated. It will be a straight line that will pass through some of the points but not necessarily all of them. There will be some points on either side of the line and not on the line.)

Time (hrs.)	Tommy Tub Temp. (˚C)	Ambient Temp (˚C)	Time (hrs.)	Tommy Tub Temp. (˚C)	Ambient Temp (˚C)
0			13		
1			14		
2			15		
3			16		
4			17		
5			18		
6			19		
7			20		
8			21		
9			22		
10			23		
11			24		
12			Average loss per hour		

Questions:

1. How does Tommy the Tub's temperature loss over the first 12 hours compare with that of a real human corpse? Explain your answer. Include data from your graph or data table to support your answer.

2. How does Tommy the Tub's temperature loss over the next 12 hours (hours 13 to 24) compare with the expected heat loss of a real human corpse? Explain your answer. Include data from your graph or data table to support your answer.

3. Explain some of the limitations of using Tommy the Tub as an appropriate model for a human body.

4. How could you design a more realistic model of a human corpse to be used in this experiment?

5. Did the ambient temperature change over the 24-hour period? If the ambient temperature did change, describe its possible impact on the loss of temperature noted on Tommy the Tub.

6. List some variables affecting the rate of heat loss from a human corpse.

CHAPTER 12

Soil Examination

WHEN SOIL IS NOT JUST DIRT

Police in California found a dead body on the platform of an oil well. Investigators noticed the rocks used to construct the platform seemed unique. Test showed that the rocks were a special type that came from a site 300 miles for the well's location. They knew finding matching samples of the rocks on a suspect would be key to solving the case.

Police arrested an acquaintance of the victim. The suspect kept repeating, "I wasn't anywhere near the crime scene!" The police suspected otherwise. Physical evidence linking the suspect to the crime scene was needed. After they obtained a search warrant, the police took possession of work boots that had been worn by the suspect. They took the boots to a laboratory. Once in the lab, forensic scientists attempted to match soil samples from the suspect's boots to soil from the area where the body was found. The suspect was not worried. To him, dirt was dirt. He was unaware of the testing that could be done to compare the soil from his boots to the soil found on the body.

Forensic geologists performed chemical and microscopic tests on the soil on the suspect's boots, as well as samples taken from his car.

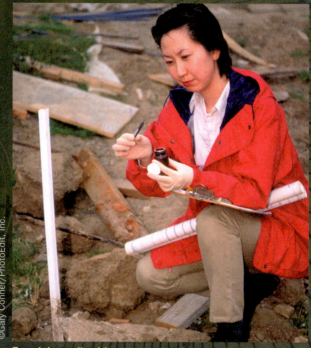

©Gary Conner/PhotoEdit, Inc.

Examining soil evidence

The test proved that the samples from the suspect came from the site where the body was found. Faced with convincing evidence, the suspect admitted his guilt and was sentenced for the crime.

OBJECTIVES

By the end of this chapter you will be able to

✔ Recognize various soil types and describe some methods for examining soil samples.

✔ Distinguish sand samples by size, color, and composition.

✔ Perform a soil analysis, including macroscopic and microscopic examination, as well as chemical and physical analysis.

✔ Explain how soil evidence can link suspects to crime scenes.

BIOLOGY   EARTH SCIENCES   CHEMISTRY

PHYSICS   PSYCHOLOGY   MATHEMATICS

TOPICAL SCIENCES KEY

VOCABULARY

clay the smallest type of soil particles that have the capacity to absorb and hold water

geology the study of soil and rocks

humus material in the upper layers of soil made up of the decaying remains of plants and animals

leaching the removal of minerals and clay as water drips through the soil

mineral a naturally occurring, crystalline solid formed over time on Earth

rock a hard substance made up of minerals

sand granules of fine rock particles

silt a type of soil whose particles are larger than clay and smaller than sand

soil a mixture of minerals, water, gases, and the remains of dead organisms that covers Earth's surface

soil profile a cross section of horizontal layers, or horizons, in the soil that have distinct compositions and properties

weathering formation of soil through the action of wind and water on rock

INTRODUCTION

Did You Know?

People in some cultures eat dirt. They do it because they say it tastes good or they think it will bring them good luck. But don't try it. Eating dirt can make you sick. On the other hand, the fact that babies of a certain age seem compelled to eat dirt has been theorized to be an adaptation allowing the immature immune system to encounter pathogens in the soil.

Soil is produced by a complicated process that is influenced by factors such as temperature, rainfall, and the chemicals and minerals present in the material from which it forms. Because of all the factors that affect soil formation, soil from different locations can have different physical and chemical characteristics that are useful to forensic scientists. Soil analysis has been used to identify livestock thieves and even the location of hot air balloon construction sites. The uniqueness of soil composition has helped locate burial sites and link suspects to crime scenes.

In one case, cattle rustlers stole a herd of cattle from Missouri and took it to Montana. The rustlers changed the brands on the cattle, thinking they would not get caught. However, they did not think of everything. Forensic scientists analyzed a sample of cow manure taken from the back of a truck thought to be used in the theft. They found small fragments of a rock made from silica in the manure. This type of rock could only have come from Missouri. Police used this soil evidence to convict the rustlers.

During World War II, Japanese scientists devised a plan for delivering explosives to the United States. The explosives would be carried in the air from Japan to the United States using hydrogen-filled balloons. The balloons also carried bags of sand as counterweights. More than 9,000 of these balloons were launched. Physical evidence has shown that about 300 balloons actually reached America. The explosives carried by the balloons did little property damage. However, six people lost their lives when one of the explosive-laden balloons detonated. By analyzing the sand used in the counterweights, geologists were able to determine that it came from one particular beach in Japan. With this information, the American forces bombed that beach and destroyed the site where the balloons were constructed.

HISTORY OF FORENSIC SOIL EXAMINATION

Real and fictional investigators have been using soil samples to identify criminals since the late 1800s. Between 1887 and 1893, Sir Arthur Conan Doyle wrote about the use of **geology** in the investigation of crime in his novels. His character, Sherlock Holmes, used soil and mud samples to help link an individual to a specific location where a crime had been committed.

Digging Deeper
with Forensic Science e-Collection

Sir Arthur Conan Doyle created the character Detective Sherlock Holmes, but Doyle wasn't only a writer. Go to the Gale Forensic Science eCollection on school.centage.com/forensicscience and research Doyle's life and Sherlock Holmes. Give a short presentation on this famous author and his famous creation. Include answers to the following questions in your presentation:

How did Conan Doyle feel about Sherlock Holmes?

After more than 100 years since the first publication, why is Sherlock Holmes still so popular?

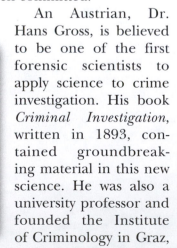

An Austrian, Dr. Hans Gross, is believed to be one of the first forensic scientists to apply science to crime investigation. His book *Criminal Investigation*, written in 1893, contained groundbreaking material in this new science. He was also a university professor and founded the Institute of Criminology in Graz,

Austria. Dr. Gross firmly believed in the value of physical evidence, including soils found at crime scenes.

A German investigator, Georg Popp, is credited with being the first forensic scientist to use soil evidence to solve a crime. During a murder investigation, he examined a handkerchief left at the crime scene and found it contained bits of coal and particles of hornblende, a greenish-black-colored mineral. Using science, Popp linked this evidence to a suspect who was known to work in a coal-burning factory and also in a gravel pit that contained hornblende. Popp also matched soil samples taken from the suspect's trousers to samples collected at the crime scene. When confronted with all of the evidence, the suspect admitted his crime.

Did You Know?

In the 1880s, a railroad company laying tracks in northwest Colorado gave the town Silt its name because of the nature of the soil in the area.

SOIL COMPOSITION

Soil is part of the top layer of Earth's crust, where most plants grow. Soil contains **minerals**, decaying organisms, water, and air, all in varying amounts. Soil texture describes the size of the mineral particles that make up soil. There are three main soil grain sizes: **sand**, **silt**, and **clay**. Sand describes the largest grain size and clay the smallest grain size.

There are also three subcategories of soil—loam, peat, and chalk. Loam is a type of soil that is made up of sand, silt, and clay. Along with mineral particles, soil can also contain organic material, such as decaying plants and animals. Soil with more than 20 percent decaying organic material is called peat soil. Chalk soil is alkaline and contains various-sized pieces of a solid, but soft, **rock** called chalk. Figure 12-1 summarizes information about these soil types.

Figure 12-1. *Soil Type Comparison*

Soil Type	Feel	Composed of	Location	Other Characteristics	
Sand	gritty	weathered rock	deserts, beaches, riverbeds	large visible particles, loses water quickly	
Clay	sticky	small particles adhering to each other	various	small particles, clumps, poor drainage	
Silt	crumbly, slippery like flour	medium-sized mineral particles	sediment in riverbeds	good drainage, easily farmed	
Peat	compressible	decaying organic material	bogs, areas where water is retained by organic matter failing to decompose	acidic, used with other soil types in fertilizer because of its ability to retain water	
Loam	loose	sand, silt, and clay mixture	various	best soil for agriculture	
Chalk	various colors, white to brown	alkaline (basic) soil with mineral stones	below the top soil	poor for agriculture, requires the addition of fertilizer and humus	

All photos ©Susan Van Etten

SOIL PROFILES

Soils form in horizons, or layers, that are more or less parallel to Earth's surface. The soil in each horizon has characteristic properties that differ from those in other horizons, as shown in Figure 12-2. Soil in a given area will have a unique profile or sequence of layers. Soil horizons within the profile are labeled with an uppercase letter, named as follows:

The uppermost horizon is called the O horizon. It is made up mostly of decaying organic matter, sometimes referred to as **humus**.

Beneath the O is the A horizon. The soil here is dark in color. The A horizon is also called topsoil. Topsoil is a mixture of humus and mineral particles. Seeds sprout and plant roots grow in the A horizon.

Next is the E horizon. The soil in the E horizon is light in color. It is made up mostly of sand and silt. Water dripping through the soil in this layer carries away most of the minerals and clay originally present. This process is called **leaching**.

The B horizon lies beneath the E horizon. Another name for this layer is the subsoil. The subsoil contains clay and mineral deposits that have leached out from layers above it as water drips through from the horizons above.

The C horizon is next. This layer is made of partially broken-up rock. Plant roots do not grow in this layer. Also, very little humus is found in this layer. If there is a solid rock layer underneath all of the other horizons, it is called the R horizon.

Figure 12-2. *A typical soil profile with horizons labeled.*

CHEMISTRY OF THE SOIL

The materials that make up a type of soil determine that soil's chemical properties. An important chemical property of soil is whether it is acidic or basic (alkaline). Chemists use a special scale, called the pH scale, to indicate how acidic or basic a substance is. The pH scale ranges from 0 to 14 (Figure 12-3). Anything with a pH of less than 7 is acidic. Substances with a pH greater than 7 are basic. A substance with a pH of 7, such as pure water, is considered neutral.

Figure 12-3. *The pH scale.*

	strong acid		weak acid			neutral		weak base			strong base		
pH value =	1	2	3	4	5	6	7	8	9	10	11	12	13	14

Examples	pH value (approximate)	Examples	pH value (approximate)
Acidic substances		Basic substances	
Battery acid	1	Baking soda, sea water	8.5
Lemon juice	2.5	Milk of Magnesia	10.5
Orange juice	3.0	Detergents	10.0
Vinegar	3.5	Ammonia water	11.0
Breads, pasta	5.0	Bleaches, oven cleaner	12.0
Rain (not acid)	5.5	Lye (drain cleaner)	13.5
Milk	6.5		

The pH of a soil can help determine if a plant will grow and survive. Whether certain minerals and nutrients are available to plants is partly determined by the soil's pH. Plants need nutrients like nitrogen and phosphorus to grow, but before a nutrient can be used by plants, it must be dissolved in the water contained within the soil. Most minerals and nutrients are more easily dissolved in acidic soils than in neutral or slightly alkaline soils. If the pH of the soil is above 5.5, nitrogen in the soil is made available to plants. Phosphorus in the soil is available to plants when soil pH is between 6.0 and 7.0. Figure 12-4 shows how the absence of certain nutrients can affect plants.

Figure 12-4. *Effects of nutrient deficiencies in plants.*

Nutrient Deficiency	Appearance of Plants
Nitrogen	Leaves of plants are yellow
Phosphorus	Small, frail plants with reddish leaves
Potassium	Leaves at bottom of plant dying from lack of chlorophyll (no longer green)
Calcium	Stems and leaves droop, unable to support upright position

Materials that make up a soil are not the only factors that affect its pH. Rainfall can change the pH of a soil. Water passing through the soil can carry away some basic substances, leaving the soil more acidic. Pollution, acid rain, and fertilizer use can also change the pH of soil.

SAND

Sand is formed by the action of wind and water on rocks, called **weathering**. As wind and water move rocks around, the rocks collide with other rocks. These collisions break the rock into smaller and smaller pieces until small grains, called sand, remain. As rock weathers, it breaks along boundaries between different types of mineral crystals. If the sand grain contains only one mineral, it is considered a crystal. If the grain contains two or more minerals, it is called a fragment.

The size of sand grains can be anywhere from 0.05 mm to 2 mm in diameter. Grains of sand can be rounded or angular, depending on the amount of weathering and the mineral composition of the rocks that formed the sand. Figure 12-5 on the next page shows different types of sands.

ROUNDING

As pieces of rock are moved by wind or water, they strike against one another. If the pieces hit in the center, they bounce off. If they strike along an edge, the edges may break off. As the edges break off, the rock pieces become more rounded. The entire rounding process may take millions of years to complete.

Sand grains carried by water lose their jagged edges and become rounded more slowly because the water acts as a buffer, so sand grains collide more gently. Wind-blown sand becomes rounded more quickly because the grains strike each other directly without a water buffer.

Figure 12-5. *Examples of different sands.*

a. Bermuda sand

b. Myrtle Beach, SC, sand

c. Vero Beach, FL, sand

d. Hawaii sand

e. Bahamas sand

f. Rhode Island sand

All photos ©Cengage Learning

Sand grains are classified as young, or immature, and old, or mature. Immature sand contains a large portion of clay, and the grains have a high percentage of fragments. This type of sand is found close to where it was formed, usually in areas where it is not exposed to waves or currents, such as the bottom of bays and lagoons, or in swamps or river floodplains. Mature sand does not contain clay and has fewer fragmented edges. Mature sand is found in areas, such as beaches and desert dunes, where much water and wind weathering has taken place.

MINERAL COMPOSITION OF SAND

Sand from different locations contains different combinations of minerals. The most common mineral found in sand is quartz. Many other minerals, such as feldspars, micas, and iron compounds, may be present in smaller quantities. Sand can also be made of organic material, such as coral and seashells. There are four basic sources of sand. These four types are summarized in Figures 12-6 and 12-7.

Figure 12-6. *Sand composition.*

Source	Composition	Identifying Features
Continental sand	granite, quartz, feldspar, mica, dark minerals	quartz
Volcanic sand	dark color, black basalt, green olivine, volcanic ash	dark color with green olivine, no quartz
Skeletal (biogenic) sand	broken shells, coral, coralline algae, sea urchin remains	shells indicate evidence of warm-water life
Precipitate sand	calcium carbonate	oolithic, egg-shaped, or round spheres of calcium carbonate from rock

Continental Sand

Continental sand is composed mostly of quartz, micas, feldspars, and dark-colored minerals like hornblende or magnetite (Figure 12-8). If feldspar is present, the sand probably came from a temperate or polar climate or from a high altitude. In warm, tropical climates, feldspar weathers away quickly. A high percentage of quartz means the sand is very old, because quartz weathers very slowly and often remains after other minerals have weathered away.

Figure 12-7. *Mineral components of sand.*

 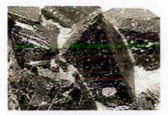 

a. *Feldspar is light in color and opaque. It comprises about 60 percent of Earth's surface.*

b. *Magnetite is opaque black or gray. It is a form of iron oxide.*

c. *Quartz can be white, gray, rose, or clear. It is a very abundant mineral.*

d. *Granite is usually found in massive deposits.*

e. *Mica appears in flakes and flat sheets.*

f. *Coral is the rocklike skeletal remains of sea life.*

Volcanic Sand

Volcanic sand is usually dark in color as a result of the presence of black basalt or green olivine (Figure 12-9). The source of this sand is from mid-ocean volcanoes, hot spot volcanoes, like those found in Hawaii. It sometimes contains volcanic cinders or other volcanic debris. Volcanic sand is very young and contains little or no quartz, except for obsidian (black) particles.

Skeletal (Biogenic) Sand

This type of sand is made of the remains of marine organisms, such as microorganisms, shells, and coral (Figure 12-10, below). It is younger than other types of sand. Skeletal sand originates in water and occurs all over the world. Skeletal sand made up of coral, however, occurs only in tropical regions. Because of the high amounts of calcium carbonate, skeletal sand gives off bubbles when mixed with a few drops of an acid.

Precipitate Sand

Water contains dissolved minerals, and when the water evaporates, these minerals precipitate, or come out of the water solution, and form crystals. Calcium carbonate will sometimes precipitate out of seawater, forming a coat of hard particles that resemble the layers of an onion. These layers eventually form small, round structures called *oolites*. Oolite formation is not an example of weathering, but rather of deposition. Sand containing these oolites is called oolithic sand and can be found in various places, including near the Great Salt Lake in Utah (Figure 12-11).

Figure 12-8. *Continental sand from Big Sur, California.*

Figure 12-9. *Volcanic sand from Hawaii.*

Figure 12-11. *Precipitate sand from the Great Salt Lake, Utah.*

Figure 12-10. *Skeletal sand from Bermuda and the Bahamas.*

SOIL COLLECTION

The following are some of the steps a crime-scene investigator uses to collect soil evidence at a crime scene. See Chapter 2 for information on correctly labeling and packaging samples. At a crime scene, investigators must follow certain procedures for collecting, labeling, and packaging soil evidence. They must:

1. Collect all samples as soon as possible before the environment of the crime scene can be changed by humans or animals walking through the scene or by weather conditions, like rain.

2. Collect at the surface baseline samples and samples that appear different or out of place from the rest of the environment of the crime scene.

3. Collect at least four tablespoons of material from several locations at the scene.

4. Sketch the crime scene and note on the map where the soil samples were obtained.

5. Be careful not to remove soil stuck to shoes, clothing, or tools found at the crime scene and package these items separately in appropriate containers.

6. Carefully remove soil from vehicles and package these soil samples separately.

7. Document all samples by taking photographs, drawing sketches, and labeling the evidence collection containers.

8. Collect additional soil samples from the four compass points (north, south, east, and west) within a few feet of the crime scene. Collect another set of samples 20 to 25 feet from the crime scene.

SOIL EXAMINATION

Soil recovered from a crime scene, victim, or suspect can be analyzed to determine if the suspect was at a particular location. Forensic scientists carefully compare the characteristics of the soil samples taken from the suspect or crime scene to those taken from a known location. The presence of soil unique to a certain area can show that a suspect or victim must have been in that area. The absence of a particular kind of soil can be used to prove or disprove an alibi. Layers of soil or mud taken from shoes or the wheels of vehicles can show that a suspect was present at a series of locations.

Scientists compare the size, shape, and color of soil by looking at samples macroscopically for soil type, amount of plant and animal material, and particle size.

Digging Deeper
with Forensic Science e-Collection

Why is it important for crime-scene investigators to properly collect and handle evidence? Go to the Gale Forensic Science eCollection on school.cengage.com/forensicscience and enter the search term "soil analysis." Click on Magazines and read about the investigation into the burial of five murder victims. Use information from the article and what you have learned about soil evidence collection procedures to write a brief essay describing the methods the investigators used to make sure their evidence was collected properly and not contaminated. Be sure to cite your resources.

Soil can also be identified and compared by studies using density columns, moisture content, and chemical testing for mineral content. Additional testing involves using specific laboratory techniques, such as X-ray diffraction. Before using X-ray diffraction, a soil sample is crushed into a very fine powder. The powder is then tested, and as the X-ray is deflected, it produces a pattern on a film. Each mineral and chemical produces a specific pattern. The pattern produced by the sample allows scientists to determine the mineral composition of the soil. Scientists compare the patterns produced by samples taken from a suspect to those taken from a specific location.

SUMMARY

©Elena Rooraid/ PhotoEdit, Inc

- Soil taken from crime scenes, victims, and suspects has been used to solve crimes since the late 1800s.

- There are three grain sizes of soil: clay, silt, and sand.

- There are three subcategories of soil: peat, loam, and chalk.

- Soil forms in layers called horizons. Starting from the layer at Earth's surface and moving downward, the horizons are labeled the O horizon, the A horizon, the E horizon, the B horizon, the C horizon, and the R horizon.

- The pH scale is used to measure how acidic or basic soil is. A pH from 0 to 6 is considered acidic. A pH from 8 to 14 is considered basic. A pH of 7 is considered neutral.

- Sand is formed when weathering breaks up rock into small grains from 0.05 mm to 2 mm in diameter.

- There are four main types of sand: continental sand, volcanic sand, skeletal sand, and precipitate sand.

- There are special procedures crime-scene investigators must follow when collecting soil evidence.

- Soil analysis can involve macroscopic and microscopic examination, chemical testing, and X-ray diffraction analysis.

CASE STUDIES

Andreas Schlicher (1908)

Georg Popp investigated the murder of Margaethe Filbert, who was found in a field in Bavaria, Germany. Neighbors considered a man named Andreas Schlicher to be a possible suspect. Schlicher denied having anything to do with the crime and denied walking in the field where the murder occurred on the day the crime was committed. Police determined that Schlicher's wife cleaned his dress shoes the night before the murder and that he had worn those shoes only on the day of the murder. The shoes were covered with particles of soil. Popp collected soil samples from the crime scene and also from the area immediately surrounding Schlicher's home. Popp also collected soil from a nearby castle where clothing and ammunition belonging to Schlicher were found.

When he examined the suspect's dress shoes, Popp noted three different layers of mud. The layers in the mud told him the sequence of Schlicher's travels on the day of the murder. In all three layers, Popp successfully compared the material on the shoe with soil from areas near the suspect's home, the crime scene, and the castle. Schlicher claimed he had walked in his own fields and was nowhere near the crime scene or the castle. His shoes told a different story. He had lied about his travels that day. The dirt on his shoes told more than any information police could obtain from an interrogation.

Enrique "Kiki" Camarena (1985)

Enrique Camarena, an undercover agent for the U.S. Drug Enforcement Agency, infiltrated drug dealings in Mexico and successfully helped destroy millions of dollars of illegal drugs. One of the drug trafficking groups under his investigation identified him as an agent. On February 7, 1985, Enrique and his pilot, Alfredo Zavala, were kidnapped in Guadalajara, Mexico, and killed. They were reported to have been killed by drug dealers during a shootout between Mexican police and the drug dealers. Soil samples taken from the two bodies were found to contain an unusual combination of minerals and volcanic particles. Investigators used this evidence to prove that the men's bodies were originally buried in the mountains, far away from the shootout. The soil evidence, along with other evidence, eventually showed that the Mexican federal police had been involved in the murders.

Janice Dodson (1995)

John B. Dodson died while on a hunting trip with his new wife, Janice. John appeared to be the victim of a hunting accident. Janice was standing in a nearby, muddy field when John was shot. She returned to their campsite and removed her muddy overalls before getting help for her wounded husband from some men who were hunting nearby.

Police determined that John was shot with a .308 caliber firearm, but no weapon was ever recovered. They did find two shell casings by a nearby fence. At the time of the shooting, Janice's ex-husband was staying at a campsite nearby. He reported that his .308 rifle and some cartridges were stolen from his tent. The ex-husband had an alibi for the time of the shooting, and the case went unsolved for three years.

In 1998, investigators returned to the area where John had been shot. They noted that the campsite where Janice's ex-husband had camped was next to a small pond. The investigators found bentonite, a mineral used to filter larger particles from the water, had been added to the pond water. This means that the mud next to the pond would also contain bentonite, and would be very different from mud taken anywhere else in the area. They reexamined the mud from Janice Dodson's overalls and found that it was a match to the sample containing bentonite taken from near the pond. This exposed Janice's lie about her location at the time of the shooting and placed her at the site of the stolen shotgun. Janice Dodson was tried and found guilty of her husband's murder. She is now serving a life sentence for her crime.

Think Critically How can layers of soil found on a suspect's shoes show a sequence of where the suspect has traveled?

Forensic Geologists

Forensic geologists use earth science and geologic materials, such as soil and rocks, to help solve criminal and civil cases. To become a forensic geologist, you would have to take college-level courses in geology, mathematics, chemistry, law, and forensic science. Some laboratories that employ forensic geologists include the FBI Laboratory in the United States, La Polizia Scientifica in Italy, the Centre of Forensic Sciences in Toronto, and the National Institute of Police Science in Japan.

Forensic geologists work in areas other than law enforcement. They can also help authenticate a work of art,

Forensic geologist working in the field.

©Catherine Ursillo/Photo Researchers, Inc.

such as a painting, by identifying the amount of mineral or organic material used to make the paints. This information can be used to tell when the painting was painted and possibly by whom.

Forensic geologists are hired to test the soil and rocks in an area being sold as a mine. The results of their tests provide information on whether that area will be a good source of gold, silver, oil, and so forth. A forensic geologist can also examine precious stones to determine their worth before purchase.

Forensic geologists also play a critical role in criminal investigations. They can tell if the soil on a body matches the soil at the location where the body is found. If it does not, they can help identify the area from where the body was moved. Forensic geologists can compare two soil samples, one collected from the suspect and the other collected from the crime scene, to

see if they came from the same source. For example, how does the soil on the suspect's shoe compare with the soil type collected at the crime scene?

A new, developing area of forensic geology is in intelligence work. A person may claim to have never been to a particular location, but soil and rock evidence from that area is found at the person's home. This evidence would link the person to a specific geographic location. Remember the rocky background behind Osama bin Laden in a tape played on television after September 11? Where was that tape filmed? If that question could be answered, bin Laden's whereabouts could be found. John Shroder, a geologist who had done field work in the area, was able to identify the region where bin Laden had been sighted in Afghanistan in 2001.

Learn More About It
To learn more about careers in forensic geology, go to
school.cengage.com/forensicscience.

CHAPTER 12 REVIEW

True or False

1. Clay particles are larger than sand.

2. Humus is decayed organic material.

3. The R horizon is composed of solid rock.

4. Acidic soil has a pH between 8 and 14.

5. Small, round grains are considered a characteristic of immature sand.

Multiple Choice

6. Oolites are characteristic of
   a) continental sand
   b) volcanic sand
   c) skeletal sand
   d) precipitate sand

7. In which soil horizon would you expect to find a carrot growing?
   a) A horizon
   b) E horizon
   c) B horizon
   d) C horizon

8. Which soil type would be best for gardening?
   a) sand
   b) loam
   c) clay
   d) chalk

Short Answer

9. A forensic scientist examines a soil sample using a microscope. She finds small, rounded grains made up of quartz crystals. There are also small bits of coral present. From this information, what can you conclude about the origin of the soil?

   _____

   _____

10. You are sent to collect soil evidence from a house that has been burglarized. Briefly describe how you would go about collecting samples.

   _____

   _____

11. Write a short story of a fictitious crime and describe how soil evidence helped link a suspect to the crime scene or victim.

_____

_____

12. Describe physical characteristics of soil that could help distinguish three different soil samples.

_____

_____

Bibliography

Books and Journals

Lee, Henry. *Physical Evidence in Forensic Science*. Tucson, AZ: Lawyers & Judges Publishing, 2000.

Lerner, K. Lee, and Brenda Wilmot Lerner. *World of Forensic Science*. Belmont, CA: Thomson Gale, 2005.

Murray, R. C., Tedwow, J. C. F. *Forensic Geology: Earth Science and Criminal Investigation*. New Brunswick: Rutgers University Press, 1975.

Murray, R. C., Tedwow, J. C. F. *Forensic Geology*. Englewood Cliffs: Prentice Hall, 1992.

_____. "The Geologist as Private Eye." *Natural History Magazine*, February 1975.

_____. *Soil in Trace Evidence Analysis*. Proceedings of the International Symposium on the Forensic Aspects of Trace Evidence 1991: 75-78.

_____. "Devil in the Details, The Science of Forensic Geology." *Geotimes* February 2000; 14-17.

Murray, Raymond. *Evidence from the Earth*. Missoula, MT: Mountain Press Publishing, 2004.

Stewart, Melissa. *Soil*. Portsmouth, NH: Heinemann Library, 2002.

Tarbuck, Edward J., Frederick K. Lutgens, and Dennis Tasa. *Earth: An Introduction to Physical Geology*, 8th ed. Upper Saddle River, NJ: Prentice Hall, 2005.

Web sites

Gale Forensic Sciences eCollection, school.cengage.com/forensicscience.

FBI Handbook of Forensic Services, Revised 2003, www.fbi.gov/hq/lab/handbook/intro.htm.

http://query.nytimes.com/gst/fullpage.html?res=9402E5DC123BF930A25751C0A962948260&sec=health&spon=&pagewanted=1

Hayes, Robert A. Forensic Geologists Uncover Evidence In Soil And Water http://www.geoforensics.com/geoforensics/art-1101a.html.

ACTIVITY 12-1
EXAMINATION OF SAND
(Modified from original lab by Mary Farina)

Objectives:

By the end of this activity, you will be able to:
1. Examine and compare various samples of sand.
2. Compare and analyze samples and determine if they have a common composition and origin.

Time Required to Complete Activity: 45 minutes

Materials:

(students working in pairs)
stereo microscope or hand lens
sand samples from five suspects and the crime scene
microscope slides
sieve set (to share with other groups)
dropper bottle of dilute hydrochloric acid (0.1 M)
small paintbrush
teaspoon or other small measuring device

Safety Precautions:

Goggles should be worn when working with dilute hydrochloric acid. Materials should be discarded as directed by your instructor.

Background:

Sand from Vero Beach, Florida.

©Cengage Learning

Five suspects were identified in a theft involving stolen property from a beach house. The suspects' shoes were confiscated, and the sand found was compared to that at the beachfront property. Sand from differing locations may be unique in appearance. Microscopic examination can tell a great deal about the origin, composition, and age of the sample of sand.

Procedure:

1. Using a stereo microscope or hand lens and slide, examine 50 grains of sand of Sample 1.
2. Describe your results on Table 1. Include the source of the sample indicated on the container and the type of crystals, fragments, or both if present. Crystals consist of one type of mineral only. Fragments consist of two or more minerals.
3. While viewing the 50 grains under the microscope, determine how many quartz, feldspar, black minerals, and others are in the count of 50 grains. Multiply each number by two to determine the percentage.
4. Examine the sand for degree of rounding. Degree of rounding can be either very round, rounded, or angular edges.

5. Use the sieve stack to separate a teaspoon of the sample and determine which sieve has collected the fragments. Record as either most, some, least, or none. This will give you an approximate idea of the percentage of particle size in the sample. View each sieve and compare which sieve has collected the most, some, least, or no particles.
6. Add a drop of hydrochloric acid to your slide of sand grains. Bubbling indicates the presence of carbonates, the shells and skeletons of dead marine organisms. Record if bubbling occurred and to what degree.
7. Wash off the slide.
8. Repeat the process with each of the other Samples 2 through 5.
9. Repeat the process for the Crime Scene sample (beachfront area). Determine if the soil from the crime scene matches any of the samples taken from the suspects.

Data Table 1: Comparison of Sand Analysis

Characteristics	Sample 1	Sample 2	Sample 3	Sample 4	Sample 5	Crime Scene
Source of sand						
Crystals or fragments or both						
Minerals found per 50 grains: Quartz———— Feldspar———— Black minerals— Other————	% of total ____% ____% ____% ____%	% of total ____% ____% ____% ____%	% of total ____% ____% ____% ____%	% of total ____% ____% ____% ____%	% of total ____% ____% ____% ____%	% of total ____% ____% ____% ____%
Rounding: Very rounded, rounded, or angular edges						
Particle size: record as most, some, least, or none *Top sieve (4 mm)* 4th sieve (2–3.99 mm) 3rd sieve (0.25–1.99 mm) 2nd sieve (0.062–0.249 mm) Bottom sieve (≤0.062 mm)						
Reaction with hydrochloric acid: Bubbles and sand totally dissolves or bubbles slightly or does not bubble at all						

Questions:

1. Your unknown sample appears closest to which sand sample? Is it an absolute match? Why or why not? Justify your answer from the data collected from your soil examination.
2. Would matching sand evidence from a crime scene to a suspect be enough evidence to convict a suspect of a crime? Explain your answer.

ACTIVITY 12-2
SOIL PROFILE EXAMINATION

Objective:

By the end of this activity, you will be able to:
Study soil characteristics and match samples taken from a suspect to samples found at a crime scene.

Time Required to Complete Activity:

60 minutes over two consecutive days

Materials:

(to share by teams of four)
5 hand lenses or compound microscopes or stereomicroscope with 40x magnification
5 graduated cylinders, 250 mL
5 soil samples, four from suspects and a crime scene sample
5 rubber bands
5 beakers, 250 mL
5 pieces of cheesecloth approximately 8" x 8"
5 teaspoons
5 pieces universal range pH paper
5 watch glasses
5 ultraviolet lights
5 flat toothpicks

Safety Precautions:

Proper eye protection is needed for the UV light.

Background:

A robbery has occurred, and four suspects have been questioned. Soil samples were taken from the wheel wells of each of the suspects' vehicles. Do any of the soil samples match soil from the crime scene?

Procedure:

Part A: Microscopic Examination
Obtain four soil samples (1 to 4) and a crime scene sample.

1. Examine a dry Soil Sample 1 (from Suspect 1) using a hand lens or low power (40x) on a microscope. Because soil changes somewhat in appearance as it dries out, make sure each sample is dry before beginning your examination.
2. Record all information requested on Table 1 on page 355.
3. Describe or sketch any organisms found in the soil.
4. Describe the color, texture, odor, and overall appearance of the soil.

Data Table 1: Soil Analysis

Soil Sample	Describe or sketch any organisms or foreign objects found	Color of Dry Sample (black, brown, gray, etc.)	Sample Texture (crumbly, gritty, loose, sticky)	Odor of Sample (smells like ? or no odor)	Overall Appearance of Sample (sand-like, organic, rocks, minerals)	Appearance under UV Light (glows, no glow)
1						
2						
3						
4						
Crime Scene						

Part B: UV Examination

5. View the sample in a darkened room using a UV light and describe what you see.
6. Repeat the procedure and examine each of the other suspect samples and the crime scene sample.

Part C: Determination of pH of the Soil Sample

The pH value is a measure of whether something is acidic, neutral, or basic.

7. Place a piece of cheesecloth on a clean surface. Place two spoonfuls of soil from Sample 1 in the center of the cheesecloth. Gather the sides of the cheesecloth and place a rubber band around the cheesecloth, capturing the soil sample in a ball within the cloth.
8. Place 50 mL of distilled water in a 250-mL beaker and label it Suspect 1.
9. Place the ball of soil in the water and leave it undisturbed for 10 minutes.
10. Repeat steps 7 to 9 for suspect samples 2, 3, 4, and the crime scene sample. Let samples sit for 10 minutes.
11. Using the pH paper, determine the pH of the water for each of the soil samples and complete the chart. If the pH value is less than 7, the soil is acidic. The lower the number, the more acidic. If the pH equals 7, the soil is neutral. If the pH is greater than 7, the soil is basic. The higher the number, the more basic the sample.
12. Record your results on Table 2 on page 356.

Part D: Sedimentation of Soil Samples

Soil samples that seem identical can be further examined by creating a sedimentation column. As shaken materials separate, they will layer, with the densest materials settling to the bottom first.

13. Into a 100-mL graduated cylinder, add one heaping teaspoon of Soil Sample 1.
14. Cover the top of the cylinder; shake the contents for 30 seconds.
15. Repeat steps 13 and 14 for each of the other samples.

Data Table 2: pH Determination

Soil Sample	pH	Acidic or Basic?	pH Paper
1			
2			
3			
4			
Crime Scene			

Data Table 3: Soil Sedimentation Results

Sample	1	2	3	4	Crime Scene
Using colored pencils, draw a sketch to scale of the layers in the column					
Number of Distinct Layers					
Description of Floating Material					

16. Allow all samples to settle overnight and compare the overall appearance of each sample. Note any floating material in your description.
17. Record your results in Table 3.

Questions:

Compare your results from Parts A to D of this activity.

1. The crime scene sample and sample _____ is a match. Justify your results.
2. How could they be used in a court of law?
3. Research what other types of analysis might be performed to compare soil samples.

ACTIVITY 12-3
CHEMICAL AND PHYSICAL ANALYSIS OF SAND

Objectives:

By the end of this activity, you will be able to:

1. Examine and compare chemical reactions in various samples of sand.
2. Examine and compare the magnetic and fluorescent properties of sand.
3. Compare samples and determine if there is a match between sand found on a crate of narcotics and sand from four possible ports of origin.

Time Required to Complete Activity:

two periods, 45 minutes each

Materials:

(students working in pairs)
1 dropper bottle of acetic acid (0.1 M CH_3COOH)
1 dropper bottle of dilute hydrochloric acid (0.1 M HCl)
1 dropper bottle of silver nitrate (0.1 M $AgNO_3$)
5 sand samples numbered location 1 to 4 and a crime scene narcotics crate
5 microscope depression slides (if available), or 5 watch glasses, or a well plate tray with 15 wells
magnet
5 flat toothpicks
UV light (to share with other groups)
5 squares of black paper, 3" x 3"

Safety Precautions:

Goggles are to be worn to avoid contact between harmful solutions and your eyes. All materials are to be handled as described by your instructor and discarded as directed. Silver nitrate solution will stain clothing and skin temporarily. Avoid looking directly into a UV light source. Proper eye protection is needed for the UV light source.

Background:

Customs officials from New York City noted the presence of sand in a crate containing narcotics. The sand was analyzed and found to be composed of quartz, feldspar, and shell fragments from a high-energy beach (a beach with lots of wave action). In an attempt to trace the route of the narcotics, a list of all ports visited by the ship was obtained. Samples of sand were obtained from each of the four possible ports and compared to the sample carried with the narcotics. The smugglers were later apprehended and arrested.

Reactions to Be Examined:

1. Test for sulfates:

   sample sulfates + $BaCl_2$ + $CH_3COOH \rightarrow BaSO_4 \downarrow$ (a white precipitate) + dissolved materials

   sample sulfates + barium chloride + acetic acid react to form $\rightarrow$ a white solid

2. Test for chlorides:

   sample chlorides + $AgNO_3$ + $CH_3COOH \rightarrow AgCl_2 \downarrow$ (a white precipitate) + dissolved materials

   sample chlorides + silver nitrate + acetic acid react to form $\rightarrow$ a white solid

3. Test for carbonates (living material):

   sample carbonates + $HCl \rightarrow CO_2 \uparrow$ (gas bubbles) + H_2O + dissolved materials

   sample carbonates + hydrochloric acid react to form $\rightarrow$ gas bubbles of carbon dioxide

Procedure:

Sulfate Test

1. Place about 50 grains of the Location 1 sample in a depression slide and add two drops of the 0.1 M barium chloride solution and two drops of the 0.1 M acetic acid solution.

2. Stir with a toothpick and observe under a hand lens or microscope at low power. If a white precipitate forms, sulfates are present in the sand sample.

3. Record your results on Table 1.

4. Repeat steps 1 to 3 for each of the other sample locations and record the results on Table 1.

5. Wash off all of the slides and prepare for the chloride test.

A small scoop of soil is all that is needed for an analysis.

©J. Beam Photography

Chloride Test

6. Place about 50 grains of the Location 1 sample in a depression slide and add two drops of the 0.1 M silver nitrate solution and two drops of the 0.1 M acetic acid solution.

7. Stir with a toothpick and observe under a hand lens or microscope on low power. If a white precipitate forms, chlorides are present in the sand sample.

8. Record your results on Table 1.

9. Repeat steps 6 to 8 for each of the other sample locations and record the results on Table 1.

10. Wash off all of the slides and prepare for the carbonate test.

Carbonate Test

11. Place about 50 grains of the Location 1 sample in a depression slide and add two drops of the 0.1 hydrochloric acid solution.

12. Observe under a hand lens or microscope. If bubbles form, carbonates are present in the sample.
13. Record your results on Table 1.
14. Repeat steps 11 to 12 for each of the other sample locations and record the results on Table 1.

Magnet Test
15. Using a magnet, determine if any of the samples contain any magnetic components.
16. Record on Table 1.

Fluorescence Test
17. In a darkened area, determine the fluorescence of each sample of sand. The UV light will cause fluorescent material to glow.
18. Sprinkle a small amount of Location 1 sand on a piece of black paper and observe it under UV light.
19. Record the size, shape, and percentage of fluorescent particles on Table 1.
20. Repeat steps 18- 19 for the other samples.

Data Table 1: Sand Testing Results

Beach Sand Sample Location	Sulfate Test (white ppt, yes or no?)	Chloride Test (white ppt, yes or no?)	Carbonate Test (bubbles, yes or no?)	Magnetic Particles (yes or no?)	UV Reaction: Fluorescence		
					Size	Shape	% of Particles
1							
2							
3							
4							
Narcotics Crate							

Questions:

1. Based on the comparison of results in Table 1, can you match a location to the beach from which the drugs were transported?
2. Describe additional testing of the sand samples that would provide further evidence for a possible match. Justify your answer.

CHAPTER 13

Forensic Anthropology: What We Learn from Bones

BURN BARREL EVIDENCE LINKS SUSPECT TO MURDER

Four days passed before 25-year-old Teresa Halbach was reported missing. Teresa, a photographer working for the *Auto Trader Magazine*, spent much of her time driving across eastern Wisconsin in her 1999 Toyota RAV 4 taking pictures of old cars. On Friday, October 31, 2005, her last stop was at the Avery Auto Salvage yard in Gibson, near Lake Michigan. She was there to meet co-owner Steven Avery and to take pictures of a Plymouth Voyager minivan he had for sale.

The police knew Steven and his brothers from earlier encounters. Recently in a very public trial, Steven was convicted of rape and attempted murder and then released as innocent when DNA evidence pointed to another man. He filed a $36 million lawsuit against the state for wrongful conviction. When the police showed up Monday afternoon after tracing Teresa's movement to his salvage yard, he announced he was "being set up because of my lawsuit."

In the yard, officers found a "burn barrel" with remains of a camera, cell phone, clothes, teeth, and bones. A team of forensic anthropologists were called in to investigate, and they determined that the remains were of an adult

©AP Photo/The Post Crescent, Patrick Ferron

Steven Avery.

human female. Damage to some of the bones also suggested the body had been mutilated.

When Steven's nephew confessed to participating in the crime, Steven was arrested on numerous charges involving Teresa's death. Do the bones in the barrel and the account of a 16-year-old tell the same story? Will they convict a man who still maintains his innocence?

OBJECTIVES

By the end of this chapter you will be able to

- ✔ Describe how bone is formed.
- ✔ Distinguish between male and female skeletal remains based on skull, jaw, brow ridge, pelvis, and femur.
- ✔ Describe how bones contain a record of injuries and disease.
- ✔ Describe how a person's approximate age could be determined by examining his or her bones.
- ✔ Explain the differences in facial structures among different races.
- ✔ Describe the role of mitochondrial DNA in bone identification.

TOPICAL SCIENCES KEY

BIOLOGY EARTH SCIENCES CHEMISTRY

PHYSICS PSYCHOLOGY MATHEMATICS

VOCABULARY

anthropology the scientific study of the origins and behavior as well as the physical, social, and cultural development of humans.

epiphysis the presence of a visible line that marks the place where cartilage is being replaced by bone

forensic anthropology the study of physical anthropology as it applies to human skeletal remains in a legal setting

joints locations where bones meet

mitochondrial DNA DNA found in the mitochondria that is inherited only through mothers

ossification the process that replaces soft cartilage with hard bone by the deposition of minerals

osteobiography the physical record of a person's life as told by his or her bones

osteoblast a type of cell capable of migrating and depositing new bone

osteoclast a bone cell involved in the breaking down of bone and the removal of wastes

osteocyte an osteoblast that becomes trapped in the construction of bone; also known as a living bone cell

osteoporosis weakening of bone, which may happen if there is not enough calcium in the diet

skeletal trauma analysis the investigation of bones and the marks on them to uncover a potential cause of death

INTRODUCTION

In forensics, analyzing bones is important for identification of a possible victim or suspect. If the remains of bones are found in association with a suspect, being able to identify the bones can be a critical step in linking the suspect to the crime (Figure 13-1). This chapter will examine how someone's identity, sex, age, height, race, and background can be revealed through an analysis of his or her bones.

Figure 13-1. *People of different races have differently shaped facial bones.*

©Pixland/Jupiter Images   ©ImageSource   ©ImageSource

Studying bones may also reveal what happened to a person before or after death. Bone evidence can help an investigator reconstruct a crime.

HISTORICAL DEVELOPMENT

Anthropology is the scientific study of all aspects of human development and interaction. It studies tools, language, traditions, and social interactions and how we relate to other societies. Physical anthropology studies human differences, especially those by which we can be identified. **Forensic anthropology** studies these identifying characteristics on the remains of an individual. These unique characteristics can be used to demonstrate the sex, race, height, and physical health of a victim from his or her remains.

Did You Know?

The founder of modern criminology, Dr. Cesare Lombroso, claimed to be able to identify people with criminal tendencies based on physical characteristics, including head size and the distribution and abundance of facial wrinkles and eye defects. His theories were later proved wrong.

- In Europe in the 1800s, the origins of the races of humans were heatedly discussed. Scientists began using skull measurements to differentiate among individuals. The differences between male and female anatomy, and the formation, aging, and fusing of bones were also examined, laying the framework for today's knowledge.

- The Luetgert murder case of 1897 accused a sausage maker of killing his wife and boiling down her corpse. Remains found in the factory appeared to be fragments of his wife's skull, finger, and arm.

- In 1932, the FBI announced the opening of its first crime lab. The Smithsonian Institution became a working partner, aiding in the identification of human remains.

- In 1939, William Krogman published the *Guide to the Identification of Human Skeletal Material.*

- The remains of soldiers killed during World War II were identified using anthropologic techniques.

- More recently new techniques in DNA found in the mitochondria of cells has been used in identification, such as the analysis of the skeletons of Nicholas and Alexandria Romanov.

CHARACTERISTICS OF BONE

Our bones are alive; they may not move or appear to have any obvious function besides making our bodies rigid, but they do have other purposes. Bones carry on a type of respiration called cellular respiration and consume energy like any other living cells. Inside of bones is a tissue called marrow, where blood cells are made (Figure 13-2). Bones are regulated by hormones that affect the amount of calcium in the blood and in the hard part of the bone. Since bones are alive, they are capable of growth and repair.

Did You Know?

Bones can reveal if a person had tuberculosis, arthritis, and leprosy, as well as iron and vitamin D deficiency. Although long healed, a record of any broken bones can be detected.

DEVELOPMENT OF BONE

Bones originate from living cells called **osteoblasts.** During the development of the fetus, bones begin as soft cartilage, the same flexible material that makes up our ears. Osteoblasts migrate to the centers of cartilage production and deposit minerals, such as calcium phosphate, that harden to form bone. This process is called **ossification** and begins during the first few weeks of pregnancy. By the eighth week of pregnancy, the outline of the skeleton has formed and is visible in an X-ray. As bone develops, a protective membrane layer that contains nerves and blood vessels covers the surface of the bone. This membrane, called the perisoteum, serves an important role in keeping bones moist and aiding in the repair of injuries.

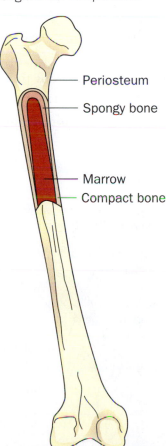

Figure 13-2. *A cross section of bone, showing its different longitudinal components.*

Periosteum

Spongy bone

Marrow

Compact bone

Throughout our lives, bone is deposited, broken down, and replaced. When an arm or leg is broken, the blood vessels at the area have the ability to increase calcium phosphate deposition to help heal the break. Newly trapped osteoblasts, called **osteocytes,** form the new bone framework. These cells can no longer produce new bone and become the basic framework for the new bone.

Osteoclasts, the second type of bone cell, are specialized to dissolve bone. As bones grow, they need to be reshaped. Simply adding layer upon layer of calcium phosphate would not maintain the proper shape of the bone. Therefore, as bones grow, the osteoclasts secrete enzymes that help dissolve certain areas of the bone.

Osteoclasts also aid in maintaining homeostasis within the body. Calcium, a mineral that is vital to normal metabolism, may be borrowed from bone when levels in the blood are low. Osteoclasts dissolve the bone and release calcium into the blood. Continued failure to eat enough calcium can result in a weakening of bones. This condition is called **osteoporosis.** A third function of the osteoclasts is to remove cellular wastes and debris from the bones. When bones become injured, the osteoclasts secrete enzymes that dissolve the injured or damaged part of the bone so that new, healthy bone can be laid down.

NUMBER OF BONES

How many bones are in the human body? Most medical students will tell you 206. That answer is only partially correct. An adult has 206 bones after all bones have become fully developed (Figure 13-3). A baby has 450 bones!

Figure 13-3. As we grow older, bones in our body fuse together.

©ImageSource ©Creative/Getty Images ©ImageSource ©ImageSource

HOW BONES CONNECT

A **joint** is the location where bones meet (articulate). Joints contain basically three kinds of connective tissue:

- *Cartilage.* Wraps the ends of the bones for protection and keeps them from scraping against one another (Figure 13-4).

- *Ligaments.* Bands of tissue connecting together two or more bones (Figure 13-5).

- *Tendons.* Connect muscle to bone (Figure 13-6).

Figure 13-4. *Cartilage*

Femur (thigh bone)
Patella (knee cap)
Articular cartilage
Media meniscus cartilage
ACL
Lateral (outer) meniscus cartilage
Tibia
Fibula

Figure 13-5. *Ligaments*

Patella (knee cap)
Anterior cruciate ligament (ACL)
Lateral collateral ligament (LCL)
Femur (thigh bone)
Posterior cruciate ligament (PCL)
Medial collateral ligament (MCL)
Fibula
Tibia

Figure 13-6. *Tendons*

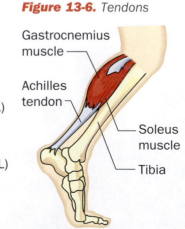

Gastrocnemius muscle
Achilles tendon
Soleus muscle
Tibia

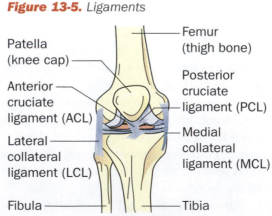

AGING OF BONE

Throughout our lifetime, bones are being produced and being broken down. Children build more bones at a faster rate than the rate of bones being broken down. As a result, bones increase in size. After 30 years, the process begins to reverse; bones deteriorate faster than they are built. This deterioration can be slowed with exercise. Without exercise, bones can become frail and less dense and are easily broken later in life.

People with osteoporosis are at risk of breaking bones because their bones have lost calcium and tend to be porous. As the vertebrae lose calcium, they begin to collapse and can give someone a hunched appearance. Some elderly people do, in fact, shrink; the loss of height is caused by the vertebrae collapsing.

The number of bones and their condition can tell an investigator about a person's age, health, and whether they had enough calcium in their food.

Did You Know?

The smallest bone in your body is 2.5 to 3.3 mm long. It is the stirrup bone, located behind your eardrum.

WHAT BONES CAN TELL US

So much about a person is revealed by examination of his or her bones (Figure 13-7). The term **osteobiography** literally translates as *the story of a life told by the bones*. Bones contain a record of the physical life. Forensic scientists know that analyzing the bones reveals clues to one's age, sex, race, approximate height, and health. For example, a loss of bone density, poor teeth, or signs of arthritis can point to nutritional deficiencies and disease. The bones of a right-handed person's arm would be slightly larger than the bones of the left arm. If someone lifted heavy objects regularly, the bones would be denser than someone who did not work physically hard. The type of sports one plays could be detected by the extra wear and tear on different joints and the sizes of the bones in general. An X ray of the bones taken during an autopsy would show previous fractures, artificial joints, and pins.

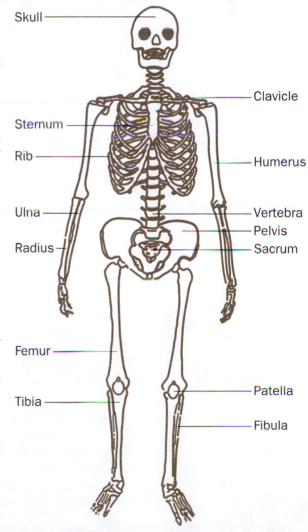

Figure 13-7. *Our skeletons reveal information about us.*

Skull — Clavicle — Sternum — Rib — Humerus — Ulna — Vertebra — Pelvis — Sacrum — Radius — Femur — Tibia — Patella — Fibula

HOW TO DISTINGUISH MALES FROM FEMALES

Often, a detective's first question to a forensic anthropologist is whether the skeleton belongs to a male or female. How can one differentiate sex from bone fragments? The overall appearance of the female's skeleton tends to be much smoother (gracile) and less knobby than that of a male's skeleton (robust). A man's skeleton is usually thicker, rougher, and appears quite bumpy. Because of male hormones, muscles are more developed in the male. When muscles are larger, they require a stronger attachment site on the bones. To accommodate the larger muscles and their tendons, the surface of the bone where a muscle and tendon attach is thicker, creating the appearance of a rough or bumpy area. One place this is especially noticeable is in the knees, because the bones of the knees are more obvious than other areas of the body.

Skull

Generally, the male skull is more massive and bumpier than the female skull. There are many specific differences, but the first step is to review Figures 13-8 and 13-9 depicting the major bones of the skull.

Figure 13-8. Front view of skull with major bones labeled.

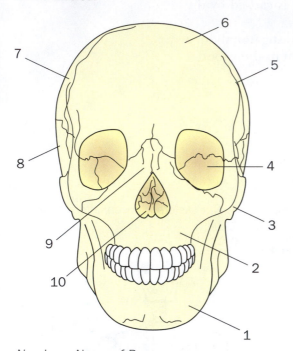

Figure 13-9. Side view of skull with major bones labeled.

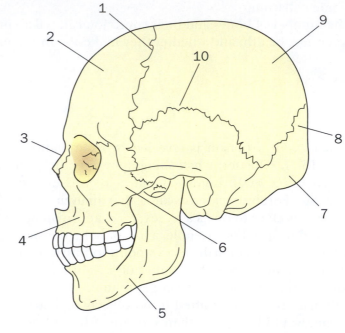

Number	Name of Bone
1	Mandible
2	Maxilla
3	Zygomatic
4	Orbits of the eye; Sphenoid
5	Coronal suture
6	Frontal
7	Parietal
8	Temporal
9	Nasal
10	Vomer

Number	Name of Bone or Suture
1	Coronal suture
2	Frontal bone
3	Nasal bone
4	Maxilla
5	Mandible
6	Zygomatic complex
7	Occipital protuberance
8	Lambdoidal suture
9	Parietal bone
10	Squamous suture

In Figures 13-10 and 13-11, note the differences between the male and female skulls. The male's frontal bone is low and sloping, whereas the female's frontal bone is higher and more rounded. The male eye orbits tend to be square, whereas the female's eye orbits are more circular. The male's lower jaw is square, with an angle of about 90 degrees. The female's lower jaw is sloped, with an angle greater than 90 degrees. Males also have squarer chins; females' chins are rounder or more V-shaped.

The occipital protuberance, a bony knob on the male skull, serves as an attachment site for the many muscles and tendons of the neck. Because the muscles in a man's neck are larger than the muscles in a woman's neck, the area of attachment needs to be thicker, creating the protuberance on the male skull.

Figure 13-10. *A side view of male and female skulls, noting the differences.*

Male; note the low sloping frontal bone and a jawbone set at 90 degrees

Female; note the rounded frontal bone and jawbone greater than 90 degrees

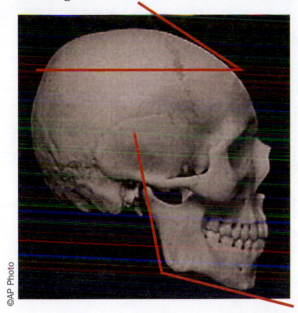

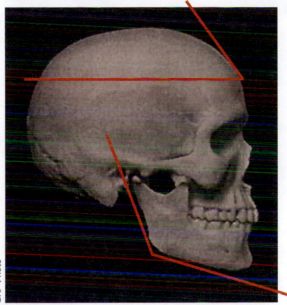

©AP Photo

©AP Photo

Figure 13-11 *Summary of male and female skull differences.*

Male Front View	Male characteristics	Trait	Female Characteristics	Female Front View
	More square	Shape of eye	More rounded	
	More square	Mandible shape from underside	More V-shaped	
	Thick and larger	Upper brow ridge	Thin and smaller	
Male Side View	**Male Characteristics**	**Trait**	**Female Characteristics**	**Female Side View**
	Present	Occipital protuberance	Absent	
	Low and sloping	Frontal bone	Higher and more rounded	
	Rough and bumpy	Surface of skull	Smooth	
	Straight	Ramus of mandible	Slanting	
	Rough and bumpy	Nuchal crest	Smooth	

Pelvis

One of the easiest methods of determining the sex of a skeleton is to examine the pelvis. Because of the anatomical differences needed for child-bearing, this region of the body exhibits many differences. The surface of a woman's pelvis is engraved with scars if she has borne children. During the fourth month of pregnancy, hormones are released that soften the tendons in the pelvic area to help accommodate the developing fetus. These scars can be detected on the pubic symphysis, a cartilaginous area where the bones meet. Review the different bones of the pelvis in Figure 13-12, on the next page.

To distinguish between the male and female pelvis, compare the following:

- Subpubic angle (Figure 13-13)
- Length, width, shape, and angle of the sacrum (Figure 13-14)
- Width of the ileum
- Angle of the sciatic notch

Figure 13-15 has a summary of these differences.

Figure 13-12. *The major bones of the pelvis.*

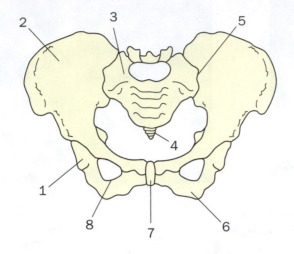

Number	Name of Bone
1	Joint
2	Ileum
3	Sacrum
4	Coccyx
5	Joint
6	Ischium
7	Pubis symphysis
8	Obturator foramen

Figure 13-13. *The subpubic angle is greater than 90 degrees on the female and less than 90 degrees on the male.*

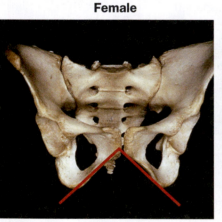
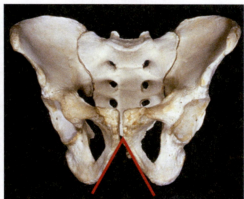

Female   Male

©VideoSurgery/Photo Researchers, Inc.

Figure 13-14. *The female's pelvic cavity is more opened than the male's.*

Female pelvic cavity (oval shaped)

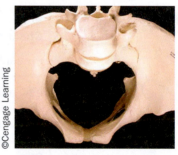

Male pelvic cavity (heart shaped)

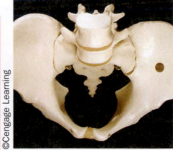

©Cengage Learning

Figure 13-15. *Summary of Male and Female Pelvis Differences*

Region	Bone	Male	Female
Pelvic	Subpubic angle	50–82 degrees	90 degrees
	Shape of pubis	Triangular pubis	Rectangular pubis
	Shape of pelvic cavity	Heart-shaped	Oval-shaped
Sacral	Sacrum	Longer, narrower, curved inward	Shorter, broader, curved outward

Thigh Bones

The thigh bone, or femur, also provides information about sex (Figure 13-16). The angle of the femur in relation to the pelvis is greater in females and straighter in males. The male femur is also thicker than that of a female.

HOW TO DISTINGUISH AGE

The age of a person can be determined by examining particular bones and by looking for the presence or absence of cartilage. Because bones do not reach maturity at the same time, it is possible to estimate the age of a person by looking for the absence or presence of specific characteristics on a range of bones (e.g., suture marks on the skull or the presence or absence of cartilage lines).

Suture Marks

Suture marks with a zigzag appearance are found on the skull where bones meet. In an immature skull, areas of softer tissue, such as the soft spot of the baby's skull (fontanel), gradually become ossified (harden). The suture marks slowly disappear as the bones mature, giving the skull a smoother appearance. There are three main areas of suture marks, marking three main areas where skull bones meet and grow together (Figure 13-17).

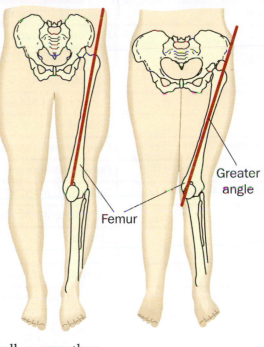

Figure 13-16. *The male femur is thicker and joins the pelvis at a straighter angle than the female femur.*

Greater angle

Femur

Figure 13-17. *The main suture marks on a skull, marking where the bones are growing to join together.*

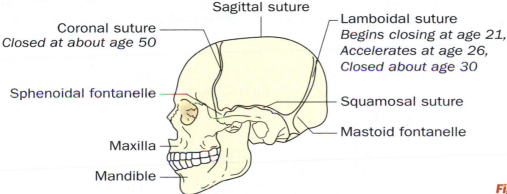

Sagittal suture

Coronal suture
Closed at about age 50

Lamboidal suture
*Begins closing at age 21,
Accelerates at age 26,
Closed about age 30*

Sphenoidal fontanelle

Squamosal suture

Mastoid fontanelle

Maxilla

Mandible

Cartilaginous Lines

Recall that we are born with more than 450 bones that later join together to form 206 bones. As the cartilage is slowly replaced with hard, compact bone, a cartilaginous line is visible, called an **epiphysis** (Figure 13-18). When the cartilage is fully replaced, a line is no longer visible. The age for the completion of growth for each bone varies. The presence or absence of these cartilaginous lines can therefore be used to approximate someone's age.

Long Bones

When the head of a long bone, like the thigh or upper arm bone (femur and humerus, respectively), has totally fused to its shaft, it is another indication of age. Various charts have been developed to help in this determination (Figure 13-19, on the next page). Because this fusing occurs at different times with different bones, this information can be used to approximate age.

Figure 13-18. *During development, a visible line occurs as the bone replaces cartilage.*

Epiphyseal plate (hyaline cartilage)

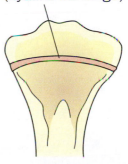

Epiphysis

Figure 13-19. *Estimation of age using bones.*

Region of the Body	Bone	Age
Arm	Humerus bones in the head fused	4–6
	Humerus bones in the head fused to shaft	18–20
Leg	Femur: Greater trochanter first appears	4
	Less trochanter first appears	13–14
	Femur head fused to shaft	16–18
	Condyles join shaft	20
Shoulder	Clavicle and sternum close	18–24
Pelvis	Pubis, ischium are almost completely united	7–8
	Ileum, ischium, and pubic bones fully ossified	20–25
	All segments of sacrum united	25–30
Skull	Lambdoidal suture close	Begins 21 ends 30
	Sagittal suture close	32
	Coronal suture close	50

HOW TO ESTIMATE HEIGHT

Did You Know?

By examining Roman skeletons, archaeologists determined that Roman males were 5'7" on average and Roman females were 5'3" on average. The average height in the United States today is 5'9" for males and a little less than 5'4" for females.

Measuring bones like the humerus or femur can help determine the approximate height of an individual. Many databases have been established that use mathematical relationships to calculate the overall height of an individual from one of the long bones of the body. There are separate tables for males, females, and different races (Figure 13-20). The mathematical formula between bone length and estimated height varies depending on the race and the bone used. If the race and sex of an individual are known, the calculation of height will be more accurate.

Here is an example of the formula: A femur measuring 40 cm belonging to an African American male is found. Use the formula on the next page to estimate his height:

$$\text{Height (cm)} = 2.10 \text{ femur} + 72.22 \text{ cm} \quad (\pm 3.91)$$
$$= 2.10 \ (49 \text{ cm}) + 72.22 \text{ cm}$$
$$= 102.9 \text{ cm} + 72.22 \text{ cm}$$
$$= 175.12 \text{ cm or } 69 \text{ inches (5 ft 9 inches)}$$

HOW TO DISTINGUISH RACE

Determination of race from skeletal remains is often difficult because through years of intermarriages, physical traits have blended and this distinction is losing its significance. Race is probably best indicated by the bones of the skull and the femur. Characteristics of the skull that differ with race include the following:

- Shape of the eye sockets

- Absence or presence of a nasal spine

- Measurements of the nasal index (the ratio of the width of the nasal opening to the height of the opening, multiplied by 100)

- Prognathism (the projection of the upper jaw, or maxilla, beyond the lower jaw) (*continued on page 372*)

Figure 13-20. Height estimation formula.

Bone length for American Caucasian males.

Factor × bone length	plus	Accuracy
Height (cm) = 2.89 × humerus	+ 78.10 cm	± 4.57
Height (cm) = 3.79 × radius	+ 79.42 cm	± 4.66
Height (cm) = 3.76 × ulna	+ 75.55 cm	± 4.72
Height (cm) = 2.32 × femur	+ 65.53 cm	± 3.94
Height (cm) = 2.60 × fibula	+ 75.50 cm	± 3.86
Height (cm) = 1.82 × (humerus + radius)	+ 67.97 cm	± 4.31
Height (cm) = 1.78 × (humerus + ulna)	+ 66.98 cm	± 4.37
Height (cm) = 1.31 × (femur + fibula)	+ 63.05 cm	± 3.62

Bone length for American Caucasian females.

Factor × bone length	plus	Accuracy
Stature (cm) = 3.36 × humerus	+ 57.97 cm	± 4.45
Stature (cm) = 4.74 × radius	+ 54.93 cm	± 4.24
Stature (cm) = 4.27 × ulna	+ 57.76 cm	± 4.30
Stature (cm) = 2.47 × femur	+ 54.10 cm	± 3.72
Stature (cm) = 2.93 × fibula	+ 59.61 cm	± 3.57

Bone length for Caucasians, both sexes.

Factor × bone length	plus	Accuracy
Stature = 4.74 × humerus	+ 15.26 cm	± 4.94
Stature = 4.03 × radius	+ 69.96 cm	± 4.98
Stature = 4.65 × ulna	+ 47.96 cm	± 4.96
Stature = 3.10 × femur	+ 28.82 cm	± 3.85
Stature = 3.02 × tibia	+ 58.94 cm	± 4.11
Stature = 3.78 × fibula	+ 30.15 cm	± 4.06

Bone length for African-American and African males.

Factor × bone length	plus	Accuracy
Height (cm) = 2.88 × humerus	+ 75.48 cm	± 4.23
Height (cm) = 3.32 × radius	+ 85.43 cm	± 4.57
Height (cm) = 3.20 × ulna	+ 82.77 cm	± 4.74
Height (cm) = 2.10 × femur	+ 72.22 cm	± 3.91
Height (cm) = 2.34 × fibula	+ 80.07 cm	± 4.02
Height (cm) = 1.66 × (humerus + radius)	+ 73.08 cm	± 4.18
Height (cm) = 1.65 × (humerus + ulna)	+ 70.67 cm	± 4.23
Height (cm) = 1.20 × (femur + fibula)	+ 67.77 cm	± 3.63

Bone length for African-American and African females.

Factor × bone length	plus	Accuracy
Stature = 3.08 × humerus	+ 64.67 cm	± 4.25
Stature = 3.67 × radius	+ 71.79 cm	± 4.59a
Stature = 3.31 × ulna	+ 75.38 cm	± 4.83
Stature = 2.28 × femur	+ 59.76 cm	± 3.41
Stature = 2.49 × fibula	+ 70.90 cm	± 3.80

Bone length for All ethnic groups or, if ethnicity is unknown, both sexes.

Factor × bone length	plus	Accuracy
Stature = 4.62 × humerus	+ 19.00 cm	± 4.89
Stature = 3.78 × radius	+ 74.70 cm	± 5.01
Stature = 4.61 × ulna	+ 46.83 cm	± 4.97
Stature = 2.71 × femur	+ 45.86 cm	± 4.49
Stature = 3.01 × femur	+ 32.52 cm	± 3.96
Stature = 3.29 × tibia	+ 47.34 cm	± 4.15
Stature = 3.59 × fibula	+ 36.31 cm	± 4.10

Digging Deeper
with *Forensic Science e-Collection*

Can the size of your head compared to your height determine if you are a thief? Can you determine a motive for crime by the size of the eye sockets? For more than a century, scientists have been measuring our bones to derive useful information from them. Compare and contrast the modern science of *anthropometry* and the traditional practice of *Bertillonage*. Search the Gale Forensic Science eCollection on school.cengage.com/forensicscience for the following article: K. Lerner and Brenda Lerner, eds. "Anthropometry," *World of Forensic Science*, Vol. 1. Detroit, MI: Gale, 2005; pp. 33–34. 2 vols.

- Width of the face
- Angulation of the jaw and face

Other racial comparative characteristics are listed in Figure 13-21.

Many other characteristics of a person can be determined by examining bones, such as whether the person was right- or left-handed; diet and nutritional diary, especially the lack of Vitamin D or calcium; diseases and genetic disorders, such as osteoporosis, arthritis, scoliosis, and osteogenesis imperfecta; previous fractures; type of work or sports based on bone structure; surgical implants, such as artificial joints (with code number stamped on them) and pins, and, in women, childbirth.

Figure 13-21. *Comparing the racial characteristics of bones.*

	Caucasoid	Negroid	Mongoloid
Shape of Eye Orbits	Rounded, somewhat square	Rectangular	Rounded, somewhat circular
Nasal Spine	Prominent spine	Very small spine	Somewhat prominent spine
Nasal Index	<.48	>.53	.48 – .53
Prognathism	Straight	Prognathic	Variable
Femur	Fingers fit under curvature of femur	Fingers will not fit under curvature of femur	Fingers will fit under curvature of femur

FACIAL RECONSTRUCTION

The exact size and shape of bones not only vary from person to person, but are also related to the overall shape and size of the muscles and tissues that lay on top of bones. Theoretically, it should be possible to rebuild a face from the skeleton up. The use of bones to reconstruct faces has been helpful in some crime investigations.

In 1895, Wilhelm His used the skull of Johann Sebastian Bach in an attempt to reconstruct his face in clay (Figure 13-22). His measurements of tissue depth taken from cadavers are the basis for the system of facial reconstruction used today. Victims of explosions or blunt force trauma often do not have enough bone structure in place to facilitate identification. Today, facial markers are positioned at critical locations on the face, and the clay is contoured to follow the height of the markers.

Reconstruction of the faces of famous historical figures has been attempted, including the reconstruction of King Tutankhamen (King Tut) (Figure 13-23). Notice the differences in the results.

The computer program Faces® (Interquest) performs a similar function today, allowing a facial manipulation and reconstruction in seconds. Investigators generate an image of the skull on a computer screen based on actual measurement and can manipulate the facial reconstruction. Features can be added, deleted, and easily modified. Nose and jaw length can be adjusted, as well as hairline, hairstyle, and the color of the skin and shape and size of the eyes.

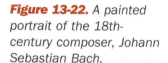

Figure 13-22. *A painted portrait of the 18th-century composer, Johann Sebastian Bach.*

©Mary Evans Picture Library/The Image Works

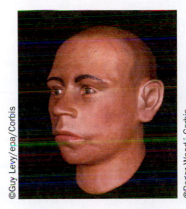

©Guy Levy/epa/Corbis

©Roger Wood/ Corbis

This technique, known as video-superimposition, can be used to match an existing photograph to someone's remains for the purpose of identification. The facial landmarks are measured and overlaid on a photograph of the skull for comparison.

These same techniques can be used to age missing persons and criminals who are still at large. Video cameras installed in banks and businesses can use this technology to apprehend criminals. Facial recognition systems may be used in the future to recognize terrorists and other criminals who attempt to superficially disguise their appearance.

DNA EVIDENCE

DNA profiling usually uses nuclear DNA, which is found in the nucleus of white blood cells and other body tissues. Bone contains little nuclear DNA, but it does contain **mitochondrial DNA.** Mitochondria are organelles found in all cells that contain DNA inherited only from the mother. There is no genetic information in mitochondria from our fathers. Long after nuclear DNA had been lost through tissue degeneration, mitochondrial DNA can be extracted from bone and profiled. The results can then be compared with living relatives on the mother's side of the family to determine the identity of skeletal remains.

BIOLOGY

SKELETAL TRAUMA ANALYSIS

Weathering and animals often damage bones that are exposed to the elements for long periods. Forensic anthropologists are trained to recognize these marks (Figure 13-24). A knife wound may leave parallel scoring on a rib, but mice and rodent chew marks can look very similar. **Skeletal trauma analysis** attempts to

Digging Deeper
with Forensic Science e-Collection

Some forensic scientists are experts on the condition of old bones. Others can easily see the difference between a blow from a hammer and one from a bullet by the scarring on a rib. Others can draw information from the soil and environment in which a body is buried to provide evidence for a crime. Discuss the three different subspecialties of forensic anthropology and crime scenes where these three skills would be required to solve the crime. Go to the Gale Forensic Science eCollection on school.cengage.com/forensicscience to find the following article: K. Lerner and Brenda Lerner, eds. "Skeletal Analysis," *World of Forensic Science*, Vol. 2. Detroit, MI: Gale, 2005; pp. 621–23. 2 vols.

Figure 13-24. *Forensic anthropologists are often required to determine if damage to bones occurred before or after death.*

make distinctions between the patterns caused by weapons and the damage and wear caused by the environment after death to decipher what happened to a body before and after death.

Generally, forensic anthropologists try to determine the weapon that caused death. Sharp-force and blunt-force trauma, gunshots, and knife wounds all have distinctive patterns. For blunt-force and sharp-force trauma, there is a difference in the shatter pattern and the amount of impact damage to the bone. Blunt objects generally have more cracks radiating from the site of impact, as well as causing more damage to the surface of the bone.

The strength of bone decreases as it ages and dries out. Living bone is actually quite flexible when compared to dry and brittle bones, and therefore they break in different ways. Living bones will usually shatter in a spiral pattern parallel to the length; old bones often break perpendicular to the length.

SUMMARY

- Bones are alive and carry on all life functions. Living cells replace the cartilage of our skeleton at birth by depositing calcium phosphate, creating a hard, compact material. This process is called ossification.

- The condition of bones can tell an investigator about a person's health and nutrition during life.

- Male and female skeletons differ in many ways, including roughness and thickness of bones, size and shape of frontal bone in the skull and the shape of the eye cavity, the angle of the pelvis, and the presence or absence of childbirth scars.

- The age of a person at death can be estimated by the number of bones, the sutures that mark bone joints, and the presence and location of cartilaginous lines.

- The height of a person can be estimated by the length of the long bones in the arms and legs. (Estimates are most accurate when the sex and race of the skeleton is known.)

- Facial reconstruction is possible using the physical measurements of the skull. Forensic investigators can match a skull's size and shape to a photograph of a person's head to make a positive identification.

- Mitochondrial DNA can be extracted from bone and used to help identify skeletal remains by comparing to maternal DNA.

- Skeletal trauma analysis examines the bones for evidence of damage. This damage may provide clues to injuries sustained when the person was alive or damage to bones after death.

- X-rays are a critical tool during an autopsy to reveal skeletal features, number of bones, conditions or bones, previous fractures, implants, disease, and disorders of the bone.

CASE STUDIES

Alfred Packer (1874)

After serving as a soldier in the Civil War, Alfred Packer worked as a guide and prospector in Colorado. During the winter of 1874, he and five other prospectors went into the mountains and disappeared. In April, Packer returned with a horrifying tale. The six men had been forced by cold, starvation, and sickness to resort to cannibalism, and only Alfred survived. He was accused of murder, but he professed innocence, saying he had acted in self-defense when he killed the last of the five prospectors. When their bodies were found two weeks later, significant amounts of flesh had been removed from them, and one body was headless.

Packer (known as the Colorado Cannibal) escaped from custody, but was recaptured nine years later, tried, and found guilty of murder. The remains of the five prospectors allegedly killed by Packer were exhumed in 1989 and re-examined. The bones showed distinctive marks of filleting, with defensive wounds on the arms and head trauma. Four of the victims had been bludgeoned to death, three with a hatchet-like implement and the fourth with perhaps a rifle butt. Distinctive scraping marks to the bone implied each body had been carefully stripped of flesh. The question of Packer's guilt still remains.

Elmer McCurdy (1911)

Train robber Elmer McCurdy was cornered in a hayloft and killed by gunshot. His body was embalmed with arsenic by the undertaker but remained unclaimed. The undertaker put Elmer on display, charging five cents to view the "Bandit Who Wouldn't Give Up." Several years later, his body was claimed by someone claiming to be his brother, and taken to California, where it was coated in wax and displayed in several circuses and amusement parks, and then forgotten.

Recently, during the filming of a TV show at the Nu-Pike Amusement Park, someone tried to move what they thought was a dummy and accidentally pulled off his mummified arm. The body was well preserved, yet no medical or dental records existed to give the body a name. Forensic anthropologists measured his facial dimensions, the lengths of his arms and legs, and the symphysis of his pubic bone. These details provided identification evidence for the state to put Elmer McCurdy to rest, once again.

The Romanovs (1918)

On July 16, 1918, the last royal family of Russia—Tsar Nicholas II, his wife Alexandra, four daughters, one son, and their servants—died at the hands of a firing squad (Figure 13-25). Bolshevik Jacob Yurosky, who commanded the death squad, boasted that the world would never know what had happened to the royal family.

That was true for the next 75 years, until a team of specialists including Michael Baden, William Maples, and forensic odontologist Lowell Levine examined the skeletons discovered in a shallow grave outside of Ekaterinburg, Russia (Figure 13-26). The team was able to determine the age and sex of all nine skeletons. Five were identified as females and four as males. The skulls had all been crushed, making identification difficult.

The bones and teeth helped. One female had poor dental work and calcification of knee joints, indicating a person who had spent time scrubbing floors and doing manual labor. One male skeleton was mature, probably the remains of the royal family physician, Dr. Botkin. The recovered dental plate and skull similarities to a photograph provided evidence to the doctor's identity. Expensive dental repairs and dental records identified the rest of the royal party. Because some of the leg bones were crushed, height estimations were calculated using arm length. The remains of Anastasia and Alexei, who were 17 and 14, respectively, were not found.

Figure 13-25. *The last royal family of Russia.*

©AP Photo/File

Figure 13-26. *Location of bones from the mass grave of the Romanovs.*

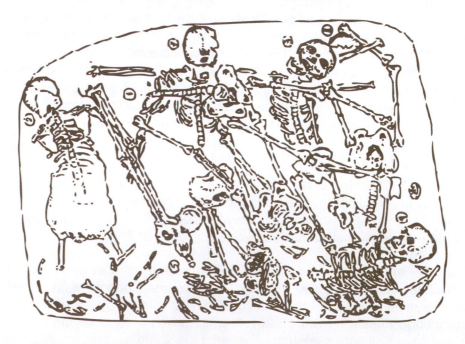

 Think Critically **Select one of the Case Studies and explain what forensic anthropology techniques were used for identification.**

Clyde Snow: The Bone Digger

"The bones don't lie and they don't forget. And they're hard to cross-examine." So says Clyde Snow, one of the world's leading forensic anthropologists, as he explains why it is so important to present the evidence of skeletal remains in court.

Clyde Snow

©AP Photo/Victor Ruiz C

Clyde has studied thousands of skeletons all over the world, revealing their secrets. Over the last two decades, Clyde has been heavily involved in international human rights. He served on the United Nations Human Rights Commission, working in Argentina, Guatemala, the Philippines, Ethiopia, Bosnia, and Iraq. Closer to home, he's worked on several important cases, including those of mass murderers John Wayne Gacy and Jeffrey Dahmer, as well as the victims of the 1995 Oklahoma City bombing. He also participated in some historic investigations, searching for the remains of Butch Cassidy and the Sundance Kid in Bolivia, digging up the bones at the site of Custer's Last Stand, and examining King Tut's mummy.

Clyde Snow's accomplishments are great— much greater than his early experiences in school might have suggested. Born in 1928 in Texas, Clyde was expelled from high school and transferred to a military school, where his grades dropped. When he finally made his way to college, Clyde flunked out on his first attempt, but then went on to achieve a Ph.D. in anthropology.

He began his career working for the Civil Aeromedical Institute examining the bodies of victims of air crashes. It wasn't until 1979 that he decided to focus entirely on forensics.

Only a few years later, Clyde found himself on a plane to Argentina to see if it was possible to investigate— and ultimately hold accountable—those responsible for the genocide committed by the previous Argentinean government. It is believed the Junta militia killed tens of thousands of civilians. Mass killings on a similar scale have been investigated by Snow in Guatemala and Iraq. In 2006, Snow testified against Saddam Hussein in the trial involving the mass murder of Kurdish people. Snow has dug up the remains of victims all over the world, many who were killed by their own or a neighboring government.

Why does he do it? He's forthright about his reasons. One is to identify the remains of victims and return them to their families. Another is to try to bring about some justice. A third is to let the governing people worldwide who have power over others know that they cannot kill their citizens without anyone trying to do something about it. His final reason is to provide a historical record. It is pretty emotional work. Clyde tells his students to "do the work in the daytime and cry at night."

Learn More About It

To learn more about careers in forensic anthropology, go to school.cengage.com/forensicscience.

CHAPTER 13 REVIEW

True or False

1. Osteoblasts are bone-building cells.

2. Bones can help us determine the age, sex, and health of a person.

3. Female hips have a subpubic angle less greater than 90 degrees.

4. Male skulls tend to have a lower, more sloping frontal bone than female skulls.

5. It is possible to estimate someone's height based on the length of a humerus.

6. Approximate age can be determined by studying the suture lines of the skull.

7. A woman's skull is usually bumpy compared to a man's skull.

8. A man's jawline usually forms a 90-degree angle.

9. If a person were right-handed, then his or her skeleton on the left side would be slightly larger than the skeleton on the right side.

10. Mitochondrial DNA contains no genetic information from the father.

Short Answer

11. Describe the features you would expect to find in a female skeleton, age 40.

_____

_____

12. Describe the process of ossification of the skull using each of the following terms:
    a. Cartilage

    _____

    b. Blood vessels

    _____

    c. Osteoblasts

    _____

    d. Osteocytes

    _____

    e. Osteoclasts

    _____

    f. Enzymes

    _____

g. Calcium

_____

h. Phosphates

_____

13. Calculate the approximate height of a Causasian male if one of the following bones is found:

   a. radius bone equal to 25 cm

   _____

   b. humerus bone equal to 30 cm

   _____

   c. ulna bone equal to 21 cm

   _____

14. Suppose that the two bones found belonging to the male in questions 13b and 13c were his ulna, which is 21 cm long, as well as his humerus, which is 30 cm. Calculate his height using both the ulna and the humerus bone measurements. Show your work.

   _____

   _____

15. Refer to your answers for questions 13b and 13c and 14. Explain which value for height should be more accurate (the two separate values or the combined values).

   _____

   _____

Bibliography

Books and Journals

Evans, C. *A Question of Evidence*. New York: Wiley, 2002.

Ferlini, R. *Silent Witness: How Forensic Anthropology Is Used to Solve the World's Toughest Crimes*. Ontario, Canada: Firefly Books, 2002.

Lerner, K., and Brenda Lerner, eds. *World of Forensic Science*, Volume 1. Detroit: Gale, 2005; 33–34.

Maples, W., and M. Browning. *Dead Men Do Tell Tales*, Main Street Books, 1995.

Massie, R. *Romanovs*. New York: Ballantine Books, 1996.

Snyder Sachs, J. *Corpse: Nature, Forensics, and the Struggle to Pinpoint Time of Death*. New York: Perseus Books Group, 2002.

Turek, S. L. *Orthopaedics: Principles and Their Application*, Volume 2, 4th ed. Philadelphia: J.B. Lippincott Co., 1984.

Ubelaker, D., *Human Skeletal Remains, Excavation, Analysis and Interpretation*, 2nd ed. Washington, DC: Taraxacum, 1989.

Web sites

Gale Forensic Sciences eCollection, school.cengage.com/forensicscience.

American Society of Bone and Miner al Reconstruction, http://depts.washington.edu/bonebio/ASBMRed/ASBMRed.html.

http://www.dnai.org

ACTIVITY 13-1
DETERMINING THE AGE OF A SKULL

Objectives:

By the end of this activity, you will be able to:
Estimate the age of a skull by studying the cranial suture marks.

Safety Precautions:

None

Time Required to Complete Activity:

15 minutes (groups of two students)

Materials:

textbook (Figure 13-17)
access to the Internet or reference books

Procedure:

Part A:

Using Figure 13-17 in your textbook showing the relationship between age and skull sutures, determine the approximate age of a skull with the following features:

1. Lambdoidal and sagittal sutures fused.   Age _____
   Coronal sutures not fused.
2. Lambdoidal sutures almost fused.   Age _____
   Sagittal and coronal sutures not fused.
3. All sutures fused.   Age _____
4. Lambdoidal wide open.   Age _____
   Sagittal and coronal sutures open.

Part B:

1. Research and then compare the infant to the adult skull with respect to:
   a. similarities
   b. differences in numbers of bones, composition
   c. percentage of body length
2. Describe the process of ossification of the skull using each of the following terms:
   > Cartilage
   > Blood vessels
   > Osteoblasts
   > Osteocytes
   > Osteoclasts
   > Enzymes
   > Calcium
   > Phosphates

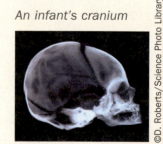

An infant's cranium

©D. Roberts/Science Photo Library

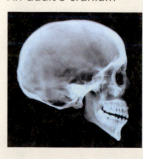

An adult's cranium

©D. Roberts/Science Photo Library

ACTIVITY 13-2
BONES: MALE OR FEMALE?

Objectives:

By the end of this activity, you will be able to:
Determine if the remains of a skeleton belong to a male or female

Safety Precautions:

None

Time Required to Complete Activity:

15 minutes (groups of two students)

Materials:

textbook figures throughout the chapter
pencil or pen

Procedure:

Refer to the figures in Chapter 13 of your textbook to help you determine if the skeletal remains listed below belong to a male or female skeleton. Complete the following questions:

Case # 1 _____
Round eye orbits, subpubic angle of 103 degrees, rectangular-shaped pubis, smooth skull. Explain your answer.

Case # 2 _____
Pelvis narrow, protuberance on occipital bone, sloping forehead. Explain your answer.

Case # 3 _____
The skull was found to be smooth with small brow ridges; would you expect to find a subpubic angle larger or smaller than 90 degrees? Explain your answer.

Case # 4 _____
A long, narrow sacrum with triangular pubis; would you expect to find the subpubic angle larger or smaller than 90 degrees? Explain your answer.

Further Research

A smooth (gracile) skull of a female appears quite different from the bumpy (robust) appearance of the male skull. What causes the male skull to be thicker with more dense bone? Research the effect on bones of each of the following:

a. XX or XY chromosomes
b. Production of higher levels of testosterone in males
c. Effect of larger muscle mass on bones

After researching these factors, form a hypothesis to account for the differences in the adult male and female skulls.

ACTIVITY 13-3
THE ROMANOVS AND DNA:
AN INTERNET ACTIVITY

Objectives:

By the end of this activity, you will be able to:
1. Examine how the Romanov family was identified from their remains.
2. Describe what the remains revealed about the family's fate.
3. Describe how DNA technology was used to help identify the skeletal remains of the Romanov family.

Safety Precautions:

None

Time Required to Complete Activity:

Part A: 40 minutes
Part B: 30 minutes
Part C: 60 minutes

Materials:

access to the Internet
pen and paper

Part A: Romanov Family
Procedure:

1. Open the Internet and type in the Internet site: http://www.dnai.org/
2. Refer to the title: DNA Interactive.
3. Refer to the menu bar at the far left of the page and double-click on the fifth heading: Applications.
4. Use the scroll bar to the right to pull the screen down. At the bottom of the screen, double-click on the second module entitled "Recovering the Romanovs."
5. Move the scroll bar up to the top of the screen to double-click on the Romanov family.
6. Move the scroll bar down to the bottom of the screen. You will see 10 circles. Click on each numbered circle and answer the questions as you progress through the material. Note the number of the circle corresponds to the numbered questions below.

Questions:

1. Refer to the pedigree of Tsar Nicholas II. What was his mother's name?
2. Alexandria was born in what country?
3. Refer to the pedigree of Tsar Nicholas II's family.
   a. How many daughters did he have?
   b. How many sons did he have?
   c. Who were the two youngest children?
4. Prince Alexei suffered from blood disorder.

a. What is the name of the blood disorder?

b. What are the symptoms of this disorder?

c. From whom did Alexei inherit this genetic disorder?

5. What other family members of Alexandra also suffered from this disorder?

6. View the four movie clips of the Tsar's family.

7. In what year did Nicholas II abdicate his throne?

8. During World War I, the leadership of Russia changed.

a. Who became the leader?

b. What was the political party?

c. What was the reason for the overthrow of the government?

9. a. What was the significance of the Ipatiev House?

b. What happened there in 1918?

10. In addition to the Tsar's family, who else was executed?

a. What was inserted into the corsets of the women that prevented some of the bullets from entering their bodies?

Part B: DNA Science Solves a Mystery

Procedure:

Go to: http://www.dnai.org

1. Applications (left side of screen)
2. Recovering the Romanovs (bottom of screen)
3. Science Solves a Mystery (top of the screen)

Questions:

1. In 1991, where was the supposed burial site of the Tsar and his family?

2. Listen to the 2 video clips narrated by Dr. Michael Baden.

a. Why was an American team called in to help identify the bodies?

b. Review the video Remains in Yekaterinburg.
What type of information can be gained by a study of the following?

1. Ridges and thick muscular insertions
2. Orbits of the eye and mandible and maxilla of the jaw
3. Pelvic girdle measurements
4. Ridges in the pubic bone
5. Leg and arm bones

3. Click on the box: Count the Skeletons.

a. How many skeletons were recovered?

b. How many people died in the massacre?

4. What can be determined by examination of:

a. Wisdom teeth

b. Vertebrae

c. Pelvic regions

5. Click on the box Analyzing the Skeletons.
Follow the directions to determine the age and sex of each of the skeletal remains.

6. What 2 people were determined to be missing from the gravesite?

a. How was this determined?

7. The bodies were buried for over 75 years. What type of evidence was preserved that enabled scientists to determine who was buried in the grave?

8. a. Click on Nuclear DNA box.
Click on the next button in the upper right-hand corner.

8. b. Click on the Mitochondrial DNA (mDNA).
Click on the next button located in the upper right-hand corner.

9. Compare and contrast mitochondrial DNA (mDNA) with nuclear DNA.
   a. Compare (How are they alike?)
   b. Contrast (How are they different?)

10. From whom do we inherit our entire mitochondrial mDNA?
Click on maternal inheritance.

11. Test yourself on Mitochondrial DNA:
   a. From whom did the Romanov children inherit their mDNA?
   b. Where did that person get his or her mDNA?
   c. Does Nicholas II have the same mDNA as his children?
      Explain your answer.

12. Double click on Tsarina's Pedigrees located in the lower right-hand corner. Find Nicholas II and Alexandria in the Pedigree chart.
   a. All mitochondrial DNA of the Tsar's children can be traced back to whom?
   b. According to this pedigree, who is the relative still alive today to have the same mDNA as the Romanov children?

13. Go to Bioserver Sequence Server.
Recall that differences in the mDNA sequence is highlighted in yellow. You will need to close out of this window by clicking on the X in the upper right hand corner.

14. Whose skeleton was #9?

15. Why was James, Duke of Fife, selected to have his DNA examined?

16. Go to Bioserver's sequence server.
Which of the male skeleton's matched with the mDNA of James, Duke of Fife?
(You will need to close out of this box by clicking on the X in the upper right hand corner.)

17. It is believed the identity of the other male skeletons were:

18. Why was Anna Anderson's mDNA being compared to the mDNA of Prince Philip?

19. Why was Carl Maucher's mDNA being examined?

20. Double click on the Hair sample video.

21. Listen to the two videos by Michael Baden and Syd Mandelbaum.
   a. If Anna Anderson was cremated, then how did they obtain a sample of her cells?
   b. Based on Michael Baden's and Syd Mandelbaum's findings, were Anna Anderson and Anastasia, the daughter of the royal family, the same person? Explain your answer.
   c. What is believed to have happened to the bodies of the two youngest children, Anastasia and Alexei?

ACTIVITY 13-4
ESTIMATION OF BODY SIZE FROM INDIVIDUAL BONES

Objectives:

By the end of this activity, you will be able to:
1. Determine the approximate height of a person from one of the long bones of the body.
2. Explain how it is possible to estimate someone's height from a single bone.

Safety Precautions:

None

Time Required to Complete Activity: 40 minutes

Materials:

textbook Figure 13-20
pencil or pen
calculator (optional)

Procedure:

1. Refer to the bone length tables in your textbook (Figure 13-20).
2. For each problem, locate the appropriate formula to calculate a person's height based on the size of the recovered bone.

Questions:

Calculate the approximate height of the person if a humerus bone was found in each of the following situations. Show your work.
  a. Caucasian male femur of 50.6 cm
  b. African-American female femur of 49.5 cm
  c. A Caucasian person, sex unknown, tibia of 34.2 cm
  d. Caucasian female humerus of 33.4 cm
  e. African-American male humerus of 41.1 cm
  f. Person of unknown sex or ethnic group, humerus of 31.6 cm

ACTIVITY 13-5
WHAT THE BONES TELL US

Objectives:

By the end of this activity, you will be able to:
Apply knowledge of bone and teeth analysis to several case studies in an effort to describe the person based upon bone and teeth examination.

Safety Precautions:

None

Time Required to Complete Activity: 40 minutes

Materials:

textbook
pen and paper

Procedure:

Using all of the information in your textbook, describe as much as you can about a person from their bones and teeth, as described in the following case studies.

Questions:

1. A skull has been found: What can you tell about the skull described below?
    a. Lambdoidal suture nearly closed
    b. Large brow ridge
    c. Large supraorbital ridge
2. Only the jaw has been found: What can you tell about the jaw from this information only?
    a. Angle = 100 degrees
    b. Wisdom teeth have emerged
    c. No fillings or bridgework
3. Femur only: What can you tell about the femur from this description?
    a. Thin
    b. Osteoporosis present
    c. Length of femur = 47 cm
4. A mass grave is found: What can you tell about the remains?
    a. Jaw angle = 105 degrees
    b. Subpubic angle = 80 degrees
    c. Left femur = 49 cm
    d. Right femur = 49.1 cm
    e. Left femur = 45.5 cm
    f. Right femur = 45.3 cm
    g. Left femur = 48 cm
    h. Skull: one partial skull found with rectangular orbits (eye sockets)
    i. How many individuals (minimum) are buried at this one site?

j. How did you determine this number?

k. Two of the femurs, one right and one left, had thick bones. What can you infer from this about this person? Explain your answer.

l. Two femurs showed very large attachment sites for tendons. What clue might this provide about the owner? Explain your answer.

5. A female was reported missing in the area. The family of the missing woman was wealthy, and their slightly built, missing daughter never did any strenuous work but did bear three children. How could you verify if the woman buried in the gravesite:

a. had three children

b. did not engage in much manual labor

c. was over the age of 21

d. was slightly built

e. was from a wealthy family

ACTIVITY 13-6
MEDICAL EXAMINER'S FINDINGS

Objectives:

By the end of this activity, you will be able to:
1. Apply knowledge of bone analysis to describe an individual from skeletal remains in several case studies.
2. Apply knowledge of lividity, rigor mortis, and algor mortis to estimate time of death
3. Apply knowledge of insect development to estimate time of death
4. Apply knowledge of digestion to estimate time of death

Safety Precautions:

None

Time Required to Complete Activity: 40 minutes

Materials:

textbook (Chapters 11 and 13)
pen and paper

Procedure:

As the medical examiner, you have collected information from the crime scene investigator as well as from autopsies. You need to provide information regarding the deceased person, his or her death, and the time of death to the police. Using your textbook for a reference answer the police's questions as completely as possible.

Questions:

Case 1: Bones found still partially clothed at bottom of a ravine.
   a. Subpubic angle = 87 degrees
   b. Length of femur = 50 cm
   c. Many fillings in teeth, no wisdom teeth
   d. Skull shows fracture marks

Based on the information available, determine the:
1. Age of the individual
2. Sex of the individual
3. Height of the individual
4. Possible cause of death

Case 2: A body was found in a wooded area. There was bruising on the neck, defensive wounds on the arms and hands, and a broken finger.
   a. Subpubic angle = 106 degrees
   b. Length of tibia = 34 cm
   c. No fillings in teeth
   d. Suture lines of skull have not begun to close

Based on the information available, determine the:
1. Age of the individual. Explain your answer.
2. Sex of the individual. Explain your answer.
3. Height of the individual. Explain your answer.
4. Possible cause of death

Estimate the time of death for each of the following. Refer to your notes on time of death from Chapter 11.

Case 3: Body found in apartment house bedroom at 9 a.m. on Monday.
    a. Male, overweight, found dead on bed. No evidence of foul play.
    b. Body temperature = 93°F (64°C)
    c. Room temperature = 83°F (58°C)
    d. Lividity showed no movement of body after death
    e. Rigor had begun

Based on the information available, can you determine the:
1. Time of death
2. What factors led to the determination of the time of death?
3. Possible cause of death
4. What tests could have been done if a drug overdose was suspected?

Case 4: Body found in dumpster at noon on Tuesday. Dumpster had been emptied on Monday.
    a. Permanent lividity was found along abdomen, chest, and front of the legs.
    b. Body temperature found to be 80°F (27°C), with the average ambient temperature being 70°F (22°C) for the past three days.
    c. Stomach was empty, but food was found in small intestine.
    d. Rigor present throughout body.
    e. Lungs filled with fresh water.
1. What is your estimated time of death?
2. Explain how you determined the time of death from the data provided.
3. Can you provide a cause of death? Explain.
4. Can you state a mechanism of death? Why or why not?
5. Can you provide the manner of death? Why or why not?
6. What other tests should be performed during the autopsy to provide more information to help identify this person?

Case 5: Woman found dead lying on her back on kitchen floor at 4 p.m. on Thursday.
    a. No blood loss. No signs of struggle.
    b. No lividity visible. Flies and eggs found on the body.
    c. Stomach contents show some undigested food, with some digested food in small intestine.
    d. Body temperature: 97°F; ambient temperature: 87°F

Provide as much information as possible from the above information regarding the death of this woman. Explain your calculations.

Case 6: Woman found in freezer at 6 p.m. on Thursday.
    a. Ligature marks on neck
    b. Body temperature: 30°F, same as freezer
    c. Undigested food in stomach

Provide as much information as possible from the above information regarding the death of this woman. Explain your calculations.

ACTIVITY 13-7
HEIGHT AND BODY PROPORTIONS

Background:

Leonard da Vinci drew the "Canons of Proportions" around 1492 and provided a text to describe what the ideal proportions of a perfect man should be. The drawing was based on the earlier writings of Vitruvius, a Roman architect. Some of the relationships described include:

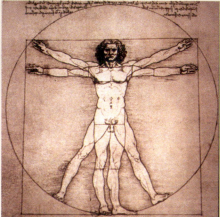

Leonardo da Vinci's "Canons of Proportions"

Photo by Susan Van Etten

- A man's height is 24 times the width of his palm.
- The length of the hand is one-tenth of a man's height.
- The distance from the elbow to the armpit is one-eighth of a man's height.
- The maximum width at the shoulders is one-half of a man's height.
- The distance from the top of the head to the bottom of the chin is one-eighth of a man's height.
- The length of a man's outstretched arms is equal to his height.

Objectives:

By the end of this activity, you will be able to:

1. Determine which of these relationships most accurately parallels your body proportions in estimating height.
2. Describe how to apply the Canons of Proportions to forensics by estimating someone's height from a limited number different body parts.

Safety Precautions:

None

Time Required to Complete Activity:

40 minutes

Materials:

(students working in pairs)
metric ruler
pen and paper
calculator (optional)
graphing paper

Part A:
Procedure:

1. Standing in your stocking feet with your back to a wall, have your partner carefully measure your height to the nearest tenth of a centimeter. Keep the top of your head level (parallel to the floor).
2. Record your results on Data Table 1.

3. Have your partner measure to the nearest .1 cm and record each of the following measurements of your body:
   a. width of your palm at the widest point
   b. length of the hand from first wrist crease nearest your hand to the tip of the longest finger
   c. distance from elbow to highest point in the armpit
   d. maximum width of shoulders
   e. the distance from the top of the head to the bottom of the chin
   f. the length of outstretched arms
4. Repeat steps 1 to 3, taking the body measurements of your partner and record in Data Table 2.
5. Your partner records your data in his or her Data Table 2.
6. Calculate and record your and your partner's estimated height using the proportions given on the data tables.
7. Determine and record the difference between your actual height and your calculated height on data tables 1 and 2. Use + and – symbols.

Data Table 1: Your Body Relationships
All measurements recorded in centimeters
Gender of person measured _____

A	B	C	D	E
Trait	Size (cm)	Multiply by	Calculated Total (cm)	Difference between actual and calculated height (cm)
Height		x 1 =		
Palm width		x 24 =		
Hand length		x 10 =		
Distance from armpit to elbow		x 8 =		
Width of shoulders		x 4 =		
Head to chin length		x 8 =		
Outstretched arms		x 1 =		

Data Table 2: Your Partner's Body Relationships
All measurements recorded in centimeters
Gender of person measured _____

A Trait	B Size (cm)	C Multiply by	D Calculated Total (cm)	E Difference between actual and calculated height (cm)
Height		x 1 =		
Palm width		x 24 =		
Hand length		x 10 =		
Distance from armpit to elbow		x 8 =		
Width of shoulders		x 4 =		
Head to chin length		x 8 =		
Outstretched arms		x 1 =		

Questions:

1. Which measurement and relationship most accurately reflected your height?
2. Was this the same measurement that most people of your gender found to most accurately estimate their actual height? Explain.
3. Which measurement and relationship most accurately reflected your partner's height?
4. Which measurement was the least accurate in estimating your height?
5. Explain why using the Canons of Proportions on teenagers to estimate height would provide less accurate data than using the canons of proportions on adults.
6. Describe a crime scene that could use the Canons of Proportions to help estimate the height of a person.

Part B:
Procedure:

1. The distance from your elbow to armpit is roughly the length of your humerus. Record the humerus length and actual length from everyone in your class and complete Data Table 3.
2. Graph the length of the humerus (*x* axis) vs. height (*y* axis). Be sure to include on your graph the following:
   - Appropriate title for graph
   - Set up an appropriate scale on each axis
   - Label units (cm) on each of the *x* and *y* axes
   - Circle each data point

Data Table 3: Comparison of Humerus to Actual Height

Name	Length of Humerus (cm)	Actual Height (cm)
1		
2		
3		
4		
5		
6		
7		
8		
9		
10		
11		
12		
13		
14		

Questions:

1. Plot the data and create the best-fit line.
2. Suppose a humerus bone was discovered at a construction site. From the graph, explain how you could estimate the person's height from the length of the humerus.
3. List the variables that would need to be considered when trying to estimate someone's height from a single bone.

CHAPTER 14

Glass Evidence

IF YOU BREAK IT . . .

The boy quickly slipped the crystal glass vase into his jacket. He thought he could easily hide it and leave the store without getting caught. However, the vase slipped and fell to the floor, shattering into many small pieces. The boy quickly ran out of the store. A store clerk remembered seeing a young man looking at the vase just before the incident occurred. The clerk gave a detailed description to the police, who were able to find the youth in a nearby music store.

When questioned by the police, the boy claimed he had not been anywhere near the store where the incident happened. Upon further investigation, the police found evidence of broken glass wedged between the grooves of his sneakers. Was the glass in his sneakers the same as the glass from the broken vase? What type of testing could be done on the glass to try to match the glass in the sneaker with the broken crystal vase in the store?

©Susan Van Etten

Glass evidence.

If the glass did match, is that enough evidence to convict the boy?

In this chapter, we will explore the formation, characteristics, history, and types of glass. Various methods of glass analysis used in forensics will follow in the activities at the end of the chapter.

OBJECTIVES

By the end of this chapter you will be able to

✔ Explain how glass is formed.

✔ List some of the characteristics of glass.

✔ Provide examples of different types of glass.

✔ Calculate the density of glass.

✔ Use the refractive index to identify different types of glass.

✔ Describe how glass fractures.

✔ Analyze glass fracture patterns to determine how glass was broken.

✔ Explain how glass is used as evidence.

TOPICAL SCIENCES KEY

VOCABULARY

amorphous without shape or form; applied to glass, it refers to having particles that are arranged randomly instead of in a definite pattern

Becke line line created as refracted light becomes concentrated around the edges of the glass fragment

density the ratio of the mass of an object to its volume, expressed by the equation:

$$density = \frac{mass}{volume}$$

glass a hard, amorphous, transparent material made by heating a mixture of sand and other additives

leaded glass glass containing lead oxide

normal line a line drawn perpendicular to the interface surface of two different media

obsidian volcanic glass

refraction the change in the direction of light as it changes speed when moving from one substance into another

refractive index a measure of how light bends as it passes from one substance to another

silicon dioxide (SiO_2) the chemical name for silica

INTRODUCTION

Glass evidence can be found at many crime scenes. Automobile accident sites may be littered with broken headlight or windshield glass. The site of a store break-in may contain shards of window glass with fibers or blood on them. If shots are fired into a window, the sequence and direction of the bullets can often be determined by examining the glass. Minute particles of glass may be transferred to a suspect's shoes or clothing and can provide a source of trace evidence linking a suspect to a crime scene.

THE HISTORY OF GLASS

Long before humans began making **glass**, glass formed naturally. When certain types of rock are exposed to extremely high temperatures, such as lightning strikes or erupting volcanoes, glass can form. Humans have used and made glass for centuries.

- Prehistoric humans used **obsidian**, a type of glass formed by volcanoes, as a cutting tool.

- Pliny, a Roman historian, described how glass was first made by accident in ancient Syria. Blocks of rock containing a chemical compound made of nitrogen and oxygen, called nitrates, were used as cooking surfaces. When these blocks were placed on top of sand and heated, the sand and nitrates melted and fused together to form a type of glass.

- The earliest man-made glass objects (glass beads) were found in Egypt dating back to 2500 BC.

- Glass blowing began sometime during the first century BC.

- In 1291, one center of glass making was moved from the city of Venice, Italy, to the island of Murano, off the coast of Venice. This happened for two reasons. First, the high temperatures needed to make glass caused frequent fires that were a danger to the large population of Venice. Moving glass production to an island with a much smaller population reduced the danger. Second, the secrets of glass making could be more closely guarded on an isolated island than in a big city on the mainland.

- By the 14th century, the knowledge of glass making spread throughout Europe.

- The Industrial Revolution brought the mass production of many kinds of glass.

WHAT IS GLASS?

Glass is a hard, **amorphous** material made by melting sand, lime—also called calcium oxide (CaO)—and sodium oxide (Na_2O) at very high temperatures. Its primary ingredient is **silicon dioxide** (SiO_2), also called silica. Sodium oxide is added to reduce the melting point of silica, or sand, and calcium oxide is added to prevent the glass from being soluble in water.

This type of glass is known as soda-lime glass because it contains sodium compounds and lime. Once it cools, the glass can be polished, ground, or cut for useful or decorative purposes. Glass blowers can form glass into many different shapes by blowing hot air through a long tube into the hot, molten, semi-liquid glass.

Glass is called an amorphous solid because its atoms are arranged in a random fashion (Figure 14-1). Because of its irregular atomic structure, when glass is broken, it produces a variety of fracture patterns.

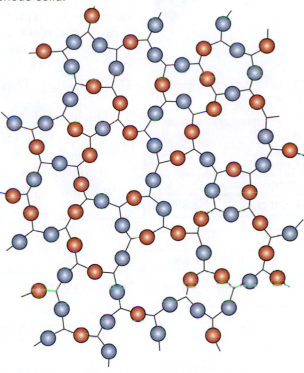

Figure 14-1. The arrangement of atoms in an amorphous solid.

TYPES OF GLASS

Because glass is a stable material that does not deteriorate over time, it has a variety of uses. The most common type of glass, soda-lime glass, is inexpensive, easy to melt and shape, and is reasonably strong. Manufacturers of most glass containers use the same basic soda-lime composition, making recycling easier.

Fine glassware and decorative art glass, called crystal or **leaded glass**, substitutes lead oxide (PbO) for calcium oxide. The addition of lead oxide makes the glass denser. As light passes through the more-dense glass, the light waves are bent, giving the glass a sparkling effect.

Ovenware and laboratory glassware is often sold as Pyrex® or Kimax® glass. These types of glass contain compounds that improve the ability of the glass to withstand a wide range of temperatures needed for cooling or heating glassware in a kitchen, laboratory, or in headlights. Adding certain metal oxides to the glass mixture produces different colors of glass. For example, nickel oxide produces colors ranging from yellow to purple, depending on the type of glass to which it is added. Cobalt oxide makes a purple-blue glass, whereas oxides of selenium make red glass.

PROPERTIES OF GLASS

Altering the compounds used to make glass changes the composition and produces different types of glass. The composition of a particular piece of glass may be unique and therefore identifiable. Because glass is made of a variety of compounds, it is possible to distinguish one type of glass from another by examining the different physical and chemical properties. We will examine some of the properties of glass, such as density, refractive index, and fracture patterns that are used in the forensic examination.

DENSITY

Each type of glass has a **density** that is specific to that glass. One method of matching glass fragments is by a density comparison.

Did You Know?

Some people say that glass is a liquid because it "flows" over time. They note that glass in very old church windows is thicker at the bottom of the pane than at the top, claiming that the glass has flowed to the bottom. In actuality, when placed into windows, the panes were installed with the heavier, thicker side at the bottom.

Density (D) is calculated by dividing the mass (m) of a substance by its volume (V). The formula for calculating density can be written as $D = \frac{m}{V}$. Once you determine the mass of a fragment of glass and the volume of water it displaces, you can calculate the density. Some of the more common glass densities are shown Figure 14-2.

Calculating the Density of a Piece of Glass

Figure 14.2. *Common glass densities.*

Type of Glass	Density (g/mL)
Bottle glass	2.50
Window glass	2.53
Lead crystal	2.98–3.01
Pyrex®	2.27
Tempered (auto)	2.98
Flint	3.70
Crown	2.50

1. The mass of a piece of glass (in grams) can be found using a balance.

2. In order to determine the volume (in milliliters) of water displaced by a piece of glass, fill a beaker completely with water, so that adding a single drop will cause spillage.

3. Position a graduated cylinder under the spout of the beaker (Figure 14-3).

4. Carefully place the piece of glass into the beaker and collect the overflow water in the graduated cylinder.

5. Measure the volume of water displaced by the glass.

6. Divide the mass (g) by the volume displaced (mL) to find the density.

Figure 14-3. *Using a beaker and graduated cylinder to determine the volume of a piece of glass.*

REFRACTIVE INDEX

Have you ever stood in a pool and looked down at your legs in the water? Did your legs appear to be where you expected them to be located? The distortion of your legs' position illustrates how a beam of light bends or refracts. When the beam of light moves from one medium (air) to another (water), its speed changes. The change in speed causes the beam to change direction, or bend. Another name for the bending of light is refraction. **Refraction** is the change in the direction of light as it speeds up or slows down when moving from one medium into another. The direction and amount the light bends varies with the densities of the two mediums.

The **refractive index** is a tool used to study how light bends as it passes through one substance and into another. Any substance through which light can pass has its own characteristic refractive index. The refractive index of a substance is calculated by dividing the speed of light in a vacuum—a space empty of all matter—by the speed of light through that particular substance.

Light in a vacuum travels at a speed of about 300,000 kilometers per second (km/s). The refractive index of a vacuum is 1 because the ratio of the speed of light in a vacuum divided by the speed of light in a vacuum is 300,000 km/s divided by 300,000 km/s equals 1.

When light travels through a vacuum, nothing interferes with the light to slow it down as it moves. When light travels through any other medium, the particles in that medium slow the light down. As the density of the medium increases, the speed of light passing through that material decreases.

The speed of light passing through air is slightly slower than the speed of light passing through a vacuum, because air is slightly denser than a vacuum. The refractive index of air is 1.0008 and is so close (for our purposes) to that of the refractive index of a vacuum that we will use 1 as the refractive index for air as well.

As light passes through air, the light travels in a straight line at a speed slightly slower than 300,000 km/s. When that same beam of light enters water, it slows down to approximately 225,000 km/s. If the beam were to pass through a piece of window glass, it would slow down even more to a speed of approximately 200,000 km/s.

If the light travels from a less-dense medium to a denser medium, the beam of light will slow down and bend toward the normal, as shown in Figure 14-4. The **normal** is a line perpendicular to the surface where the two different mediums meet. The red line in the figure indicates the normal line. The incoming beam of light passing through the first medium is called the *incident ray,* and the beam of light as it passes through the second medium is called the *refracted ray.* The angle the incident ray in medium 1 forms with the normal is called the *angle of incidence,* labeled here as "Angle 1." The angle the refracted ray in medium 2 forms with the normal is called the *angle of refraction,* labeled here as "Angle 2."

If medium 1 is denser than medium 2, the light will bend away from the normal. You can see this in in Figure 14-5.

Snell's Law

Snell's law describes the behavior of light as it travels from one medium into a different medium. Snell's law can be written as:

n_1 (sine angle 1) = n_2 (sine angle 2)

In this equation, n_1 is the refractive index of medium 1 and n_2 is the refractive index of medium 2. Angle 1 is the angle of incidence and angle 2 is the angle of refraction. The sine of an angle (abbreviated *sin*) is a trigonometric function. In this text, we will use a scientific calculator or a sine table to find the sine of an angle.

Did You Know?

Scientists have found a way to slow down the speed of light from 186,000 miles per second to about 38 miles per hour, the speed of a car in rush-hour traffic.

Figure 14-4. *Light travels through air and enters the glass of a windowpane. As it moves from a less-dense medium into a denser medium, the light slows down and changes direction.*

Figure 14-5. *As a beam of laser light travels from oil, medium 1, into air, medium 2, it speeds up and bends away from the normal.*

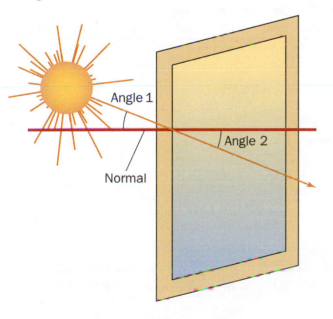

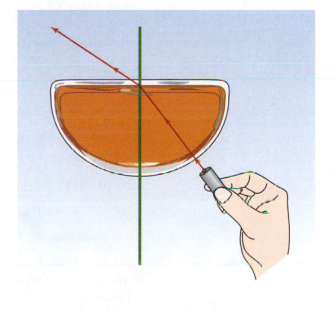

Example 1: A beam of light travels in air (medium 1) and then passes through a piece of glass (medium 2). As the light passes from the air into the piece of glass, the light ray is bent (Figure 14-6). What is the angle of refraction measured from the normal?

Here is what we know so far:

refractive index of air (medium 1) = 1.00

refractive index of glass (medium 2) = 1.50

angle 1 = 45°

angle 2 = ?

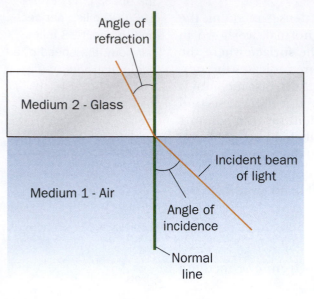

Figure 14-6. *Light refraction passing from air through glass.*

Using a scientific calculator to solve for the measure of angle 2

n_1 (sin of angle 1) = n_2 (sin of angle 2)
Substituting what we know into Snell's law:

1.00 (sin of 45°) = 1.50 (sin of angle 2)
To find the sine of an angle on most scientific calculators, enter the angle measurement, and then press the [sin] key.

The sine of 45° is 0.7071.

1.00 (0.7071) = 1.50 (sin of angle 2)

(0.7071)/1.50 = sin of angle 2

0.4714 = sin of angle 2

Now, you must find the measure of angle 2. On most scientific calculators, enter 0.4714, and then press the SHIFT key or the second function key, and then press the [sin–1] key.
angle 2 = 28.1° ≈ 28°

This answer makes sense because the light is moving from a less-dense substance in medium 1 (air) to a denser substance in medium 2 (glass). The light will slow down and bend toward the normal. Angle 1 is 45° and angle 2 is smaller at 28°, indicating that the light did bend toward the normal.

Using a sine table to solve for the measure of angle 2
It is possible to calculate the measure of an angle using a sine table (see Appendix A) if a scientific calculator is not available.

n_1 (sin angle 1) = n_2 (sin angle 2)
Substituting what we know into Snell's law:

1.00 (sin 45°) = 1.50 (sin angle 2)

Look up sine 45° in the sine table.

sin 45° = 0.7071

1.00 (0.7071) = 1.50 (sin angle 2)

0.7071/1.50 = sin angle 2

0.4714 = sin angle 2

Find the value closest to 0.4714 in the table. The sine value closest to 0.4714 in the table is 0.4695. According to the table, the angle measure with a sine value of 0.4695 is 28°.

So, sin 28° ≈ 0.470

angle 2 ≈ 28°

Angle 2 is smaller than angle 1. This means the beam of light will bend toward the normal. This makes sense because the glass is denser than the air and will slow down the light and bend it toward the normal.

Example 2: As light travels from water (medium 1) to air (medium 2), it bends (see Figure 14-7). Does it slow down or speed up as it passes from medium 1 through medium 2? Will the light bend toward the normal or away from the normal? Use Snell's Law to determine the size of angle 2.

Here is what we know so far:

refractive index of water (medium 1) = 1.33

refractive index of air (medium 2) = 1.00

angle 1 = 30°

angle 2 = ?

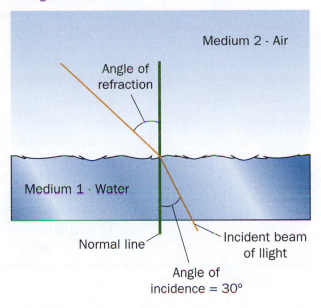

Figure 14-7. Light refraction passing from air through water.

Medium 2 - Air

Angle of refraction

Medium 1 - Water

Normal line

Incident beam of llight

Angle of incidence = 30°

Using a scientific calculator to solve for the measure of angle 2

n_1 (sin angle 1) = n_2 (sin angle 2)

1.33 (sin 30°) = 1.00 (sin angle 2)

To find the sine of an angle on most scientific calculators, enter the angle measure, and then press the [sin] key. The sin of 30° is 0.5.

1.33 (0.5) = 1.00 (sin angle 2)

0.665 = 1.00 (sin angle 2)

Now, you must determine the size of angle 2.
On most scientific calculators, enter 0.665, press the SHIFT key, and then press the [sin−1] key.

angle 2 = 41.6° ≈ 42°

Angle 2 is larger than angle 1. The beam of light will bend away from the normal. This makes sense because water is denser than air. When passing from water into air, light will speed up and bend away from the normal.

Using a sine table to solve for the measure of angle 2

n_1 (sin angle 1) = n_2 (sin angle 2)
Substituting what we know into Snell's law:

1.33 (sin 30°) = 1.00 (sin angle 2)
Look up sin 30° in the sine table.

sin 30° = 0.5000

1.33 (0.5000) = 1.00 (sin angle 2)

0.665 = sin angle 2

Find the value closest to 0.665 in the table. The sine value closest to 0.665 in the table is 0.6691. According to the table, the angle measure with a sine value of 0.6691 is 42°.

So, sin 42° ≈ 0.6691

angle 2 ≈ 42°

This means the beam of light will bend away from the normal. This makes sense because water is denser than the air. When the light passes from water into air, it should speed up and bend away from the normal.

APPLICATION OF REFRACTIVE INDEX TO FORENSICS

Glass fragments from car headlights or broken windows can link a suspect to a crime scene. Forensic scientists can use the refractive index of glass fragments and determine if the evidence at a crime scene is consistent or remove suspicion from a suspect. How is it possible to match glass from a crime scene to glass collected as evidence?

One method of determining if the evidence glass matches the glass from the crime scene is to compare the refractive index of the evidence glass to the refractive index of the glass from the crime scene. Car dealers and glass manufacturers have databases containing the refractive indexes of their products. However, the evidence glass obtained from the clothing or shoes of a suspect is often too small to easily check consistency. Therefore, the *submersion method* can be used (Figure 14-8). This method involves placing the glass fragment into different liquids of known refractive indexes. If a piece of glass and a liquid have the same refractive index, the glass fragment will seem to disappear when placed in the liquid.

If the refractive indexes of several different liquids are known, the submersion method can be used to estimate the refractive index of the glass. Notice in Figure 14-9 that the piece of glass is totally invisible in test tube 5. This means the glass fragment and the refractive index of the liquid in test tube 5 are very similar.

The refractive indexes of some of the more common liquids used in determining the refractive index of glass are listed in Figure 14-10. The second half of the table lists examples of different types of glass and their refractive indexes.

In which liquid would headlight glass seem to disappear? Which liquid would you select to determine if the evidence glass was quartz glass?

Figure 14-8. *The submersion method of determining the refractive index of a piece of glass.*

Piece of glass

Level of liquid

Figure 14-9. *Comparing the known refractive indexes of different liquids to an unknown refractive index of a piece of glass.*

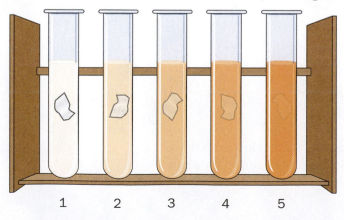

1   2   3   4   5

Index of refraction increasing ⟶

BECKE LINES

You have learned that a piece of glass will seem to disappear when submerged in a transparent liquid that has the same refractive index. Another technique involves submerging a fragment of glass in a liquid and then viewing it under a low power using a compound microscope. If the refractive index (*n*) of the liquid medium is different from the refractive index of the piece of glass, a halo-like ring appears around the edge of the glass. This halo-like effect is called a **Becke line**. It appears because the refracted light becomes concentrated around the edges of the glass fragment. A Becke line is visible under a microscope when the glass and liquid have different refractive indexes. The position of the Becke line surrounding the glass fragment is significant. The Becke line is located in the medium that has the higher refractive index.

If the Becke line is located inside the perimeter of the glass fragment, then the refractive index of the glass is higher than the refractive index of the surrounding liquid. If the Becke line is located on the outside of the perimeter of the glass fragment, then the refractive index of the surrounding medium is higher than the refractive index of the glass. Look at Figure 14-11. The arrow shows the position of the Becke line in each photograph.

Figure 14-10. *Refractive indexes of liquids and glasses.*

Liquid	Refractive Index
Methanol	1.33
Water	1.33
Isopropyl alcohol	1.37
Olive oil	1.47
Glycerin	1.47
Castor oil	1.48
Clove oil	1.54
Cinnamon oil	1.62

Type of Glass	Refractive Index
Pyrex	1.47
Automotive headlight glass	1.47–1.49
Television glass	1.49–1.51
Pane window glass	1.49–1.51
Bottle glass	1.51–1.52
Eyeglass lenses	1.52–1.53
Quartz glass	1.54–1.55
Lead glass	1.56–1.61

Figure 14-11. *The position of the Becke line indicates whether the refractive index of the glass sample is higher or lower than the liquid in which it is placed.*

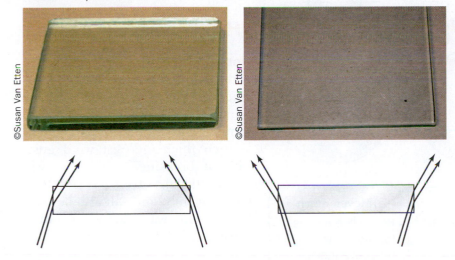

©Susan Van Etten
©Susan Van Etten

· Glass has higher refractive index
· Becke line seen inside
· Rays converge

· Glass has lower refractive index
· Becke line seen outside
· Rays converge

Estimating the Refractive Index of Glass Using a Microscope and Becke Lines

The refractive index of glass collected as evidence can be compared to the refractive index of glass collected at the crime scene. This is accomplished by following these steps:

1. Place the glass on a slide surrounded by a liquid medium of known refractive index.

2. Place the slide on the microscope stage and focus the lens on the glass.

   • If the glass seems to disappear when focused, the glass and the surrounding medium have the same refractive index.

- If the glass did not disappear, the surrounding medium and the glass have different refractive indexes. In the next step, you will use the position of Becke lines to estimate which medium, the glass or the surrounding liquid, has the higher refractive index.

3. Increase the distance between the stage and the lens. Look for the appearance of a Becke line.

- If a Becke line appears inside the perimeter of the glass, then the glass has a higher refractive index than the surrounding liquid.

- If a Becke line appears outside the perimeter of the glass, then the surrounding liquid has a higher refractive index than the glass.

By using several different types of liquids with different refractive indexes, it is possible to arrive at a good estimate of the refractive index of the glass fragment (Figure 14-12).

Figure 14-12. *Estimating the refractive index of a glass sample using a compound microscope and Becke lines.*

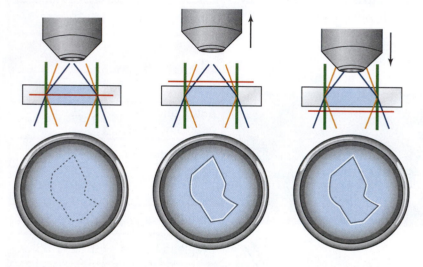

Match Point	Moved up	Moved down
No Becke line.	Becke line moves towards higher refractive index.	Becke line moves towards lower refractive index.
Glass and liquid refractive indices match.		
Glass seems to disappear.	Glass piece has a higher refractive index than the liquid.	Glass piece has a lower refractive index than the liquid.

There are automated instruments, such as the GRIM 2 (Glass Refractive Index Measurement), that measure the refractive index of glass. However, these instruments are too expensive for most forensic laboratories to own.

THICKNESS OF GLASS

Not all glass is the same thickness, and this difference provides another clue for identifying glass. Picture frame glass is $\frac{1}{8}$ inch thick, while window glass must be $\frac{3}{32}$ inch to $\frac{1}{8}$ inch thick to resist wind gusts without breaking. Door glass will vary in thickness from $\frac{3}{16}$ inch to $\frac{1}{4}$ inch thickness and can also be reinforced with wire threads running through it.

Glass as a Source of Trace Evidence

The trained examiner will be able to determine the composition, type, and perhaps the manufacturer from a sample of glass found on a victim, suspect, or at a crime scene. By determining the thickness, refractive index, and density of the glass collected, glass fragments can be matched assuming a large enough piece can be recovered.

Digging Deeper
with Forensic Science e-Collection

How can recent technological developments help investigators detect trace evidence? Go to the Gale Forensic Science eCollection at school.cengage.com/forensicscience and enter the search terms "trace evidence." Click on "Magazines" and read about remote information-sharing techniques. Write a short report on the new field of teleforensics. What is teleforensics? How is it helping investigators? What do you think are the advantages and disadvantages of this technology?

FRACTURE PATTERNS IN BROKEN GLASS

Glass has some flexibility. When glass is hit, it can stretch slightly. However, when the glass is forced to stretch too far, fracture lines appear and the glass may break. Recall that glass is an amorphous solid. Its atoms are not arranged in a pattern, but have a random structure. Therefore, glass will break into fragments, not into regular pieces with straight lines at the edges. The fracture patterns formed on broken glass can provide clues about the direction and rate of impact.

When glass breaks, fracture patterns form on the surface. Breaks, called *primary radial fractures*, are produced. These fractures start at the point of impact and radiate, or move outward, from there. Radial fractures form on the side opposite the point of impact. Secondary fractures may also form. These fractures take the form of concentric circles around the point of impact. *Concentric circles* are circles that have the same center. Concentric circle fractures form on the same side of the glass as the point of impact. By examining these glass fracture patterns, it is possible to determine which side of the glass was hit (Figure 14-13).

Why Radial and Concentric Fractures Form

When an object such as a bullet or rock hits glass, the glass stretches. On the side where the impact takes place, the glass surface is compressed, or squeezed together. The opposite side of the glass (the side away from the impact) stretches and is under tension. Glass is weaker under tension than under compression. It will break first on the weaker side, the side opposite the strike, producing radial fractures. The radial fracture lines will start at the center of the spot where the object hit the glass and move outward.

After the primary radial fractures form, the secondary or concentric fractures form. These are in the shape of concentric circles. They are formed on the same side as the impact or force on the glass. Knowing how and where glass fractures form, it is possible to determine where the force came from that broke the glass. Figures 14-14 and 14-15 (on next page) compare radial and concentric glass fractures.

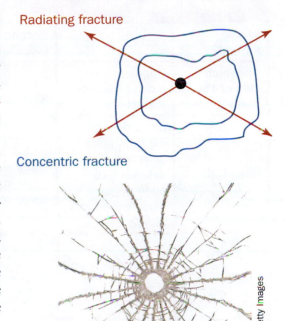

Figure 14-13. Glass after an impact shows radial fractures (red) and concentric circle fractures (blue).

Radiating fracture

Concentric fracture

©Photodisc/Getty Images

Figure 14-14. How radial and concentric circle fractures form when glass is hit.

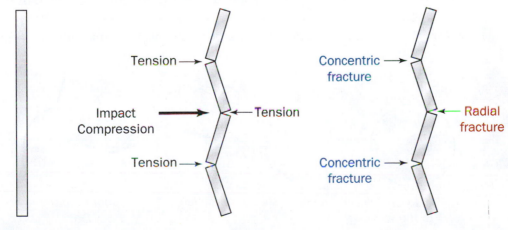

Unbroken glass

Tension
Impact Compression
Tension
Tension

Concentric fracture
Radial fracture
Concentric fracture

Figure 14-15. *Comparison of radial and concentric fractures in glass.*

	Radial Fracture	**Concentric Fracture**
When formed?	First (primary)	Second (secondary)
On which side of the glass?	Opposite the side of force or impact	Same side as force or impact
Description	Lines originating from point of impact and extending to edge of glass (like spokes in a bicycle tire)	Series of circles one inside the other sharing the same center
Diagram		

Bullet Fractures

The direction of a single bullet fired through glass can be easily determined (Figure 14-16). As the bullet passes through the glass, it pushes some glass ahead of it, causing a cone-shaped piece of glass to exit along with the bullet. This cone of glass makes the exit hole larger than the entrance hole of the bullet.

If several shots are fired through the glass, the order in which the shots were fired can be determined if enough of the glass is available or can be reconstructed. The first shot produces the first set of fracture lines. These lines set the boundaries for further *fracturing by following shots.* Radiating fracture lines from a second shot stop at the edge of fracture lines already present in the glass. Notice in Figure 14-17 that the fracture pattern on the right occurred before the fracture pattern on the left. The first shot made fracture lines that blocked the fracture lines made by the second shot. The second shot produced radiating fractures that stopped when they reached the first set of fractures.

Figure 14-16. *A bullet passing through glass.*

©Digital Vision/Getty Images

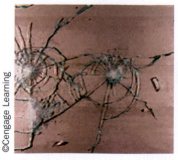

Figure 14-17. *Three bullet holes and the fracture patterns they produced in glass.*

©Cengage Learning

Path of a Bullet Passing through Window Glass

The angle at which a bullet enters a piece of window glass can help locate the position of the shooter (Figure 14-18). If the bullet was fired perpendicular to the windowpane, the entry hole of the bullet will be round. If the bullet was fired into the window at an angle, fracture patterns in the glass left by the bullet can be used to help locate the shooter's position.

If the shooter was firing at an angle coming from the left, glass pieces will be forced out to the right. The bullet's exit hole will form an irregular oval as it exits to the right. If the shot originated at an angle coming from the right, glass pieces will be forced out to the left, leaving an irregular oval hole to the left.

Ammunition type may be determined from the size and characteristics of the bullet hole. The distance from the shooter to the window can be estimated based on knowledge of the type of ammunition and its effect on the window. However, a high-speed bullet fired from a great distance will often exhibit characteristics of a slower-speed bullet fired from a closer range.

Figure 14-18. *The glass pattern left by a bullet can help determine the position of the shooter.*

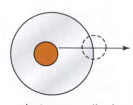

shot perpendicular to the window

shot from the left

shot from the right

Bulletproof Glass

Bulletproof glass is a combination of two or more types of glass, one hard and one soft. The softer layer makes the glass more elastic, so it can flex instead of shatter (Figure 14-19). The index of refraction for both of the glasses used in the bulletproof layers must be almost the same to keep the glass transparent and allow a clear view through the glass. Bulletproof glass varies in thickness from three-quarter inch to three inches.

Tempered Glass

The point of impact is more difficult to determine on windshield glass. Safety glass, also known as tempered glass, used in windshields is composed of two layers of glass bonded together by a layer of plastic in the middle. When hit, tempered glass is supposed to crack, but not break apart (Figure 14-20). Tempered glass is designed to protect passengers in the vehicle from being showered with broken pieces of glass following an impact. Tempered glass tends to produce a pattern of large pieces with fewer concentric circle fractures than other types of glass. Additionally, the pieces break into small, nearly cubic pieces, so they do not cut deeply.

Figure 14-19. *A bullet hole in bulletproof glass.*

Figure 14-20. *Fracture patterns on tempered glass.*

©AFP/Getty Images

©Getty Images News

Backscatter

When a window breaks, most of the fragments will be carried forward. However, some of the fragments, known as *backscatter*, can be projected backward because as the glass shatters, fragments collide and tumble in various directions. Backscatter is a form of trace evidence that may link a suspect to the crime scene.

Heat Fractures

During a fire, glass may break as a result of heat fracturing. Heat fracturing produces breakage patterns on glass that are

Digging Deeper
with *Forensic Science e-Collection*

Not all bulletproof glass will stop bullets. Go to the Gale Forensic Science eCollection eCollection at school.cengage.com/forensic science and research safety glass. Prepare a short presentation on what you find. Include these topics in your presentation: What is the rating system used to determine the level of protection given by each material? What other materials besides glass are used to protect people and property?

Figure 14-21. *(a) Wiper marks in a windshield.*

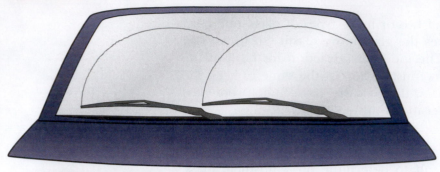

(b) Scratch marks on a car window

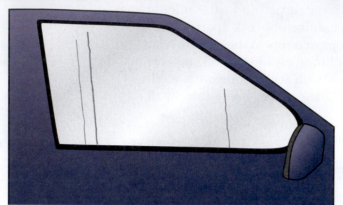

different from breakage patterns caused by an impact. Wavy fracture lines develop in glass that has been exposed to high heat. Also, glass will tend to break toward the region of higher temperature. If the glass was not broken before the fire, there will be no radial or concentric circle fracture patterns in glass broken by high heat.

Other Scratch Patterns

Patterns or scratches other than those already mentioned may be found on evidence glass. For example, windshield wipers may leave marks on the windshield (Figure 14-21a).

Any dirt or other particles embedded in the rubber insulation of a car window may leave scratch marks when the window is opened or closed (Figure 14-21b).

The speed of an object when it hits a piece of glass influences the number of concentric circles in the fracture pattern. An object moving at a high rate of speed at impact, such as a bullet striking the glass, produces fewer concentric circles. An object moving at a slower rate of speed at impact, such as a rock thrown at a window, produces a greater number of circles.

HANDLING OF CRIME-SCENE GLASS SAMPLES

When collecting glass evidence, it is important to follow the correct evidence collection procedures to avoid the loss or contamination of any evidence samples. Here are some general rules to follow when collecting glass evidence.

- Identify and photograph any glass samples before moving them.
- Collect the largest fragments that can be reasonably collected.
- Identify the outside and inside surfaces of any glass.
- If multiple window panes are involved, indicate their relative position in a diagram.
- Note any other trace evidence found on or embedded in the glass, such as skin, hair, blood, or fibers.
- Properly package all materials collected to maintain the proper chain of custody.

Glass collected by crime-scene investigators should be initially separated by physical properties, such as size, color, and texture. Samples should be carefully catalogued and kept separate to avoid contamination between two different sources.

Once in the lab, other trace evidence (e.g., hair, fibers, blood) can be separated from glass fragments. Any clothing related to the crime scene should be examined for glass fragments and other trace evidence. Any objects that might have been used to break the glass should also be collected and examined for glass fragments.

CLEANING AND PREPARING THE GLASS FRAGMENTS

After glass fragments have been documented and examined, they should be cleaned. Any surface debris (e.g., grease, soil particles), which might serve as additional evidence, should be noted and collected before cleaning. Cleaning solvents may be used as well as ultrasound cleaners to clean the glass.

It is important to avoid destroying any glass samples collected as evidence, if possible. If glass fragments are needed for further analysis, non-destructive methods, such as examination under a compound microscope, should be considered first. However, sometimes techniques must be used that do destroy the evidence sample during analysis. If this is the case, it is important to try to avoid using all of the evidence in the testing procedures. Additional material may be needed for duplicate lab tests.

SUMMARY

- Humans have been making glass for centuries. Glass is also produced by natural forces, such as volcanoes and lightning.

- Glass is an amorphous solid usually made from silica, calcium oxide, and sodium oxide.

- Glass fragments can be identified by their density, thickness, and refractive index.

- The density of an object is calculated by dividing its mass (usually in grams) by its volume (usually in milliliters).

- The refractive index of a material is a measure of how much light bends, or refracts, as it travels through that material.

- Snell's law can be used to calculate the refractive index of a piece of glass.

- The submersion method is a laboratory method used to estimate the refractive index of a glass sample.

- The position of a Becke line is another laboratory method used to estimate the refractive index of a piece of glass.

- When glass is hit, it first stretches, then breaks, forming radial fracture and concentric fracture patterns. Radial fracture patterns occur on the side of the glass opposite the point of impact. Concentric fracture patterns occur on the same side as the point of impact.

- When a bullet is shot through glass, the exit hole is larger than the entry hole.

- The angle of entrance of a bullet into glass can be determined by the pattern of glass left by the bullet's entry and exit.

- When glass breaks from exposure to a fire, the fracture pattern formed is different from the fracture pattern caused by an impact. In a fire, glass tends to break toward the area of higher temperature.

CASE STUDY

Susan Nutt (1987)

At 9:30 P.M. on a cloudy, dark night in February, 19-year-old Craig Elliott Kalani went for a walk in his neighborhood in northwest Oregon but never returned home. A hit-and-run driver killed him. Crime-scene investigators collected pieces of glass embedded in Craig's jacket and other glass fragments found on the ground near his body.

Police searched for a vehicle that had damages consistent with a hit-and-run accident. They found a car with those types of damages that belonged to a woman named Susan Nutt. In order to connect Ms. Nutt and her car with the crime, the police had to match the glass from the crime scene to the glass in her car. Researchers at Oregon State University's Radiation Center compared the glass from both sources. The scientists found that windshield glass from the crime scene contained the same 22 chemical elements as those used to make the glass in Ms. Nutt's car. The scientists considered both samples of glass to be a definite match.

The glass evidence helped convict Susan Nutt of failure to perform the duties of a driver for an injured person. She was sentenced to up to five years in prison and five years' probation.

©Nonstock/Jupiter Images

 Think Critically Describe a scenario in which glass evidence is important in solving a crime.

Neutron Activation Analyst

A neutron activation analyst applies techniques that can be used in many different fields, such as medicine, geology, engineering, and forensics. Given a sample of trace evidence, a nuclear activation analyst bombards the sample with neutrons produced by a nuclear reactor. Some atoms in the sample may absorb one of these neutrons and become radioactive.

The radioactive sample gives off radiation of different wavelengths, which the analyst measures to determine the elements contained in the sample. Each chemical element gives off radiation of a specific wavelength. The amount of radiation at a certain wavelength can indicate the amount of the element in the sample. Instrumental neutron activation analysis is a very sensitive technique that can detect chemicals that other methods cannot. In addition to being very sensitive, neutron activity analysis requires a very small sample, does not destroy the sample, and is very expensive.

Although neutron activation analysis is a powerful technique that would be useful in helping to solve many crimes, most laboratories cannot perform this type of analysis, because it requires a nuclear reactor. One laboratory that does have the facilities to perform neutron activation analysis is the Radiation Center at Oregon State University. Police departments have

Photo courtesy of Oregon State University Radiation Center

called on analysts there to help solve crimes by analyzing trace evidence, such as glass fragments.

Neutron activation analysis does not work as well on human hair, because hair is porous and can absorb contaminants from the environment. Besides glass evidence, this analysis also works well on bullets. A neutron activation analyst can use this technique to match a bullet found at a crime scene to a bullet found in a suspect's possession, even when the police have not found a gun connected to the crime. Like paint chips, the material used to make bullets contains varying amounts of several elements, such as arsenic, copper, silver, tin, mercury, or gold. The exact chemical composition of a bullet made in one batch is different from that of a bullet made in a different batch. According to one analyst there is only one out of 10,000 chance that a bullet from one batch will match a bullet from a different batch of bullets.

An analyst from the Radiation Center at Oregon State University was able to determine the exact chemical composition of bullets found at the scene of a murder and link the bullets to bullets found at a suspect's home. These findings helped convict the suspect on three counts of aggravated murder. The suspect was sentenced to life in prison without parole.

Learn More About It

To learn more about a career as a neutron activation analyst, go to school.cengage.com/forensicscience.

CHAPTER 14 REVIEW

True or False

1. Obsidian is a type of glass-like sedimentary rock formed in the ground under great pressure.

2. One of the earliest forms of man-made glass was accidentally produced in Syria during food preparation.

3. Metal oxides added to glass produce glass of different colors.

4. Safety glass is made so it will not crack and not break apart into pieces upon impact.

5. When it is hit, glass bends and stretches before breaking.

6. Secondary fracture lines radiate out from the center of impact as glass shatters.

7. If glass is located near a fire within a building, it will tend to shatter outward.

Multiple Choice

8. The refractive index refers to the ability of a substance to
   a) bend light
   b) reflect light
   c) absorb light
   d) convert light to heat energy

9. The substance added to glass that makes crystal glasses seem to sparkle is
   a) copper
   b) silver
   c) lead
   d) aluminum

10. The speed of light in a vacuum is approximately
   a) 30,000 kilometers/hour
   b) 300,000 kilometers/hour
   c) 30,000 miles/hour
   d) 300,000 meters/hour

11. Which refractive index would indicate the densest material?
   a) 1.2
   b) 1.3
   c) 1.4
   d) 1.5

12. As the density of a medium increases, the refractive index should
   a) increase
   b) decrease
   c) stay the same

13. The normal is the line that is
    a) parallel to the surface where two different mediums meet
    b) moving in the same direction as the beam of light through the first medium
    c) perpendicular to the surface where two different mediums meet
    d) the line of light passing through a vacuum

14. What are the correct units when measuring density?
    a) milliliters/gram
    b) cubic centimeters/milliliter
    c) grams/milliliter
    d) millimeters/second

Short Answer

15. Describe how to use the submersion method to determine the refractive index of a piece of glass found at a crime scene.

    _____

    _____

    _____

16. In order to determine the refractive index of a small piece of glass, the glass is submerged in different liquids and viewed under a compound microscope. The appearance of a Becke line is used to determine the refractive index.

    a) What is a Becke line?

    _____

    _____

    _____

    b) Why does a Becke line form?

    _____

    _____

    _____

    c) What does the location of the Becke line tell you about the refractive index of the piece of glass and the surrounding liquid in which it is placed?

    _____

    _____

    _____

Refer to Figure 14-10 on page 403 to answer question 17.

17. You are testifying as an expert in glass evidence. You want to demonstrate that the evidence glass found embedded in a blanket came from a broken headlight of a vehicle suspected to be the vehicle involved in a hit-and-run accident.

   a. What liquid would you use to demonstrate that the glass fragment was obtained from a broken car headlight?

   _____

   _____

   b. Describe the demonstration that you would show the jury.

   _____

   _____

   c. What explanation would you provide to the jury to convince them that the glass evidence has to come from a glass like the glass found in a car headlight?

   _____

   _____

18. A beam of light passes through air into a second medium. Angle 1 (angle of incidence) is 33°. Angle 2 (angle of refraction) is 48°. Calculate the refractive index. Show all your work.

   _____

   _____

19. Compare and contrast radial and concentric glass fractures. Include in your answer:

   a. Description of each type of fracture

   _____

   _____

   b. On which side of the glass will they form

   _____

   _____

   c. Which type of fracture will form first and why

   _____

   _____

20. A window is broken. The group of vandals who were standing behind the broken window run away. If the broken glass projected inward, how is it possible that a small amount of trace broken glass evidence was found on their clothes?

_____

_____

21. List characteristics of glass that can be used to compare suspect glass samples to glass found at a crime scene.

_____

_____

22. Using a light source and a protractor, explain how to calculate the refractive index of a liquid.

_____

_____

Bibliography

Books and Journals
Giancoli, Douglas C. *Physics, Principles and Applications,* 5th ed. Englewood Cliffs, NJ: Prentice Hall, 1998.
Saferstein, Richard, ed. "Criminalistics," in *Introduction to Forensic Science*, 9th ed. Englewood Cliffs, NJ: Prentice Hall, 2006.
Stratton, David R. "Reading the Clues in Fractured Glass." *Security Management* 38(1): 56, January 2004.

Web sites
Gale Forensic Science eCollection, school.cengage.com/forensicscience.
http://hypertextbook.com/facts/2004/ShayeStorm.shtml
http://www.glassonline.com/infoserv/history.html
http://tpub.com/neets/book10/39h.htm
http://scienceworld.wolfram.com/physics/SnellsLaw.html
http://www.gwu.edu/~forchem/BeckeLine/BeckeLinePage.htm
http://www.fbi.gov/hq/lab/fsc/backissu/jan2005/standards/2005standards9.htm
http://www.newton.dep.anl.gov/askasci/chem00/chem00135.htm
http://www.matter.org.uk/schools/SchoolsGlossary/refractive_index.html
http://www.gazettetimes.com/articles/2005/10/09/news/top_story/news01.txt
http://radiationcenter.oregonstate.edu/Research/Research%20home.html
http://www.gazettetimes.com/articles/2005/10/09/news/top_story/news01.txt
http://www.haines.com.au/Gee_Store/Scripts/prodList.asp?idCategory=1136

ACTIVITY 14-1
GLASS FRACTURE PATTERNS

Objectives:

By the end of this activity, you will be able to:

1. Use glass fracture patterns to explain how to sequence events that occurred to form the broken glass.
2. Analyze glass fracture patterns and determine the order of the breaks in the glass.
3. Distinguish the differences between fractures formed in tempered or safety glass and the fractures formed in window glass.

Time Required to Complete Activity: 45 minutes

Materials:

(per group of two students)
diagrams included with lab
pencil
ruler
piece of broken window glass (demonstration table)
piece of broken tempered glass or safety glass (demonstration table)

Safety Precautions:

This activity involves paper exercises only. If your teacher chooses to include additional hands-on work with fractured glass, wear gloves and goggles while handling glass. Spread newspaper or construction paper in your work area. Dispose of all materials as directed by your teacher.

Background:

Forensic examiners need to be able to look at evidence left at a crime scene and try to determine what happened. If witnesses or suspects are at the crime scene, they may describe their version of what happened. Evidence can either corroborate their story or present a new version of what actually occurred. In this activity, you will examine glass fracture patterns to determine the sequence of events that lead to the breaking of the glass. The fracture patterns may also indicate where force was applied to break the window, and if there are a series of impacts, which impact occurred first.

Procedure:

1. Examine the diagrams to the right, which show a side view of a window both before and after impact. Determine the point of impact and direction of force.

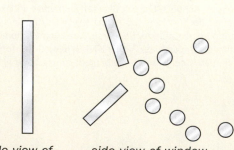

side view of window before impact

side view of window after impact

2. Draw an arrow showing the direction in which the force was applied to the window. Explain your answer using the terms *tension* and *compression*.

3. Tempered glass is also known as safety glass. One of the main uses of this type of glass is for car windshields. When impacted, safety glass fractures differently than window glass. Analyze the pictures of safety and window glass. (If you have actual pieces of broken safety glass and window glass, refer to them.)

    a. Record some of the differences you observed between the two types of glass.

    b. Explain how safety glass is better suited for windshield glass than window glass.

4. Police responded to an incident involving gunshots. When they arrived at the scene, they found two men arguing. One man named Henry was inside the home, and the other man, Ralph, was outside near a shed. Two bullets went through the large living-room window. The police did not recover the bullets. Henry claimed that he was firing in self-defense and that Ralph had fired the first shot noted by A in the diagram. In self-defense, Henry shot the second bullet from inside the house, noted by letter B in the diagram. Ralph claimed that he did not fire any guns and that the man in the house fired both bullets at him. You have been called in as a glass expert to analyze the glass. Based on the glass fracture patterns, can you determine if either of the two men is telling the truth or if both men are lying.

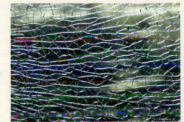

(a) Safety glass

(b) Window glass

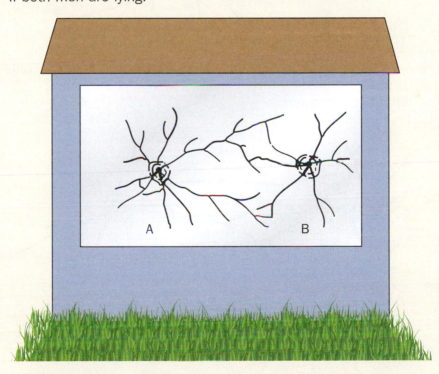

a. On diagram B, label the primary and secondary fracture lines. Explain your answer.

b. Based on the fracture lines, which impact, A or B, occurred first? Explain your answer.

c. As a glass expert, you told the police that if you could examine the actual broken glass, it would be possible for you to determine the direction in which the bullets were fired. You explain that it is possible to determine if they were fired from inside the house or from the outside. Explain two different methods you could use to determine the direction of the bullets based on glass analysis.

d. If you examined the actual glass, could you determine who was telling the truth: Henry or Ralph? Explain your answer.

5. Examine the diagram of the glass fracture patterns. Three different impacts resulted in the breaking of the glass. Which impact occurred first, second, and third? Justify your answer in writing and by labeling the diagram using words like "boundary" and "radial fracture."

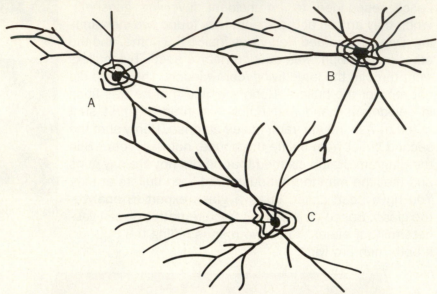

6. Review the diagram showing four different impacts from four different rocks striking glass.
   a. What is the sequence?
   b. Justify your answer.

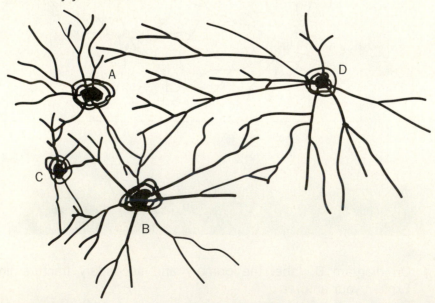

ACTIVITY 14-2
GLASS DENSITY

Objectives:

By the end of this activity, you will be able to:
1. Describe how density is determined using a water displacement method.
2. Calculate the density of various samples of glass.
3. Determine if any of the glass evidence obtained from the four suspects has the same density as the glass found at the crime scene.
4. Maintain the proper chain of evidence when collecting and examining glass evidence.

Time Required to Complete Activity: 45 minutes

Materials:

(per group of three students)
evidence bags containing glass labeled Suspects 1, 2, 3, and 4
evidence bag containing crime-scene evidence labeled CS
displacement containers *or* 10 mL graduated cylinders
beaker (250 mL)
water
dropper bottle of water containing 50 mL of water
balance (accurate to at least .01 gram)
forceps
newspaper or construction paper
labeling tape
permanent marker pen

Safety Precautions:

Wear gloves and goggles while handling glass. Spread newspaper or construction paper in your work area. Dispose of all materials as directed by your teacher. Immediately report any accidents with glass to the teacher.

Background:

The density of glass fragments found at a crime scene can be compared to the glass fragments found on suspects. Keep in mind that if the densities do match, this does not prove that the suspect is guilty, because glass would be considered class evidence.

Glass fragments from a crime scene need to be matched with any glass fragments associated with the four suspects.

In this activity, you will be asked to determine the density of glass fragments found at the crime scene and the densities of glass fragments found on any of the suspects. If the densities do not match, you may be able to disqualify a suspect. If you find that the densities do match, then you will need to collect further evidence to help prove that a particular suspect was at the crime scene.

Procedure:

1. Obtain the evidence envelopes labeled:
   Suspect 1
   Suspect 2
   Suspect 3
   Suspect 4
   Crime Scene

2. Using Suspect #1 evidence bag, record your name, date, and time on the Chain of Possession form.

3. Open the envelope labeled Suspect #1. Do not disturb the signatures on the evidence envelope. Open it from a different side.

4. Remove two pieces of glass fragments from suspect's evidence bag #1. Using a balance, determine the combined mass of both pieces. Record the mass on Table 1. Leave the two pieces of glass on the balance for further testing.

5. Reseal the evidence bag. Place a piece of tape over the opened edge. Write your signature or initials across the interface of the tape and the bag to maintain chain of custody. At this point, there should be two taped areas on the bag, both containing signatures or initials on top of the tape. Refer to proper chain of command described in Chapter 2.

6. Set up a 250 mL beaker of water filled to overflowing. You may need to add the last few drops with a dropper.

7. Position a clean, dry, 10 mL graduated cylinder to receive overflow water. Several books may have to be placed under the beaker to adjust the height of the beaker.

8. Slowly add your two glass fragments of glass Sample 1 into the beaker one at a time. Water will spill over into your graduated cylinder.

9. Measure the volume of water displaced by the addition of the two glass fragments. This is determined by reading the amount of water that has overflowed into the graduated cylinder.

10. Record the combined volume for the two glass fragments in Data Table 1.

11. Calculate the density of the glass fragments from Suspect 1 evidence bag and record your answer on the data table.

12. Remove the two glass fragments from the beaker and handle as described by your teacher.

13. Refill the beaker to just overflowing.

14. Repeat the process with glass from (Suspect) 2. Be sure to properly open and reseal the evidence bag. Record your name, date, and time on the Chain of Possession form. Record all information for Suspect 2 in the Data Table.

15. Repeat the process until you have recorded all the information for the glass found on suspects 3 and 4, and the crime-scene evidence envelope.

Data Table: Density of Glass Samples

Sample	Combined Mass of Two Fragments (mass)	Combined Volume of Two Fragments (milliliters)	Density (M/V) (grams/ml)
Suspect 1			
Suspect 2			
Suspect 3			
Suspect 4			
Crime Scene			

Questions:

1. Did the density of the glass found on any of the four suspects match the density of the glass found at the crime scene? Explain your answer.
2. Check your results with your classmates. How did your results compare to those of the rest of the class?
3. Describe how you could improve your experiment to have more reliable results.
4. Based on your results, are you able to link any of the suspects to the crime scene based on your glass analysis?
5. Explain why glass is considered to be a form of class evidence.
6. If you did find that the glass density of fragments found at the crime scene matched those found on one or more of the suspects, what other additional tests could be done on the glass evidence to further link the suspect(s) to the crime scene?
7. In checking the density of the glass fragments, why did you use only two fragments of glass and not all of the glass fragments found in the evidence bag?

ACTIVITY 14-3
DETERMINING THE REFRACTIVE INDEX OF LIQUIDS USING SNELL'S LAW

Objectives:

By the end of this activity, you will be able to:
Determine the refractive index of three liquids.

Time Required to Complete Activity: 45 minutes

Materials:

(per group of two to three students)
laser pointer (that can be turned on without having to keep pressing down
   on a button)
paper (8 × 11)
protractor
calculator with sine function or Sine Table
three different liquids (Samples 1, 2, and 3)
ruler (mm or inches)
three semicircle plastic dishes
pencil

Safety Precautions:

Students should not look directly at the laser light.
Students should be careful when handling glass to avoid being cut.
Immediately report any injuries to the teacher.

Background:

Light travels through different mediums at different velocities. There is a
relationship between the density of a medium and the speed of light: the
higher the density of the medium, the slower the velocity of light through
that medium. When light passes between two different mediums of different densities, the velocity of the light will be altered. The change in the
velocity of light results in a bending of the light wave as it passes through
this medium. This bending is known as *refraction*.

   As a means of comparison, a ratio of the speed of light through a vacuum
(186,000 mi/sec or 300,000 km/sec) is compared to the speed of light
through a different medium. This ratio is referred to as a *refraction index*
(RI). For example:

Refraction index for water
speed of light in a vacuum   = 186,000 mi/sec (300,000 km/sec)
speed of light through water = 140,000 mi/sec (225,800 km/sec)

$$= \frac{186{,}000 \text{ mi/sec}}{140{,}000 \text{ mi/sec}}$$

$$= 1.33$$

Snell's Law:

Mathematically, the relationship between the refractive indices and the angle of incidence and angle of refraction is expressed as Snell's Law.

n_1 (sine angle of incidence) = n_2 (sine angle of refraction)

n_1 = refractive index of medium 1

n_2 = refractive index of medium 2

Comparison of Angle of Incidence and Angle of Refraction as light passes between two different mediums.

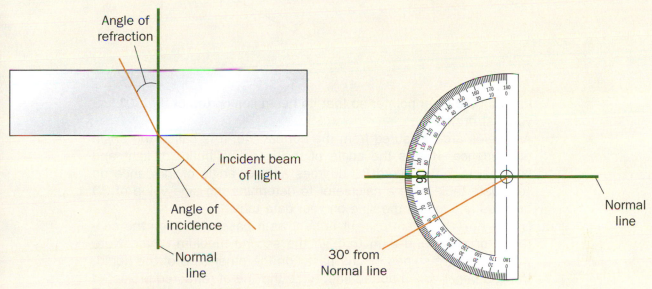

Procedure:

1. Place a piece of white paper on the table and draw a straight line down the center of the paper. Draw a second line that is perpendicular (90°) to the first line. This second line is called a normal (N).

2. Using a protractor, draw a line on the paper at 30° from the normal as shown in the diagram. This line will also be at a 60° angle from the first line you drew.

3. Obtain a semicircular plastic dish filled with Sample 1 and arrange the dish as pictured in the diagram. Position the center of the protractor at the point where the normal and 30° line meet.

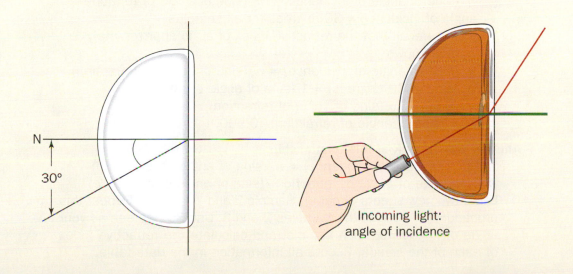

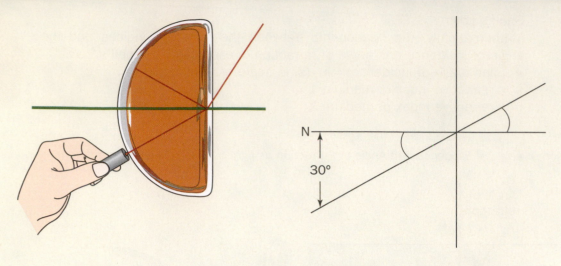

4. Position the laser pointer so that its beam lies on top of the 30° line you just drew.

5. All angles are measured from the normal line. It is called the *angle of incidence.* This is the angle of the incoming beam of light and is equal to 30°. Record 30 degrees as your angle of incidence on your data table. Use a calculator to determine the sine value of 30 degrees and record the sine on your data table.

6. Notice how the beam of light is bent as it passes through the container or liquid and then through the second medium of air. Note where the light shines on the paper on the other side of the liquid. Using a pencil and a dotted line, trace the line of refracted projected light as it exits the plastic dish and passes through the air. Label the line Sample 1.

7. Using this dotted line and the normal line and a protractor, determine R, the *angle of refraction*. Record the angle of refraction on your data table.

8. Using a calculator, determine the sine value of your angle of refraction. Alternatively, a sine table can be used and is found in the appendix at the back of your textbook.

9. Calculate the refractive index for Sample 1 using Snell's Law. Note: the angle of refraction was determined after the laser light was shown through the plastic dish.

n_1 (sine of angle of incidence) = n_2 (sine of angle of refraction)
angle of incidence = 30 degrees for all samples
n_2 = 1 (recall that the refractive index of air is approximately 1)
n_1 is unknown
n_1 (sine of angle of incidence) = n_2 (sine of angle of refraction)
n_1 (sine of 30 degrees) = 1 (sine of angle of refraction)
n_1 (0.5000) = (sine of angle of refraction)
n_1 = (sine of angle of refraction)/(0.5000)

Another way to look at this relationship is:

n_1 (sine of angle of incidence) = n_2 (sine of angle of refraction)
n_1 = n_2 (sine of angle of refraction)/(sine or angle of incidence)

10. Label two new sheets of paper, Sample 2 and Sample 3. Repeat this procedure for each of the Samples 2 and 3. Use 30 degrees as your angle of impact for each sample, and calculate the refractive index of each of the liquids. Record all information in the data table.

Refractive Index of Three Liquid Samples

Liquid Sample	Angle of Incidence (I)	Sine of Angle I	Angle of Refraction	Sine of Angle R	Refractive Index of Liquid
1	30°				
2	30°				
3	30°				

1. Compare the refractive indices of each of the three liquids. Which liquid had the highest refractive index and which one had the lowest refractive index?
2. Describe any visual correlation you can make about the refractive index and the appearance of the three different liquids.
3. Suppose you set your angle of impact at 45 degrees instead of at 30 degrees. Would you have obtained a different refractive index using the 45-degree angle of incidence instead of the 30-degree angle of incidence? Explain your answer.
4. Your two different mediums through which light passed in this experiment were the liquids and the air. In terms of density, air is less dense than any of the liquids. Would you expect the velocity of light to be faster or slower moving through the liquids than the air? Explain your answer.
5. If we assume that air is less dense than the liquids, did the angle of refraction move toward the normal, or did it move away from the normal as the light traveled from the liquid (more dense) through the air (less dense)? Support your answer with data from your table.
6. Suggest other liquids to test for refractive index that you think might give you a higher refractive index than any of the liquids selected in this experiment.
7. As the thickness of the liquid increases, what effect might that have on the bending of light? Will the bending become more pronounced or less pronounced? Explain.

ACTIVITY 14-4
DETERMINING REFRACTIVE INDEX OF GLASS USING LIQUID COMPARISONS IN A SUBMERSION TEST

Objectives:

By the end of this activity, you will be able to:

1. Perform a submersion test on glass fragments to estimate the refractive index of the glass fragment.
2. Explain how to do a submersion test on glass fragments.
3. Compare the refractive indices of the evidence glass pieces to the refractive index of the crime-scene glass.
4. Determine if the suspects can be linked to the crime scene based on the refractive indices of the evidence glass and the refractive index of the crime-scene glass.

Time Required to Complete Activity: 45 minutes

Materials

(per group of four students)
3 samples of glass fragments contained in evidence bags labeled Crime Scene, Suspect #1, Suspect #2
1 pair forceps
1 beaker of 250 mL detergent solution
250 mL of tap water
permanent marker
1 test tube rack
labeling tape
paper toweling
set of 7 small test tubes half filled with
    methanol
    water
    isopropyl alcohol
    olive oil
    cinnamon oil*
    castor oil
    clove oil*

*Because of expense and odor, these two tubes may be set up by your teacher as a demonstration for the entire class to use.

Safety Precautions:

Spread newspaper or construction paper in the work area to capture small fragments of glass. Handle glass with forceps.

Scenario:

Students at a local high school decided to steal the basketball trophy in the locked display case near the gym. They planned to steal the trophy after 7 p.m., when only a few janitors would be in the building. Once the glass case was broken, they thought they could easily run out of the back door by the gym. What they didn't plan on was that the coach came back to his office just as he heard the glass breaking and saw the two boys running out of the gym.

The coach thought he recognized one of the boys by his coat that held numerous old snowboard lift tickets from the past four years. He reported the incident and gave a description to the police, who quickly located both boys at the pizza place across the street from the school.

Did they break into the display case and steal the trophy? The police brought the boys to the police station, where they examined the bottom of their sneakers. As expected, small particles of glass were embedded in the soles of their sneakers. Did the glass in their sneakers match the glass found in the display window where the trophy case was removed? What type of testing can be used to match the glass in their sneakers to the glass found in the display case?

Background:

In this activity, we will try to match the glass found in the sneakers on the two suspects to the glass found in the broken display case. Refractive indices of the glass found in their sneakers will be compared to the refractive index of the glass in the display case.

Liquids of known refractive indices can be used to visually estimate the refractive index of glass samples. In order to estimate the refractive index of a piece of glass, submerge individual pieces of the glass in a series of different liquids. If the glass appears to disappear when submerged, then the glass has a refractive index similar to the solution. If the refractive index of the glass differs from the refractive index of the solution in which it is immersed, you will be able to see an outline of the glass. This is known as a submersion test for refractive index.

After testing the evidence glass from the sneakers and determining the refractive indices of the glass, you will test the crime-scene glass for its refractive index. If the glass from the crime scene has the same refractive index as the glass from the suspects' sneakers, it can place the suspects at the crime scene.

Procedure:

1. Obtain a test tube rack with five test tubes. Label the tops of the five test tubes in the following order:
     - methanol
     - water
     - isopropyl alcohol
     - olive oil
     - castor oil
2. Half fill each test tube with the five different liquids as numbered in the table. They are arranged in increasing order of refractive index. The table provides the refractive index of each liquid.

Data Table 1: Refractive Indices of Liquids

Liquid	Refractive Index
1. Methanol	1.33
2. Water	1.33
3. Isopropyl alcohol	1.37
4. Olive oil	1.47
5. Castor oil	1.48
6. Clove oil (teacher display)	1.54
7. Cinnamon oil (teacher display)	1.62

3. You need to maintain proper chain of custody when opening and resealing each of the evidence envelopes. Remember to:
   a. Cut open the bag in an area that does not disrupt the signature of the person who had the evidence bag before you.
   b. Sign and date the Chain of Possession form on the front of the evidence bag.
   c. When resealing the bag, tape each bag and sign your name or initials across the interface of the tape and the package.
4. Starting with the crime-scene glass, submerge a fragment of the glass into test tube 1. Hold the glass fragment with a forceps. Do not let go of the glass fragment.
5. View the submerged glass at eye level. Is the glass invisible, or can you see the glass fragment? Record your results on Table 2. Remember: If the glass and the liquid have the same refractive index, the glass will seem to disappear in the solution. The test tube containing the liquid that results in the disappearance of the glass fragment best approximates the refractive index of the glass.
6. Remove the piece of glass from test tube 1 using the forceps. Wash the glass and forceps in the soap solution, rinse the glass and forceps in a beaker of water, and dry them on the paper towel.
7. Submerge the *same piece* or a new piece of glass in test tube 2 and observe. Is the glass visible, slightly visible, or invisible? Record the information on Data Table 2.
8. Rinse the glass piece and forceps in the soapy water followed by tap water and dry before continuing with the same procedure until you have tested the crime-scene glass on all seven solutions (your five solutions and the two solutions set up by your teacher).
9. Repeat the procedure for the submersion test on a glass fragment found on Suspect #1 and Suspect #2 and record your results in Data Table 2.

Data Table 2: Submersion Test Results

Test Tube #	Refractive Index of Liquid	Visibility: Visible Slightly visible Invisible		
		Crime Scene	Suspect 1	Suspect 2
1. Methanol	1.33			
2. Water	1.33			
3. Isopropyl alcohol	1.37			
4. Olive oil	1.47			
5. Castor oil	1.48			
6. Clove oil (teacher display)	1.54			
7. Cinnamon oil (teacher display)	1.62			

Questions:

1. Based on the results of your submersion test, record the estimated refractive indices for each of the glass fragments.
   The Refractive Index (RI) of the Crime Scene glass = _____
   The Refractive Index (RI) of the glass from Suspect #1 = _____
   The Refractive Index (RI) of the glass from Suspect #2 = _____
2. What would you consider to be the experimental errors in using this technique?
3. What could you do to improve the reliability of this experiment?
4. Is your match conclusive? Why or why not?
5. Why would the match of glass from a crime scene and a suspect be considered to be class evidence?
6. Explain refraction of light. Include in your answer the following:
   - two different mediums
   - light velocity
   - density
7. A refractive index of olive oil is equal to 1.47. It is calculated as a ratio between what two numbers?

CHAPTER 15

Casts and Impressions

THE MAN IN THE BRUNO MAGLI SHOES

One morning, the residents of Los Angeles's South Bundy Drive woke to sirens and fluttering police tape. Two bodies had been found, having suffered a horrific knife attack. Bloody footprints, made by size 12 Bruno Magli shoes, tracked the victim's blood along the path. The date was June 13, 1994, and the two bodies were those of Nicole Brown Simpson and Ronald Goldman.

The ensuing court case culminated in one of the most-watched television events in U.S. history. Many people were surprised by the "not guilty" verdict against Brown's ex-husband, Orenthal James (O.J.) Simpson. There was significant evidence linking Simpson to the crime scene, but all of it was class evidence, and nothing pointed to Simpson directly. Who was the man in the Bruno Magli shoes?

The shoe impression was created when the individual wearing the size 12 shoes stepped in a pool of congealed blood. Specialists determined that the impression was made about 20 minutes after the attack. Was the man in the Bruno Magli shoes the attacker or someone who came to the scene soon after the attack? Simpson denied owning a pair of Bruno Maglis, and the prosecutor could find no way to dispute this fact. The footprint evidence was of little value in the criminal case. However, after the criminal trial ended, a photograph was

©AP Photo/LAPD

Bloody footprints at the Brown Simpson/Goldman murder scene.

found showing Simpson wearing similar shoes. In the civil case that followed, jurors believed the attacker made the footprints, and Simpson was found responsible for the deaths of Brown and Goldman.

OBJECTIVES

By the end of this chapter you will be able to

✔ Distinguish between latent, patent, and plastic impressions.

✔ Explain how various types of impressions can be used as trace evidence.

✔ Describe how to make foot, shoe, and tire impressions.

✔ Use track width and wheelbase information to identify vehicles.

✔ Prepare dental impressions and match them with bite marks.

TOPICAL SCIENCES KEY

VOCABULARY

latent impressions hidden impressions requiring special techniques to be visualized

patent impressions two-dimensional impressions that are already visible

plastic impressions three-dimensional impressions cast in soft materials, such as soil and snow or blood

sole (outsole) the pattern on the bottom of a piece of footwear

tire groove a depression in the tread pattern

tire rib an individual ridge of tread running down the tread area and around the circumference of the tire

tire ridge elevated area on the tread pattern

track width the distance from the center of the tread pattern on the left tire to the center of the tread pattern on the corresponding right tire

tread pattern the unique design of a tire's surface

turning diameter a measure of how tight a circle can be driven by a vehicle

wheelbase the distance from the center of the front axle on a vehicle to the center of the rear axle

INTRODUCTION

Crime-scene investigators do not always have the advantage of a crime being captured on camera. Often, the crime scene is devoid of all people. However, the environment has a way of recording what has happened. People, vehicles, and objects leave evidence of their presence at the crime scene in the form of a mark or an imprint. In an earlier chapter, you learned how fingerprints can identify an individual. In this chapter, you will explore impressions made by footwear, feet, teeth, and tires.

TYPES OF IMPRESSIONS

Impressions fall into three basic categories: patent, latent, and plastic (Figure 15-1). **Patent impressions** are visible, two-dimensional impressions produced as an object moves through soil, dust, paint, blood, or other fine particles and leaves a trace. By contrast, **latent impressions** are hidden to the eye but can be visualized through the use of special dusting and electrostatic techniques or chemical developers. Oils, fine soil, and other minute debris can be carried onto clean floors and be transferred as a latent impression. Even clean shoes or feet can transfer materials onto newly waxed or polished floors.

Plastic impressions are three-dimensional imprints. These impressions can be left in soft materials, such as snow, mud, soil, or soap. One difficulty in dealing with plastic prints is that they are easily lost. A strong wind or a sudden change in the weather can mean the loss of important evidence. It is critical that photographs be taken immediately before trying to make any permanent cast.

Figure 15-1. *Examples of latent (left), patent (center), and plastic (right) impressions.*

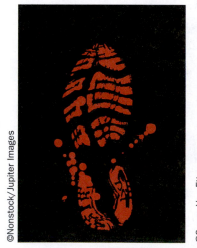
©Nonstock/Jupiter Images

©Susan Van Etten

©Stock Connection/Jupiter Images

INDIVIDUAL OR CLASS EVIDENCE?

Depending on how it is made, impression evidence may be either class evidence or individual evidence. A particular **tread pattern** in shoes or in tires may identify the brand and size, but it does not identify a specific individual or tire. Distinguishing characteristics, such as a split on a shoe sole or unusual wear on a car tire, can be used as individual evidence. Dental impressions are typically considered individual evidence and have a long history of use to identify individuals, especially during wartime to identify remains.

SHOE IMPRESSIONS

The crime-scene investigator can obtain information about the person or persons involved or about the crime itself from a shoe or tire impression found at a crime scene. For example, the size of a shoeprint can tell police the size of the person's foot. The depth of a shoe or foot impression can tell police something of the person's weight. The type of shoe (e.g., work boots versus flat dress shoes) may tell something of the person's job or personality (Figure 15-2). The brand of shoe provides information about the buyer. A retired factory worker on a small pension probably would not wear expensive, imported dress shoes, nor would a wealthy, middle-aged woman last seen in an upscale restaurant wear worn tennis shoes.

Databases contain the names of specific manufacturers and tread patterns used to identify different types of shoes. The number of manufacturers that use the same generic **sole** patterns complicates shoe pattern identification. When a shoe impression is found at a crime scene, the crime-scene investigator will search the databases to find the manufacturer that produced the sole as well as to search for the company that purchased that sole for the shoes. Once the footwear has been identified, the impression can be used as class evidence to link a suspect to the crime scene by matching their footwear to an impression. Recall that merely matching the footwear may not provide sufficient evidence to prosecute a suspect.

SHOE WEAR PATTERNS

Although two different people may purchase the same type of shoes, the wear pattern on the shoes will appear quite different (Figure 15-3). We

Figure 15-2. *The type of footwear a person chooses to wear reveals information about him or her.*

Figure 15-3. *The wear on a shoe's sole can tell you something about the owner.*

©Susan Van Etten

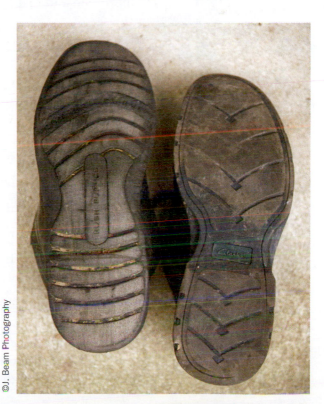

©J. Beam Photography

personalize shoes with our own characteristic way of walking and usage. A shoe tread showing strong individual character is classified as individual evidence. Some factors that personalize our footwear include:

- Whether a person walks on his or her toes or heels

- Body weight

- If the person walks straight ahead or tends to walk with toes pointed inward or outward

- The shape of the foot and the wearer's activities

- The surface on which the person usually walks

- Unique holes, cuts, or debris that may become embedded in the shoe

GAIT AND TRACKS

Figure 15-4. *Stride pattern is also individual.*

©Brand X Pictures/Jupiter Images

Numerous prints together tell an investigator about the person's *gait,* or walking habits (Figure 15-4). A limp or injury creates an asymmetrical gait; that is, when one foot is angled differently from the other or one foot makes a deeper impression than the other. This pattern is also created when someone is carrying a heavy weight. Tracks can indicate if a person was walking or running by the length of the stride and the pressure and shape of the impression. A trail of footprints can also point to the movements of their maker and possibly assist in recreating events.

Information that can be obtained from shoe impressions includes:

1. The number of people at the crime scene
2. Movements of the individuals at the crime scene

   - Did the event happen in just one room?

   - Did the event happen in several rooms?

3. The entrance and exit to the crime scene

COLLECTION OF SHOE IMPRESSION EVIDENCE

The steps necessary in collecting shoe impression evidence are (1) photographing impressions, (2) lifting latent impressions, and (3) casting plastic impressions.

Photographing Impressions

Use the following guidelines when photographing impressions:

- Take photographs before anyone touches or alters an impression!

- Fill the camera's viewfinder with the impression.

- Take photographs with the lens perpendicular to the impression to reduce distortion.

- Take multiple photographs of the impression from at least two different orientations (angles).

- If using a digital camera, check your photographs for clarity and retake them when necessary. (Forensic photographs are often taken with black-and-white film, although some departments have moved to digital color photography. Film is used to guarantee the photograph was not altered. Color film is used to photograph blood spatter.)

- Place an identifying label and a ruler in position with the impression and re-photograph, making sure to focus on the impression, not the ruler.

- Use oblique lighting when possible. Sunlight can produce a glare.

- If an additional flash is needed, position the flash at least three to four feet away to avoid reflection in the photograph.

- If the impression is faint, spray it with a *light* coat of color-contrasting paint.

In a court case, the defense tries to discredit the prosecution's evidence. This is easily done if there is a lack of photographic proof that Exhibit A was found at the crime scene. Therefore, the forensic investigator's first task at the crime scene is to document the scene and the evidence with photographs (Figure 15-5). The only exception occurs when paramedics move a body when applying their lifesaving skills.

Lifting Latent Impressions

When a shoe or bare foot walks on a smooth surface, it leaves a print that is not usually visible to the unaided eye. A bare foot leaves a thin layer of body oil, while a shoe leaves a thin film of substances from either the plastic in the sole or dirt. It takes a trained crime-scene investigator to know where to look for latent prints. If the entry and exit areas have been identified, the task of latent print identification is made easier. There are several different methods to make latent prints visible, which include:

- Luminol to make bloody footprints visible and able to be photographed.

- Dusting of the latent print, similar to dusting for fingerprints, to reveal an impression and make it visible to be photographed and lifted.

- Electrostatic lifting and gel lifting techniques to capture invisible impressions.

✗ *Electrostatic Dusting and Lifting* A dry shoe may deposit a small amount of dust with each step. Electrostatic dusting can reveal this fine dust and create an impression. It works by applying an electrostatic charge on a piece of lifting film, which is then placed over the latent print (Figure 15-6). The film picks up and holds the dust of the latent print. By viewing the film with a special light source, the impression can be made visible and photographed.

Electrostatic charges can lift impressions from paper; flooring, including carpeting, wood surfaces, and linoleum; and pavements, such as asphalt and concrete. This method can also help clean up surface debris and expose a more complete impression.

✗ *Gel Lifting* Gel lifting is another method used to visualize and recover latent impressions. A gel lifter is a layer of thick gel sandwiched between paper backing and a plastic cover sheet. It is thick and flexible to conform to uneven surfaces. Best used on oily or moist impressions, the print is first dusted with powder, which sticks to the moist residues of the latent print. The protective plastic cover is peeled off the gel lifter, and the gel is pressed firmly over the print. Because the gel is not very sticky, it can be used on surfaces, such as paper, from which it would be impossible to lift an impression

Figure 15-5. Impression evidence is documented before any attempt is made at lifting impressions.

©Cengage Learning

Did You Know?

Multiple lifts of the same print are possible using gel lifters. The second lift is fainter, but generally sharper in detail. Electrostatic lifting can also be done before gel lifting.

Figure 15-6. Electrostatic lifting uses a charge to hold the dust particles in place.

©Cengage Learning

Figure 15-7. *Plaster cast of a shoe print.*

©Cengage Learning

using normal fingerprint tape. The gel lifter is lifted off and the protective cover replaced. The gel lifter is then photographed to reveal the print. There are black, white, and clear gel lifters. The color used depends on the conditions and the color of the powder used (e.g., a white, black, gold, or silver powder).

Casting Plastic Impressions

A three-dimensional impression, such as a shoeprint in mud or snow, is called a *cast* and may be made to preserve physical evidence (Figure 15-7). The casting materials and techniques vary with the conditions at the crime scene. If the impression is in sand or dirt, a Plaster of Paris impression is made. Before pouring in the mixed Plaster of Paris, a light film of hair spray is applied to prevent the impression from collapsing under the weight of the plaster.

Snow presents a problem simply because it melts. Therefore, casting material needs to be used that will set at low temperatures and will not generate heat. An impression in snow can be cast by first using a spray wax applied in thin layers over the snow. Freezing instantly, the wax provides protection to the delicate print before the casting material is poured into the impression. The casting material used for snow is called *dental stone,* which hardens faster than Plaster of Paris.

Digging Deeper
with Forensic Science e-Collection

Go to the Gale Forensic Sciences eCollection on school.cengage .com/forensicscience and find the following article. Read the article and discuss, in not less than one page, how broadening the search for evidence beyond the obvious sources may be the future of crime detection.

Peter A. Bull, Adrian Parker, and, Ruth M. Morgan, "The forensic analysis of soils and sediment taken from the cast of a footprint," *Forensic Science International* 162.1–3 (Oct 16, 2006): pp. 6(7). From *Forensic Science Journals.*

FOOT LENGTH AND SHOE SIZE

The evidence of a shoe print is not the same as a fingerprint. A shoe print is not a direct record of the person, but it provides information about the person who left it. The size of a shoe varies by the shoe type. For example, a running shoe will give a much smaller impression than a steel-toed work-boot of the same size. The shoe model must first be identified in order to gauge the correct shoe size to obtain an estimate of the foot size. A person's height is generally related to his or her foot size, but it is impossible to predict someone's exact height from foot size. Figure 15-8 compares mens' and womens' shoe sizing. Figure 15-9 shows a rough comparison of shoe size and height of an individual.

Figure 15-8. *Comparison of foot length and U.S. shoe sizing.*

Foot Length (Inches)		9	9 ⅛	9 ¼	9 ⅜	9 ½	9 ⅝	9 ¾	9 ⅞	10	10 ⅛	10 ¼	10 ½	10 ¾	11	11 ¼	11 ½
Shoe Size	M	3 ½	4	4 ½	5	5 ½	6	6 ½	7	7 ½	8	8 ½	9	10 ½	11 ½	12 ½	14
	W	5	5 ½	6	6 ½	7	7 ½	8	8 ½	9	9 ½	10	10 ½	12	13	14	15 ½

Figure 15-9. *Comparison of shoe size and height.*

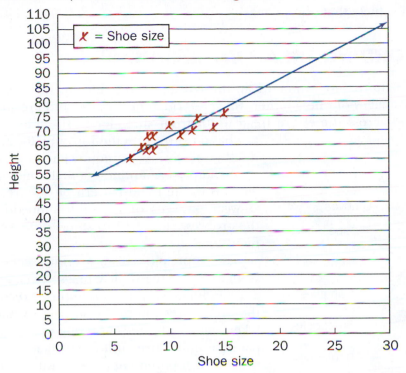

TIRE TREADS AND IMPRESSIONS

Tire evidence may be used to link a suspect or victim to a crime scene, and it can also reveal the events that took place. Tire marks may indicate the speed a car traveled when it left a road or the direction in which it traveled when fleeing a crime scene. Skid marks at the scene of a fatal accident are used to determine who may have been at fault. A forensic scientist examines tire tread and impression for two characteristics:

1. Tread pattern and measurements to identify the type of tire and perhaps the make and model of car

2. Nature of the impression to determine how the vehicle was driven

Motor vehicles can leave patent, latent, or plastic tire patterns. Patent impressions appear after the vehicle has tracked through a fluid material like oil, tar, or blood. Latent tracks may be left on asphalt or concrete roads by manufacturers' oils used to keep tires soft and pliable. Plastic or three-dimensional impressions may be left in off-road surfaces, such as mud, lawns, sand, or snow.

THE ANATOMY OF A TIRE

A tire's tread surface is divided into **ridges** (elevated regions) and **grooves** (indentations). The purpose of these ridges and grooves is to channel water away and provide traction as the surface area makes contact with the road or ground. Generally, every model of tire is unique as manufacturers continually try to improve handling qualities. The width and angle of the grooves in the tread are engineered to perform best on a specific surface. Touring tires have many smaller grooves to channel air and water at high speeds on smooth pavement, and knobby off-road tires have very wide grooves for traction in slippery mud. A single tire tread usually indicates the general type of vehicle that left the mark.

RECORDING TREAD IMPRESSIONS

Tread patterns are symmetrical; the left and right sides of the tread are mirror images of one another. **Ribs,** the ridge of the tire, and grooves are counted across the entire tread width from shoulder to shoulder. If the tire has a central rib, then it will have an odd number of ribs. Any unique characteristics in the tread pattern are noted, including imperfections in the tread pattern of the tire, such as wear patterns or a pebble embedded in the grooves. Figure 15-10 shows different elements of a tire.

The impression from a crime scene may be matched to the vehicle of a suspect, known as a candidate vehicle. To do this, a record of the tire impression of the suspect's vehicle is taken. Similar to getting a fingerprint, ink is painted onto a tire, and the vehicle is driven over smooth pavement covered with paper or cardboard. A print at least three meters long is produced to ensure that the tire rotated through one revolution. All possible individual characteristics are noted. The individual characteristics link a specific vehicle with a tread impression from a crime scene.

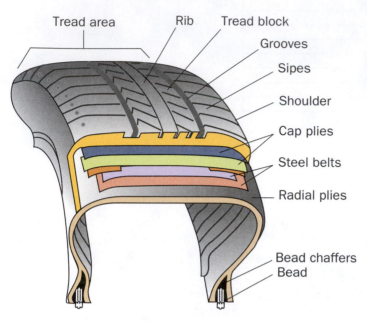

Figure 15-10. *The anatomy of a tire.*

Tread area — Rib — Tread block — Grooves — Sipes — Shoulder — Cap plies — Steel belts — Radial plies — Bead chaffers — Bead

IDENTIFYING A VEHICLE

Tire impressions can identify the make and model of vehicle that was driven at a crime scene. Because the same type of tire may be used on many vehicles, identifying the tread pattern is often not enough. The track width and the wheelbase of the treads narrow down the search. **Track width** is measured from the center of each tire to the center of the opposite tire. Note that the front and rear track width measurements may differ. The **wheelbase** of a vehicle is the distance between the center of the front axle and the center of the rear axle (Figure 15-11). Measurements should be taken to the nearest millimeter.

Figure 15-11. *Every make and model of vehicle has its own track width and wheelbase measurements.*

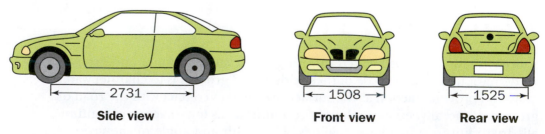

Side view — 2731

Front view — 1508

Rear view — 1525

Turning Diameter

The **turning diameter** is a measure of how tight a circle can be driven by a vehicle or, put another way, the minimal space required for a car to make a

Figure 15-12. *Tread marks revealing turning diameter can help identify a vehicle. Some samples are shown here.*

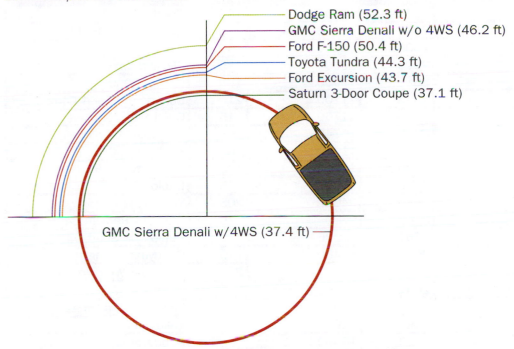

Dodge Ram (52.3 ft)
GMC Sierra Denali w/o 4WS (46.2 ft)
Ford F-150 (50.4 ft)
Toyota Tundra (44.3 ft)
Ford Excursion (43.7 ft)
Saturn 3-Door Coupe (37.1 ft)

GMC Sierra Denali w/4WS (37.4 ft)

U-turn (Figure 15-12). When a vehicle turns a sharp corner, even at moderate speeds, a track is created by the additional stress put on the front outer tire. When the driver makes a U-turn or similar sharp turn, the wheel is turned as far as possible. A longer wheelbase increases the turning diameter. This information helps the investigator determine the size of the vehicle. Sometimes turning diameter, also called turning radius, is not mathematically a diameter, but a radius, measured in millimeters, inches, or the nearest one-tenth foot.

A large database contains the track width, wheelbase, and turning diameter measurements for all makes and models of cars and can be easily checked to identify the vehicle that left the impressions (Figure 15-13).

Figure 15-13. *Examples from the database of automobile statistics by make and model.*

Make	Model	Wheelbase in mm	Turning Diameter in mm	Tire Size in mm	Tire Make
ALFA ROMEO	Alfa 146	2540	10300	175	Pirelli P4000
ALFA ROMEO	Alfa 156	2595	11600	185	Michelin Energy XH-1
AUDI	A3	2512	10900	205	Michelin Pilot HX MXM
AUDI	A4	2617	11100	195	Michelin Energy MXT
AUDI	A8	2882	12300	225	Michelin Pilot CX
BMW	3 series	2725	10500	225	Michelin Pilot HX
CADILLAC	DeVille	2890	12300	225	Goodyear Eagle RS
CADILLAC	Seville	2819	11700	225	Goodyear Eagle GA
CADILLAC	Seville	2850	12340	235	Goodyear Eagle Touring
CHEVROLET	Blazer	3122	12600	205	Uniroyal Tiger Paws
CHEVROLET	Corvette	2444	12200	285	GoodYear
CHEVROLET	Corvette	2655	11700	275	Goodyear F1 EMT
CHRYSLER	Grand Voyager	3030	12500	215	Goodyear NCT2 Touring
CHRYSLER	Neon	2642	10800	175	Goodyear Eagle NCT2
CHRYSLER	New Yorker	2870	11500	225	Michelin XGTV4
CHRYSLER	Stratus	2692	11000	215	Michelin MXV3A

Make	Model	Wheelbase in mm	Turning Diameter in mm	Tire Size in mm	Tire Make
CHRYSLER	Stratus	2743	11000	185	Goodyear NCT2
CHRYSLER	Viper	2444	12300	335	Michelin Pilot SX MXX3
FERRARI	550	2500	11600	295	Pirelli P Zero
FORD	Escort	2525	10000	185	Michelin MXV2
FORD	Focus	2615	10900	185	Pirelli P6000
FORD	Galaxy	2835	11100	215	Continental Sport Contact
FORD	Galaxy	2835	11700	215	Conti Sport Contact
HONDA	Accord	2670	10740	195	Bridgestone Potenza
HONDA	Accord	2720	11000	185	Pirelli P4000
HONDA	Civic	2620	10000	175	Bridgestone SF-321
HONDA	Civic	2620	10200	175	Dunlop SP9
HONDA	Prelude	2585	9400	205	Yokohama A085
HYUNDAI	Excel	2400	9700	175	Hankook Radial 884
INFINITI	J 30 t	2761	11000	215	Dunlop SP Sport D31
JEEP	Grand Cherokee	2690	11400	225	Goodyear Wrangler HP
JEEP	Grand Cherokee	2690	11200	225	Goodyear Wrangler HP
JEEP	Grand Cherokee	2691	11100	245	Goodyear Wrangler HP
LEXUS	GS 300	2800	11000	225	Yokohama Advan A460
LEXUS	GS 300	2780	11800	275	Yokohama
MERCEDES	A	2423	10300	175	Goodyear GT

ESTABLISHING CAR MOVEMENTS FROM TIRE MARKS

A vehicle's direction of travel can be determined by studying:

- Vegetation disturbed as a vehicle entered or left a road

- Debris patterns cast off by a moving vehicle

- Splash patterns created as a vehicle moved through a puddle of water (or some other substance) or from a wet to dry pavement

- Substance transfer, such as oil leakage from vehicle to pavement or soil (The drips would be farther apart as the vehicle accelerates.)

- Tire marks left on the pavement or ground (Figure 15-14)

Figure 15-14. *In a multiple-car accident, skid marks are used to determine the path and speed of each vehicle.*

©David Frazier/PhotoEdit

ACCIDENT RECONSTRUCTION

Just as fingerprints at the scene of a robbery can incriminate a suspect, tire marks at an accident can incriminate a driver. Sometimes the drivers may not recall the exact series of events, or the story may not be consistent with the evidence. Photographs and measurements are recorded to reconstruct the events of an accident. The goal of accident reconstruction is to analyze the accident to help determine what happened, when it happened, where it happened, why it happened, how fast the vehicles were traveling, who was involved, and ultimately, who was at fault.

In hit-and-run situations, the car is gone, but the tire marks leave clues leading to its identity. Tire marks at an accident scene can also provide clues to the speed and direction of the vehicle or vehicles involved in the accident. For example, a head-on collision between an innocent driver and a drunk driver would likely show only one set of skid marks on the correct side of the road leading to the impact site. A quick getaway is noted by a skid pattern left behind when the tires spin during acceleration and leave rubber markings on the pavement.

There are three basic types of tire marks:

1. Skid marks:

- Formed when someone brakes suddenly and lock the wheels.

- Provides evidence of the distance brakes were applied.

- Calculation of velocity can be made from skid marks.

Note the skid marks in Figure 15-15.

Figure 15-15. *Skid marks.*

©Cengage Learning

2. Yaw marks:

- Produced when a vehicle travels in a curved path faster than the vehicle can handle and skids sideways.

- Tires and road surface melt from extreme temperatures.

- Audible squeal and often smoke occurs.

3. Tire scrubs:

- Produced by a damaged or overloaded tire or tires during or immediately after impact.

- Usually curved, irregular in width

- May have striations that look like stripes

- Determine area of impact

Through experience and experimentation, investigators can also estimate speeds of vehicles using the "skid-to-stop" formula. Measuring the weight of the car, the texture of the road surface, and the length of the skid marks, investigators can calculate the approximate speed of the vehicle when the brakes were pressed. This provides information, such as if the car that caused the accident was going over the speed limit. In such a case, more charges could be filed against the driver.

DENTAL IMPRESSIONS

Locard's *principle of exchange* refers to an exchange of materials between a suspect and a victim or a suspect and a crime scene. Occasionally, a perpetrator will leave behind a bite mark. Like fingerprints, bite marks are considered to be individual evidence. Factors that contribute to the individuality of our teeth include the number, size, coloration, alignment, unique fillings, crowns, caps, the distance between teeth, and the overall condition of our teeth. In an older person, the teeth may have a unique pattern of

fillings, breakage, crowns, and caps. Certain antibiotics taken by children have been known to discolor their teeth.

STRUCTURE OF TEETH

The solid, white part of teeth is composed of two different kinds of tissue: a tough covering of *enamel* that protects the living *dentin* tissue underneath (Figure 15-16). Dentin is similar to bone and is composed largely of calcium and phosphorous. Enamel, also composed of calcium and phosphorus, is the hardest substance in the human body, which protects teeth at high temperatures.

Figure 15-16. *Cross-sectional view of a human tooth.*

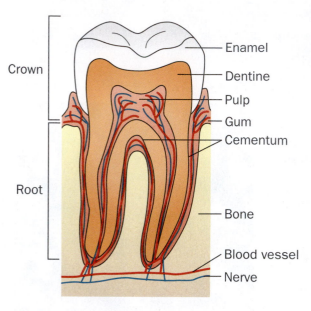

THE DEVELOPMENT OF TEETH

The appearance of 20 primary, baby teeth follows a predictable pattern beginning in the first 7 to 12 months of life. Gradually, the primary teeth are replaced by 32 permanent, adult teeth. The last teeth to develop are the wisdom teeth, which emerge between the ages of 17 and 21. The approximate age of a child can be estimated by viewing the child's teeth (Figure 15-17). An older child may have a mixture of baby and adult teeth. The presence of wisdom teeth usually indicates an age of over 17.

The complete, adult set of teeth encompasses 32 teeth, including wisdom teeth. There are eight incisors at the very front—four on the upper jaw and four on the lower jaw. These are straight teeth that work well in cutting food. The incisors are sandwiched between sharp, pointy canine teeth. There are four canines in total, one on each side of the incisors. Canines are also good for cutting and tearing. Next come eight premolars, two on each side. Premolars are flatter than canines, with ridges on them. There are 12 molars that are even flatter and wider and are involved in chewing and grinding. The shape of a set of teeth (the *dentition pattern*) varies from person to person. Differences in the size of teeth and jaws, position, and crowding make the inside of each person's mouth unique (Figure 15-18).

Figure 15-17. *Pathologists can determine the approximate age of a person from an impression of his or her teeth.*

When teeth "come in"	When teeth "fall out"
7-12 mos.	6-8 yrs.
9-13 mos.	7-8 yrs.
16-22 mos.	10-12 yrs.
13-19 mos.	9-11 yrs.
25-33 mos.	10-12 yrs.
20-31 mos.	10-12 yrs.
12-18 mos.	9-11 yrs.
16-23 mos.	9-12 yrs.
7-16 mos.	7-8 yrs.
6-10 mos.	6-8 yrs.

This chart is a guideline only, children grow at their own pace.

Baby teeth

Age tooth comes in (years)

Central Incisor	7.35
Lateral Incisor	8.45
Canine (Cuspid)	11.35
First Premolar (Bicuspid)	10.20
Second Premolar (Bicuspid)	11.05
First Molar	6.30
Second Molar	12.25
Third Molar	Variable 17 to 21
Third Molar	
Second Molar	11.90
First Molar	6.05
Second Premolar (Bicuspid)	11.20
First Premolar (Bicuspid)	10.50
Canine (Cuspid)	10.35
Lateral Incisor	7.50
Central Incisor	6.40

Adult teeth

DENTAL PATTERNS IN FORENSICS

The individual pattern of teeth is used in forensic investigations in two ways. First, teeth can be used to identify remains, such as those of Adolf Hitler, Joseph Mengele, and the victims of the Waco Branch Davidian Church disaster (1993). Teeth can also be used in profiling and identifying a suspect from unique bite patterns or bite marks left at the scene of the crime.

The bite pattern of a suspect can be matched to the bite marks associated with a crime scene, just as fingerprints of suspects can be matched to fingerprints at a crime scene (Figure 15-19). Up to 76 points of comparison may be used when comparing bite marks, including dental chipping, surface indentations, distances between teeth, individual tooth dimensions, alignment of teeth, and the angle of the mouth arch. The presence or absence of certain teeth can be an indication of age, diet, economic status, and country of origin. Dental procedures and materials may also vary from country to country. All of these factors can provide clues leading to a crime suspect.

If an assailant bites a victim, it is important that the bite marks be photographed while the impression is still visible. The photographs should include a ruler to establish a reference for size to better compare bite marks to a suspect's bite pattern. When an attacker bites a victim, saliva may be left on the victim's skin. If the bite mark is swabbed with a sterile cotton swab, DNA from the saliva may be collected and analyzed. The DNA profile can then be compared to the DNA of suspects.

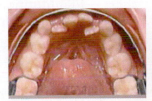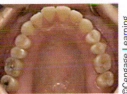

Figure 15-18. *The difference in teeth placement is used to individualize an impression.*

©Cengage Learning

Did You Know?

Dental remains have been studied since the time of Neru (66 A.D.)

Figure 15-19. *Overlaying a transparent tracing of the teeth points with the impression reveal when there is a match.*

©Cengage Learning

Dental impressions on overlay—no match

Dental impressions on overlay—match

SUMMARY

- There are three types of impressions: patent impressions, latent impressions, and plastic impressions.

- Generally, any impression evidence made by an object will be considered class evidence unless it has individualizing features, such as wear or damage.

- Tire impressions at a crime scene can lead to the identification of a vehicle and can provide evidence pertaining to events that occurred before an accident.

- Tire impressions are classified as skid, yaw, or tire scrub impressions.

- Impressions from teeth are considered individual evidence and, like fingerprints, reliability depends on the number of points of comparison and the clarity of the impression.

- Impression evidence must be carefully documented before it is moved. Photographs of the original impression always accompany the cast or record, such as a gel lift, used in court.

Did You Know?

Dental technician James Kim has patented an identification method that is based on embedding information in crowns, bridgework, or other dental work. The dental record may help identify individuals killed in plane crashes, explosions, or other disasters when other identification methods fail.

● Impressions may be used two ways: (1) to identify a person or object and (2) to determine actions that occurred in committing the crime. Identification is usually done by first matching the pattern and then the individual characteristics. A story is told by the series of footprints and/ or tire marks that when carefully observed or recreated can illustrate what actions happened before the crime or accident.

CASE STUDIES

Gordon Hay (1967)

In Scotland, the dead body of 15-year-old Linda Peacock was found with a distinct bite mark on her body. Bite mark impressions were taken from the residents of a nearby boy's detention facility, and a match was found. Gordon Hay became the first person convicted of murder based on a bite mark.

©J.R. Eyerman/Time Life Pictures/Getty Images

Theodore Bundy (1978)

A man wearing a stocking cap entered a Florida State University sorority house and attacked some of the women inside. Two women were killed and two more seriously injured. One of the women had a bite mark that was photographed as evidence. Subsequent attacks followed in other states. Ted Bundy was charged with the Florida State University attacks after his dental impressions were compared to those left on a victim. The FBI's Behavioral Science Unit had profiled Bundy as a very neat, organized, serial killer. Bundy was so meticulous that he never left fingerprints even in his own apartment. Bundy escaped from police twice, only to be recaptured. Bundy was found guilty of murder and was executed in 1989. Before his execution, he implied having committed approximately 50 murders.

Lemuel Smith (1983)

Lemuel Smith had a history of violence, which started while he was a teenager. After spending 17 years in prison for multiple violent crimes, Smith was released. Six weeks later, the bodies of two people were found murdered in the neighborhood where Smith was living. Another rape and murder occurred later the same year. Seven months later, the mutilated body of another female victim was found.

Smith was finally arrested during the kidnapping of another female. A bite mark on the nose of one of his victims matched an imprint of his teeth. In March 1978, Smith confessed to five murders. His defense included that he might be suffering from a multiple personality disorder. Based partly on the bite mark evidence, Smith was ultimately found guilty of multiple murders, kidnappings, and rapes and sentenced to more than 100 years of prison time.

In 1981, in the Green Haven Correctional Facility, a female corrections officer named Donna Payant disappeared. Her mutilated body was found in a Dumpster. This was the first time in the United States that a female corrections officer was killed on duty. On examination, Payant's body showed a bite mark. Dr. Lowell Levine, a forensic odontologist who worked on one of Smith's earlier convictions, recognized the bite mark pattern. Smith was charged, convicted, and sentenced to die.) On a legal technicality, his sentence was changed to life imprisonment. Levine still works at the crime lab in Albany, New York.

Tire Evidence Solves a Murder

In Largo, Florida, tire tracks were found next to a dead woman's body. Police sent photographs of the tracks to Pete McDonald, former Firestone Tire & Rubber Co. tire design engineer. McDonald helps police all over the United States solve crimes by analyzing photographs of tire impressions. McDonald can determine the type, size, and brand of the tire that made the impression, as well as the vehicle that was likely to be fitted with the tire.

In the Florida case, McDonald was able to identify the brand and size of the tire and the likely vehicle. Police checked purchase records for local tire dealerships and found a match. A woman who purchased the tire lived with a man who had recently served time in prison for violent crime. Using the tire impression information, police found a suspect.

The police asked the tire dealer who sold the tire for help in gathering evidence against the suspected murderer. The dealer agreed and called the suspect and told him his tires had been recalled. The dealer offered new tires to replace the old ones. The suspect came to the shop and traded for the new set of tires. The old tires were sent to McDonald for further analysis. The tire impression, combined with other evidence, convinced a jury to convict the man and send him back to prison for murder.

 Think Critically Although they might seem easy to cover up, why might footprints, tire tracks, and bite marks be hard to conceal?

CAREERS IN FORENSICS

Before *CSI*, there was *Quincy*; before *Quincy*, there was Thomas Noguchi

In 1961, Thomas Noguchi emigrated from Japan and was hired as a medical examiner with Los Angeles County. Just a year into the job, he performed the autopsy on Marilyn Monroe. Fame and fortune have followed him ever since. He did the autopsies on Senator Robert F. Kennedy, Sharon Tate, Natalie Wood, and John Belushi. If a case seemed particularly important or perplexing,

AP Photo/Damian Dovarganes

Thomas Noguchi.

Dr. Noguchi was called. Before the age of big forensic labs, Dr. Noguchi was a forensic investigator, going over every crime scene with a fine-toothed comb. With a career as a coroner spanning decades, Dr. Noguchi has numerous stories to tell about how his forensic investigations have helped the dead tell their tales.

In an interview with *Omni* magazine, Dr. Noguchi told of "one homicide I investigated, the homeowner returned early, surprising the burglar, so the burglary ended in murder. But the burglar was hungry, so he had a bite to eat before leaving. We found distinct teeth marks in the cheese!" In another case, Dr. Noguchi examined the body right at the crime scene, to determine if a truck had dumped her, as it appeared. The corpse told Dr. Noguchi otherwise. She had been brought to the site alive, and later shot. Because the murderer had to get out of the van to shoot the victim, the area was searched

for footprints. When the footprints were found, an arrest was made.

One of Dr. Noguchi's claims to fame was that he invented a unique technique to cast a stab wound. Through trial and error, Dr. Noguchi found just what he needed: a substance that was liquid at the boiling temperature of water, but quickly hardened into a solid. He used mercury, which is injected into the stab wound, and five minutes later can pull out a detailed three-dimensional replica of the weapon. You can read more about Dr. Noguchi's cases in his book, *Coroner*, published in 1983, in which he demonstrates one of his favorite sayings, "let the dead speak for himself."

The career of Thomas Noguchi has its own tales to tell. He was fired from his job as coroner for L.A. County twice because of his willingness to openly talk to the media about a case. In each instance, he was reinstated. In fact, Dr. Noguchi has been very outspoken about the consequences of drug use and speaks of drug designers as mass murderers. Dr. Noguchi is now in his eighties, but he continues to write and to educate. He feels death—and its study—is very important. He has said: "There are lessons to be learned from death. And because these death events are repeated over and over again, we must strive to understand them."

Learn More About It

To learn more about careers casting impressions at crime scenes, go to school.cengage.com/forensicscience.

True or False

1. There are two basic types of impressions—patent and latent.

2. Impressions are always considered class evidence only.

3. Before anyone touches or alters an impression, it should be photographed.

4. Electrostatic dusting is used on impressions that are dry and lack depth.

5. Footprints can be recovered from snow.

6. Tire patterns are generic and difficult to trace.

7. The turning diameter of a Volkswagen Beetle is greater than that of a Cadillac sedan.

8. The front and rear track width of a vehicle is always the same.

9. Dental work and dental impressions can be matched to a particular suspect.

10. The presence of wisdom teeth is a partial indication of age.

Short Answer

11. Distinguish between the following measurements on a car:

    a. Track width

    _____

    b. Wheelbase

    _____

    c. Turning diameter

    _____

12. List characteristics of teeth that would enable them to be used as individual evidence as opposed to class evidence.

    _____

    _____

13. What characteristics of shoes are noted when trying to match a shoeprint found at a crime scene with a shoeprint of a suspect's shoe?

    _____

    _____

14. Describe how a Plaster of Paris impression is produced. Include in your answer the role of each of the following:

a. Plaster of Paris

_____

b. Hair spray

_____

c. Ruler

_____

15. List five different characteristics to note when comparing a tire from a suspected tire to the tire mark found at the crime scene.

_____

_____

_____

_____

_____

16. Do further research on imprint or impression evidence in the O.J. Simpson or Ted Bundy cases, or research case studies not mentioned in the text that had their cases resolved as a result of impression or imprint evidence. Submit a written report on your findings.

_____

_____

17. Describe how to calculate the turning diameter of a car.

_____

_____

18. Tire marks are categorized into one of three categories:

a. Skid

_____

b. Yaw

_____

c. Tire scrub

_____

What are the differences among these tire marks?

_____

_____

19. Tire impressions of a tire must be made using the complete rotation of a tire. Explain why this is important if the tire impression is to be compared with a photograph of a tire impression found at a crime scene.

_____

_____

20. Describe how you could distinguish the teeth of an adult from the teeth of a teenager or child.

_____

_____

Bibliography

Books and Journals

Bodziak, W. J. *Footwear Impression Evidence: Detection, Recovery, and Examination,* 2nd ed. Boca Raton, FL: CRC Press, 2000.

Evans, Colin. *The Casebook of Forensic Detection.* New York: John Wiley & Sons, 1998.

Hilderbrand, Dwayne S., and Tia Kalla (illustrator). *Footwear: The Missed Evidence—A Field Guide to the Collection and Preservation of Forensic Footwear Impression Evidence.* Wildomar, CA: Staggs Publishing, 1999.

Johansen, Raymond J., and C. M. Bowers. *Digital Analysis of Bite Mark Evidence.* Indianapolis, IN: Forensic Imaging Institute, 2000.

McDonald, Peter. *Tire Imprint Evidence.* Boca Raton, FL: CRC Press, 1993.

McDonald, Peter. *Tire Imprint Evidence (Practical Aspects of Criminal and Forensic Investigations.* Boca Raton, FL: CRC Press, 1992.

Nause, Lawren. *Forensic Tire Impression Identification.* Ottawa, Canada: Canadian Police Research Center, 2001.

Noguchi, Thomas. *Coroner.* New York: Pocket Books, 1985.

Web sites

Gale Forensic Sciences eCollection, school.cengage.com/forensicscience.

www.fbi.gov/hq/lab/handbook/intro14.htm

www.fbi.gov/hq/lab/handbook/examshoe.html

www.crimelibrary.com/criminal_minds/forensics/bitemarks/4.htm

www.crimelibrary.com/notorious_murders/famous/simpson/brentwood_2.html

ACTIVITY 15-1
CASTING PLASTER OF PARIS IMPRESSIONS

Objective:

By the end of this activity, you will be able to:
Construct a Plaster of Paris impression of shoe or tire prints.

Background:

Shoeprints and other impressions made by suspects can often provide the clues that lead to solving a crime. As well as photographing the shoe or tire print, a permanent impression is produced. These impressions are used as evidence during court proceedings. In this activity, you will make Plaster of Paris impressions using two different methods.

Materials:

(per group of two to four students)

Part A:
1 cardboard box
1 large plastic garbage bag (at least kitchen-sized)
1 empty plastic, 39 oz. coffee can filled with sandbox sand
1 quart-sized resealable bag filled with Plaster of Paris (premeasured)
1 sneaker or boot that shows a tread pattern on the sole
1 empty, small coffee can (any 13 oz. metal or plastic coffee can)
1 can inexpensive aerosol hair spray (to share with other groups)
1 paintbrush
1 pencil or awl or sharpened skewer
1 stiff toothbrush (old ones are fine!)
1 wooden paint stir stick or paddle
12-inch ruler
newspapers
(optional) several tree twigs or wooden cooking skewers
(optional) digital camera

Part B:
all of the materials listed in Part A plus
1 set of Plexiglas strips
4 pieces of duct tape (or clamps)

Time Required to Complete Activity:

Two 45-minute periods: one to make the cast and one to allow casts to dry and to clean the cast impression

Safety Precautions:

1. Handle all materials as directed by your instructor.
2. Wear safety goggles when mixing Plaster of Paris and when using the hairspray. The spray can also irritate some people's throat and nasal passages.

3. Dispose of unused Plaster of Paris in a garbage container; do not dump it down the sink drain!
4. Thoroughly wash all equipment as soon as possible after use. After dumping out any excess Plaster of Paris, allow the remaining plaster to dry. The old plaster can be removed after it dries by pounding on the sides of the plastic coffee can.
5. Work outside if conditions permit. If you are working inside, use newspaper to line any desktops. The sand can scratch the surface of some countertops.

Procedure:

Part A: Preparing a Shoeprint Impression

This procedure allows you to prepare the impression inside the classroom, or you can also take an impression of a footprint left in the sand outside of the classroom.

1. Gather all materials.
2. Line the cardboard box with the plastic garbage bag and place the box on top of the newspapers.
3. Empty the sand from the large coffee container into the bottom of the cardboard box.
4. Using your hand or the paint stir stick, smooth out the surface of the sand in the cardboard box.
5. Step firmly into the sand. Do not rock your foot back and forth. Gently remove your foot from the sand by lifting your foot straight up. View your impression. If it does not seem to be a good impression, smooth out the sand and repeat the process.
6. Hold the hairspray can 12 inches above the impression and spray. If you hold the can too close, the spray will destroy your impression. Spray the entire area of the impression, along with one inch of the sand surrounding the impression. The sand will look wet.
7. Place a ruler next to the print. Use a digital picture to photograph the print and the ruler. Take your photo from directly above the print and perpendicular to it.
8. Fill the small coffee container halfway with tap water and then pour the water into the large coffee can.
9. Slowly add the Plaster of Paris in the plastic bag to the water. Do not dump the plaster into the water all at once. Stir continually as the plaster is being added to the water.
10. Stir until you have a smooth consistency (similar to pancake batter) and until the plaster is dissolved. Do not overstir, because this will cause your plaster to become very thick.
11. Hold the paint stirrer about an inch or two above the footprint. Pour the Plaster of Paris over the paint stirrer and onto the print. (This helps to keep the plaster from splashing into the print and collapsing the print.)
12. If the impression of the shoeprint is small, you will not need to use all of the Plaster of Paris. Add enough Plaster of Paris to cover the impression, but do not overload the impression.
13. Discard the unused Plaster of Paris in the garbage. *Do not dump Plaster of Paris down the sink drain.*
14. Gently smooth out the top of the Plaster of Paris with the paint stir stick, distributing the plaster evenly over the footprint. Twigs can be

added to the plaster at this point to provide additional strength while hardening.

15. Allow the shoeprint to harden for at least five minutes. Use an awl, stick, or pencil to label the shoeprint with your initials, date, and case number.

16. At this point, you can leave the plaster cast until the next day or next class. Do not attempt to lift the print until the plaster has dried. Wait at least 30 minutes. Lifting the cast impressions too early results in the plaster cracking.

17. Gently pry up the plaster cast with your fingers along its edge. Turn it over. It will probably still be damp on the side closest to the sand.

18. Gently brush away the sand with the paintbrush.

19. Use the toothbrush or awl to remove any additional loose sand.

Cleanup:

20. Wipe clean and dry the pencil, awl, stir stick, and small coffee can.

21. Save the resealable plastic bag so that it can be refilled for later use.

22. After dumping out any excess Plaster of Paris from the large coffee can, wipe it with a paper towel. Allow the remaining Plaster of Paris to air-dry. Once the plaster is dry, it is possible to knock out the remaining Plaster of Paris by pounding on the sides of the plastic can. Dry out the large coffee can thoroughly because it will be used to store the sand.

23. Mix sand by hand to break up the hair spray coating.

24. Remove and discard the residual plaster from the sand.

25. Pick up the opposite sides of the big plastic bag (filled with sand) and funnel the sand back into the now-clean, dry 39 oz. coffee can.

26. Dispose of the plastic bag and cardboard box.

27. Shake out the paintbrush and toothbrush to remove any residual sand.

28. Return all lab materials to your instructor.

29. Sweep and clean the area surrounding your setup to remove all sand.

Storage of Castings:

30. Allow the castings to air-dry.

31. Wrap in newspapers or paper towels, and store in shoeboxes.

Part B: Casting Outdoor Shoe Prints, Footprints, or Tire Prints

This procedure works especially well if you plan to cast shoe, foot, or tire impressions outside in their natural environment.

1. Obtain a set of precut Plexiglas strips long enough to cover the area you wish to cast. Tape the outside corners of the Plexiglas using five-inch pieces of tape. If you need to cast a longer section of the tire print to obtain a full rotation of the tire, you can use longer sections of Plexiglas.

The Plexiglas tray.

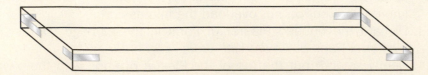

2. This type of frame prevents the tire or shoe print from being disturbed by wind, people, or animals. A second advantage is that this setup can be easily disassembled and conveniently stored. When using the Plexiglas frame, you do not need a cardboard box.
3. Place the plastic bag on level ground or a board to protect it and provide for quick cleanup.
4. Place the rectangular Plexiglas frame on top of the plastic bag and around the impression.
5. Empty the 39 oz. coffee can of sand into the frame. Level off the sand with the paint-stirring stick.
6. Follow the procedure in Part A from Step 5 onward.

Part C: Casting an Impression from a Tire Track
In this part of the activity, you will prepare a cast of a preexisting impression found in the dirt or soil outside of the building. Follow the same directions as in Part B, except:
1. Prepare your frame to fit the size of the tire track. You need a frame long enough to make an impression that represents a full rotation of the tire.
2. Place the rectangular Plexiglas frame around the tire impressions.
3. Gently spray the tire impression with hair spray holding the can at least 12 inches away from the impression.
4. Continue with the rest of the lab described in Part A, Step 6 onward. If the tire impression is long, you will need to mix more than one can of Plaster of Paris.

Questions:
1. Why do crime-scene investigators build a frame around the shoe print before casting?
2. What was the purpose of spraying with the hair spray?
3. Why is it important to brush the excess sand from the plaster casting?
4. Shoe prints can be obtained in the winter from snow, but Plaster of Paris cannot be used. What would happen if you tried to produce a cast of an imprint in snow using Plaster of Paris?
5. Why are footprints easier to cast than tire tracks?
6. Suppose two people both wear size 10, Michael Jordan sneakers. Describe characteristics that could be found in the impressions of these two sneakers that could be used to distinguish which of the two people left his or her shoe print.
7. Describe the print patterns that would distinguish between the following:
   a. an obese person and a thin person
   b. a person shuffling and a person walking normally
   c. a person who is running and a person who is walking
   d. a person walking with an injured right foot and a person walking with both feet uninjured
   e. a person walking who is in a hurry and a person walking at normal speed
8. Suppose a lawyer exhibited a small section of a tire imprint during a trial. Why would the opposing lawyer want to see the tire impression of the entire tire and not just the one section?

ACTIVITY 15-2
SHOE SIZE (FOOT SIZE) AND HEIGHT

Objective:

By the end of this activity, you will be able to:
Explore the relationship between foot length, shoe size, shoe length, and height.

Background:

Shoeprints left at a crime scene have been photographed, and casts have been made of the impressions. What information about the suspect can be obtained from this evidence?

Materials:

1 pen or pencil
1 calculator
1 yardstick
12-inch ruler

Time Required to Complete Activity:

40 minutes (student work in pairs)

Safety Precautions:

Procedure:

Part A: Shoe and Foot Size Comparison

1. Measure the length of your right shoe (or footprint cast) at its greatest length to the nearest one-eighth of an inch. To measure your shoe size, stand next to a 12-inch ruler. Place the end of the ruler next to the end of your heel and measure to the end of your big toe.
2. Record the length of your shoe in Data Table 1. This should be approximately 1 to 1 ¼ inches longer than your foot.
3. Remove your shoe. Measure the length of your right foot at its longest point. (Note that some people's second toe is longer than their big toe.) To measure your foot, stand alongside the ruler.
4. Record the length of your foot to the nearest one-eighth of an inch in Data Table 1.
5. Find your foot length in your textbook, Figure 15-8. What shoe size is given *in the table* for your foot length? Record your answer in Data Table 1.
6. When you purchase shoes, what shoe size do you buy? Record your answer in Data Table 1.
7. Is your actual shoe size and the estimated shoe size found in Figure 15-8 the same? Record your answer.
8. Record your gender in Data Table 1. Record the difference (if any) between your actual shoe size and your estimated shoe size and record your gender. Use + or − signs for shoe size difference. (Actual − Estimated = Difference)

9. Choose a partner of the opposite gender. Repeat Steps 1 to 8 for your partner. Record all answers in Data Table 1.

Part B: Estimating Height Based on Shoe Size

10. Measure your actual height. To do this, remove your shoes. Stand against a wall with your head looking straight ahead. Have your partner measure your height to the nearest half-inch. Record your answer in Data Table 2.
11. Using your actual shoe size and the information found in Figure 15-9, your height is estimated to be how many inches? Record your answer in Data Table 2.
12. Are your actual height and the height according to Figure 15-9 the same? Record your answer in Data Table 2 as "yes" or "no."
13. What is the difference between your actual height and your height according to the table? Record your answer in Data Table 2 as described in Step 8.
14. Repeat Steps 10 to 12 for your lab partner. Record the information in Data Table 2. Record your gender in Table 2.

Data Table 1: Foot Length and Shoe Size

Questions	Your Measurements (inches)	Lab Partner's Measurements (inches)
Length of shoe		
Length of foot		
Shoe size according to Figure 15-8		
Actual shoe size		
Is your actual shoe size the same as size in Figure 15-8 in your textbook? (yes or no)		
Difference between actual size and estimated size (+ or −)		
Gender (male or female?)		

Data Table 2: Shoe Size and Height Comparison

Questions	Your Data (inches)	Your Partner's Data (inches)
Actual height		
Height according to Figure 15-9 in your textbook		
Are your actual height and the graph height the same (yes or no)?		
What is the difference between your actual height and the height according to the graph?		
Gender (male or female?)		

Questions:

Part A: Comparison of Shoe Size to Foot Size

1. Compare your results in Data Table 1 to the results of other members of the class. To make it a more valid comparison, compare all of the male results separately from all of the female results.
2. If you allow a plus or minus one- to two-inch difference, did you find that a majority of the students got their actual shoe size from Figure 15-8 or was the shoe size indicated in Figure 15-8 different from their actual shoe size?
3. List some of the experimental errors that may have been made in doing this study.
4. List ways to improve the reliability of your data.
5. Based on class results, how accurate is Figure 15-8 in estimating someone's shoe size based on his or her foot size? Support your answer with data from your experiment.
6. Why would data collected from teenagers give more variable results than data collected from adults?

Part B: Comparison of Height and Shoe Size

1. Compare all of the results for the males in the class. How accurate was Figure 15-9?
2. Compare all of the results for the females in the class. How accurate was Figure 15-9?
3. List the possible experimental errors in this study.
4. List ways to improve the reliability of your results.
5. Based on class results, how accurate is Figure 15-9, the graph that estimates someone's height based on shoe size? Support your answer using data.
6. Consider another formula to estimate someone's height based on shoe size:

   Foot length (in inches) $\times$ 6.54 = height (in inches)

   Use this formula to calculate your own height:

   Your height based on this formula = _____ inches

   Your actual height = _____ inches

   Is this a more accurate correlation between foot size and height? Why or why not?
7. Based on your observations, can someone's height be estimated by measuring the size of his or her shoes? Explain your answer.

ACTIVITY 15-3
TIRE IMPRESSIONS AND ANALYSIS

Objectives:

By the end of this activity, you will be able to:
1. Produce tire tread patterns from a tire.
2. Describe how to create tire tread patterns using three different methods.
3. Analyze and measure tire treads.
4. Compare a partial tire tread pattern with two complete tire tread patterns.

Background:

Mailboxes were being knocked down and vandalized on Oak Hill Drive. Bicycle tire tracks were identified near all the damaged mailboxes. Two neighborhood teenagers were among the list of possible suspects. Tread mark impressions were made from their bike tires and compared with impressions found at the sites of vandalism. Did any of the tread patterns match?

Time Required to Complete Activity: 40 minutes

Materials:

(teams of two to four students)
petroleum jelly and fingerprint powder *or*
ink pad and small paint roller *or*
inkless pad and paper
newsprint paper or a cardboard box
marker
gloves
scissors
chalk
rulers
at least two bicycle tires on rims or two small tractor tires or lawn mower tires
unknown tire tread sample approximately 0.6 meter in length

Safety Precautions:

Make sure the surface of the tire to be examined is free from hazardous materials (e.g., glass, nails). Wear gloves when working. This activity can be very messy and is best done outside. If working inside, be sure to cover the floor with newspaper or cardboard.

Procedure:

1. Place either a piece of newsprint about 2.5 meters long or cardboard pieces at least 2.5 meters long on the floor. (Cardboard boxes cut into 1-meter lengths and a width of at least 20 cm.)

2. Put on gloves.
3. Select one of the following three methods to produce your tire impression:
   a. Using your gloved hand, smear petroleum jelly over the surface of the tread and brush the tread with fingerprint powder.
   b. Roll the small paint roller over the fingerprint ink pad. Roll the inked roller over the tire to ink the tread of the entire rotation of the tire.
   c. If your tire is small, use an inkless ink pad and roll the tire over the inkless ink pad.
4. Mark the tire with chalk and mark the paper (or cardboard) with a marker as a starting point. Be sure the tire completes at least one entire revolution as it marks the paper or cardboard. Mark the end point of the revolution on the paper.
5. Lay paper or cardboard aside to dry.
6. Label the tire impression paper or cardboard (where you plan to roll the tire) with the following information:
   a. Date
   b. Names of investigating team members
   c. Case number
   d. Tire size
   e. Tire manufacturer
   f. Serial number from tire
   g. Tire placement on vehicle (i.e., front or rear, driver or passenger side)
7. Examine the tire impression. Label all unique identifiers you find on the tire print with a pen. Measure the position of any unique identifiers on the tire print starting from the starting point. Record the distance to the nearest millimeter.
8. Record your information on Data Table 1 (Suspect 1).
9. Repeat Steps 1 to 8 using the second tire. Record your information in Data Table 2 (Suspect 2).
10. Obtain a copy of a partial crime scene tire tread pattern from your instructor; examine it as you did tire samples 1 and 2.
11. Record your answers in Data Table 3 (Crime Scene).
12. Attempt to match the unknown sample to Tire 1 or Tire 2. Six to eight unique identifiers between a tire and tread pattern are considered to be a good match.

Data Table 1: Tire 1 Data (Suspect 1)

Identifier	Location from Starting Mark (cm)	Description

Data Table 2: Tire 2 Data (Suspect 2)

Identifier	Location from Starting Mark (cm)	Description

Data Table 3: Crime Scene Data

Identifier	Location from Starting Mark (cm)	Description

Questions:

1. Did the crime scene sample match Tire 1 or Tire 2? Justify your answer.
2. Why was it necessary to produce a complete rotation of the tire?
3. When viewing tread pattern analysis, what type of information can be obtained from the tread pattern that would help identify a specific car? Explain.
4. Do an Internet search and find tire identification database that could be used to help identify a tire.

Further Study:

Obtain old tires from a garage or landfill. Prepare tire prints that would demonstrate the following:
1. Car that did not have its tires balanced properly.
2. Car that did not have its tires aligned.
3. Car that was driven with underinflated tires.
4. Car that has a damaged tire.
5. Car that was driven without sufficient tread.
6. Car that was driven with snow tires.

ACTIVITY 15-4
VEHICLE IDENTIFICATION

Objectives:

By the end of this activity, you will be able to:
1. Using an actual car, measure the tire width, track width, and wheelbase.
2. Use information in databases to identify cars based on their tire width, track width, wheelbase, and turning ratio.
3. Given information obtained from an accident report, either eliminate or link a suspect's car to the crime scene.

Time Required to Complete Activity:

40 minutes for each part of activity

Materials:

per team (three teams of students; one for each car)
3 tape measures or 3 meter sticks (or yardsticks)
3 laser measuring tools and one square meter of cardboard or wood
3 different vehicles to measure
data sheet

Safety Precautions:

Be aware of traffic patterns in the area where vehicles are being measured. Traffic should be blocked off before students begin their measurements.

Scenario:

An eyewitness said that a young, male driver in a black car ran a stop sign, striking the oncoming car in the middle of the intersection. It appeared that the young man who ran the stop sign was talking on his cell phone at the time of the incident. Brakes were applied but applied too late and the vehicles crashed. In a panic, the young driver of the black car quickly backed up, made a U-turn, and abruptly left the scene of the accident. When the police arrived, the black car was gone. The other car was struck on the driver's side just in front of the driver. All that was left at the crime scene were tire marks. The crime-scene investigators took photographs and began diagramming the crime scene. All tire marks were carefully measured and documented.

With the tire marks and measurements, the crime-scene investigator will be able to help reconstruct the accident. Information about the hit-and-run car can be obtained from the tire mark measurements. Just like fingerprints at the scene of a robbery can incriminate a suspect, tire marks at an accident can also incriminate a driver. Sometimes the drivers may not recall what happened, or the story they tell the police may be somewhat altered to try to protect themselves from getting a ticket or a jail sentence. When accident scenes are reconstructed using tire marks and measurements are made at the scene of the accident, the tire marks do indeed tell a story.

Procedure:

Part A: Measuring a Car's Track Size, Track Width, and Wheelbase

1. Review definitions for tire width, track width, and wheelbase in the chapter.
2. Three different teams will be assigned to three different cars. Each team is responsible for obtaining all of the measurements from their assigned car and sharing their information with other team members.
3. On Data Table 1, record the manufacturer, model, and year of the car that you will be measuring.
4. To measure the front track width, place a board or piece of cardboard in front of the tire. The right edge of the board or cardboard should be aligned directly in front of the tire.
5. Using the cardboard (or board) and the laser pointer or tape, measure the track width of the front tires of the assigned vehicle to the nearest millimeter (see the figure).
6. This measurement can be taken from center to center.
7. Record your information in Data Table 1.
8. Repeat the process for the rear tires, and record the rear track width in Data Table 1.
9. Measure the tire width of a front tire and a back tire on the driver's side. Record your answer in millimeters in Data Table 1.

Measuring track width.

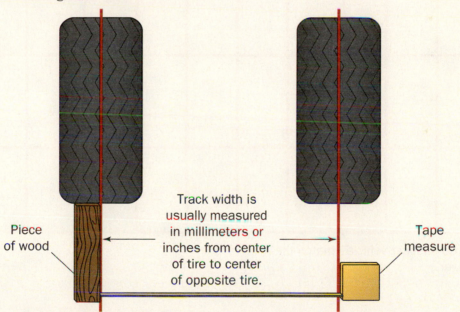

Piece of wood

Track width is usually measured in millimeters or inches from center of tire to center of opposite tire.

Tape measure

The board should be aligned with the middle of the tire.

Front edge of ruler (or laser) aligned with the middle of the tire.

10. Measure the wheelbase of the vehicle. Recall that your measurements are taken from the middle of the front tire to the middle of the rear tire. Record your answer in millimeters in Data Table 1.
11. Record information about the number and patterns of the tire ridge in Data Table 1.

12. Repeat the process for each of the other two vehicles, and record your information in Data Table 1 or obtain the information for the other two cars from the other lab teams.
13. Your instructor has information pertaining to an accident scene that was determined based on the tire marks left at the accident. Record that information in Data Table 1.
14. Compare the data from the accident to the data obtained from the measurements taken of the three suspected cars. Is it possible to eliminate any of the cars based on their measurements? Is it possible to link a particular car to that crime scene?

Data Table 1: Car Measurements

Vehicle Number	Manu-facturer	Model	Year	Front Track Width (mm)	Rear Track Width (mm)	Tire Width Front (mm)	Tire Width Rear (mm)	Wheel-base (mm)	Number of ridges (ribs)	Ridge Pattern (angular, straight, direction)	Other Visual Obser-vations
Car 1											
Car 2											
Car 3											
Evidence Data from Crime Scene											

Do any of the three suspect vehicles match the crime scene?
Explain your answer.

Part B: Linking a Car to a Crime Scene
Try to identify the car using information provided on the pavement at an accident site with the database in Figure 15-13.

1. Wheelbase = ~2689 mm
   Turning diameter = ~11,100 mm
   Tire size = 245
   Vehicle was probably a _____
   Justify your answer.
2. Wheelbase = ~2620 mm
   Turning diameter = 11,102 mm
   Tire size = 195
   Vehicle was probably a _____
   Justify your answer.
3. Wheelbase = 2689 mm
   Turning diameter = 10,739 mm
   Tire size = 195
   Vehicle was probably a _____
   Justify your answer.
4. Wheelbase = ~2835 mm
   Turning diameter = ~11,700 mm
   Tire size = 215
   Vehicle was a _____
   Justify your answer.
5. Wheelbase = ~2408 mm
   Turning diameter = ~10,800 mm
   Tire size = 215
   Vehicle was probably a _____
   Justify your answer.

Questions:

1. When evaluating evidence, which of the following is most significant, and which is the least significant, and why?
   a. Wheelbase
   b. Turning diameter
   c. Tire size
2. Is it possible to eliminate any of the cars based solely on this most significant measurement? Explain.
3. In Part A, did any of the skid marks found at the crime scene match any of the suspects' cars? Explain.
4. If any of the cars' data matched the crime scene, would this be sufficient evidence to convict? Explain.
5. What additional testing might be performed on the suspected car?

Further Study:

Accident reconstruction is an important area in forensics. This area of forensics requires knowledge of physics and math. Explain the role of each of the following:
   Friction
   Kinetic energy
   Acceleration
   Velocity

ACTIVITY 15-5
DENTAL IMPRESSIONS

Objectives:

By the end of this activity, you will be able to:

1. Create a Styrofoam impression of your own bite marks or bite marks created from professional dental castings.
2. Produce a transparency of your bite marks from your own dental impression or from the bite marks produced from the professional dental castings.
3. Match bite marks found on a victim with bite marks from a suspect.

Plaster casts are often made of a suspect's teeth to be used in the courtroom to better see the match between impressions and teeth.

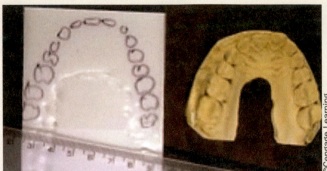

©Cengage Learning

Materials:

(students work in teams of two)
4 Styrofoam bite plates, approximately 6 cm by 7.5 cm
2 hand lenses
tissues
4 transparency sheets (8 × 8 cm)
2 permanent markers
scissors
2 transparencies of bite marks from victim
1 resealable plastic bag (for the garbage)
two sets of professional dental impressions (optional)
2 metric rulers

Time Required to Complete Activity: 45 minutes

Safety Precautions:

Use tissues to wipe any residual saliva from the Styrofoam dental impressions. When the activity is completed, all tissues should be discarded in the trash (or in a resealable plastic bag) to avoid spreading bacteria. Students should wash their hands with soap and water after wiping their Styrofoam dental impressions.

Procedure:

Part A: Making Bite Plates

1. Obtain two equally sized Styrofoam plates from your instructor. The plates must be large enough to be in contact with all of your teeth but small enough to fit into your mouth. If the pieces are too large, cut the Styrofoam plate with the scissors. You need to put in the largest size possible to get a good impression of your back teeth. It will be a tight fit and a little bit uncomfortable!

2. Label one plate *upper* and the other plate *lower*.

3. With the upper and lower plates aligned, place both plates in your mouth at once. Make sure they are placed back far enough to sit between your back molars.

4. Bite down firmly on the plates and gently grind your teeth. Do not chew on the Styrofoam or bite completely through the Styrofoam plates. Remember that all you are trying to do is to get an impression of your teeth.

5. Remove the plates and wipe off any residual saliva with a tissue. Immediately discard the used tissue and place it in a resealable bag.

6. Wash your hands with soap and water to avoid spreading bacteria.

7. Obtain two transparency sheets. Label both with your initials in the upper right-hand corner. In the upper left-hand corner, label one as upper and the other as lower.

8. Examine the bite marks on the Styrofoam plates. Place a transparency sheet labeled upper over the upper impressions.

9. With a permanent pen, outline the pattern made by each of your upper teeth. Be careful to trace each tooth individually.

10. Repeat the process for your lower teeth.

11. Compare your dental transparencies with those of your partner.

Styrofoam bite plates.

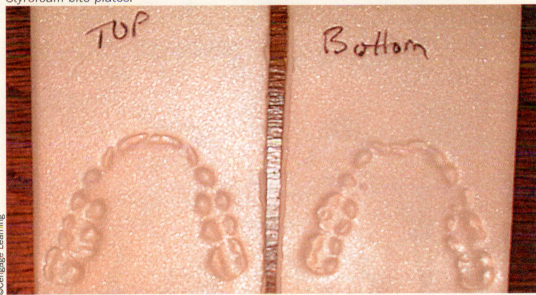

©Cengage Learning

Part B: Identifying Dental Patterns

Compare the photographs of five dental molds made from the upper and lower jaws of five different people.

1. Using a permanent pen and a half sheet of transparency, carefully trace the dental pattern of each of the five upper jaws (1 through 5). The front incisors can be drawn as dashes, while the side and rear teeth can be drawn as irregular circles. Label your transparency sheet with your initials, and label each dental pattern 1 through 5.

2. Using your tracings of the upper jaws (1 through 5), attempt to match the upper jaw with the lower jaws labeled A through E. The lower teeth should fit into the arch of the upper teeth. For most people, the upper teeth will extend beyond the lower teeth. Report

Five sample sets of teeth, upper teeth (top photo) and lower teeth (bottom photo).

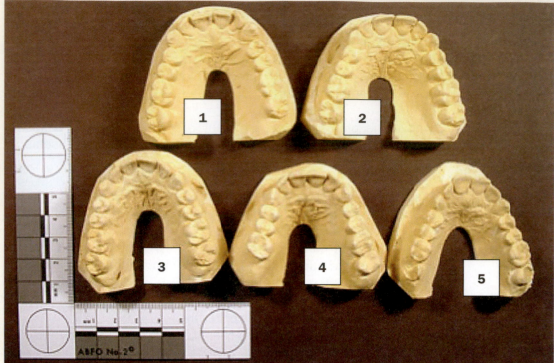

©Cengage Learning

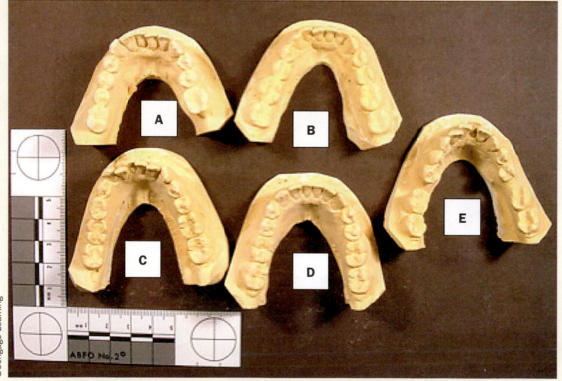

©Cengage Learning

your results on Data Table 1. Check your results with your teacher before proceeding to the next step.

Part C: Bite Mark Identification

3. Compare your five tracings of upper dental impressions with photographs X, Y, Z of bite marks made in cheese by placing the transparency over the photograph. Report your results on Table 2.

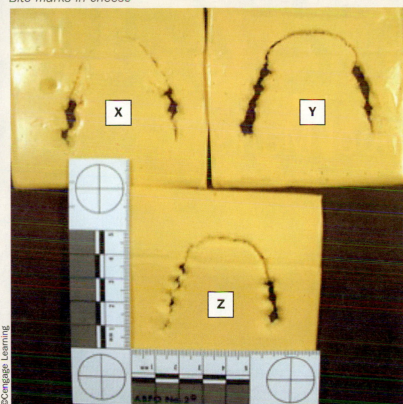

Bite marks in cheese

©Cengage Learning

Data Table 1: Matching Upper and Lower Dental Plates

Upper Teeth (number)	Matches Lower Teeth (letter)
1	
2	
3	
4	
5	

Data Table 2: Bite Marks

Bite Mark (letter)	Upper Teeth (number)
X	
Y	
Z	

Questions:

1. Measurements can be taken of dental patterns and bite marks. List what characteristics make dental patterns unique.
2. Explain how a dental impression provides clues to someone's age.
3. How would obtaining dental records help in the identification of unknown bodies?
4. Besides the use of dental impressions to identify a person, DNA analysis is also being done on teeth. What part of the tooth could contain DNA? Explain your answer.
5. Refer to the diagram of adult teeth in your text and your dental impressions. Based on the structure of the teeth, what do you suppose is the function of each of the different types of teeth: Incisors, Canines, Molars

Further Study:

1. Using the transparencies of your bite marks and those of several other classmates, design a demonstration that requires the use of dental impressions to help solve the crime. Do not actually bite anyone! Someone in your team could bite into a slice of American cheese to provide the evidence bite mark. Others in your team could become some of the suspects. Using the bite marks in cheese and the transparencies of your teammates, demonstrate how the suspect's bite mark can link that suspect to the crime.
2. During wartime, soldiers were each issued their own dog tags. Research these military dog tags and describe:
   a. What are they?
   b. How were they made?
   c. What was their function?
   d. How did they identify the individual soldier?

CHAPTER 16

Tool Marks

TOOL MARKS LINK A CHAIN OF ROBBERIES

A young man watches as the family drives away. Assured that no one is home, he circles to the back of the house, where a fence and several trees protect him from being seen. He removes a crowbar from his duffel bag. After jamming the crowbar between the doorjamb and the door, he gains entry to the home. Ten minutes later, he flees the home, taking with him money, electronics, and jewelry. The experienced burglar used gloves to avoid leaving fingerprints.

When the police are notified, the first thing they do before entering the home is photograph the pry marks on the jamb and door. After photographing and measuring the pry marks, they take a silicon cast of the tool mark impression. It is a perfect replica of the tool down to the smallest detail. The doorjamb around the pry marks is cut away, preserved, and labeled as evidence. All of the evidence is sent to the Police Department's Forensic Identification Section.

Several other burglaries occur over the next month. The same pry bar impression is found at each crime scene. Each time the burglar is able to escape. A break finally occurrs when a vigilant neighbor reports an unknown person repeatedly driving slowly around his

©Florida Images/Alamy

Tools can leave marks that provide physical evidence at a crime scene.

neighborhood. The police pull the nervous stranger over. As they began to ask him questions, they notice his duffel bag open in the back seat. His crowbar is in plain sight. When asked, the man agrees to let police examiners test his crowbar.

A forensic expert compared the crowbar with the tool marks made at each of the crime scenes. The crowbar matched the pry mark impression at several crime scenes. When faced with the evidence of the crowbar match, the man confessed to the series of burgularies.

OBJECTIVES

By the end of this chapter you will be able to

✔ Discuss the significance of tool mark impressions in criminal investigations.

✔ Describe three major types of tool mark impressions.

✔ Describe variations in tool surface characteristics that are used to identify individual tools.

✔ Summarize the steps of a tool mark examination and analysis.

✔ Summarize how technology is helping tool experts in criminal investigations.

✔ Match tool marks with the instrument that produced them.

✔ Describe how tool mark evidence is collected, preserved, and documented.

TOPICAL SCIENCES KEY

VOCABULARY

abrasion mark a mark produced when a surface slides across another surface

cutting mark a mark produced along the edge as a surface is cut

indentation mark a mark or impression made by a tool on a softer surface

tool mark any impression, scratch, or abrasion made when contact occurs between a tool and another object

INTRODUCTION

Did You Know?

Tool marks are usually found at areas of forced entry, such as a window of a house or the steering wheel of a stolen automobile.

Tool marks are a good example of physical evidence found at a crime scene. A **tool mark** is any impression, scratch, or abrasion made when contact occurs between a tool and an object. Two tools made by the same manufacturer may appear to be identical, but on microscopic examination, they can be distinguished and identified. These mass-produced tools have minor differences that can differentiate one tool from another. The impressions made by these tools could link the tool to a crime scene and ultimately to the tool's owner.

TOOLS AND CRIME SCENES

Tools simplify everyday living. They serve as extensions to our hands, multiplying our ability to do work. Simple tools such as pry bars, knives, screwdrivers, and hammers increase our ability to handle manual tasks. But the same tools that help us with everyday tasks can also be used in crimes. Criminals use tools to force their way into locked buildings and cars (Figure 16-1).

Figure 16-1. Tool marks are often found at areas of forced entry.

©Gary Roebuck/Alamy

The tool used in a crime may lead to the criminal. Ownership of the tool used in the commission of a crime, however, is only circumstantial evidence. The suspect's fingerprints on the tool make for a stronger case, but still do not eliminate the possibility of someone else using the identified tool at the crime scene. This information alone is circumstantial evidence and will not lead to a conviction, but used with other evidence, it might be enough to convict a suspect.

TOOL MARK IMPRESSIONS

Tools usually leave distinctive marks where they are used. The hardness of a tool influences the resulting marks left in the softer object. As shown in Figures 16-2, 16-3, and 16-4, there are three major categories of tool marks: indentation marks, abrasion marks, and cutting marks.

INDENTATION MARKS

Indentation marks are marks or impressions made by a tool when it is pressed against a softer surface (Figure 16-2). This mark is the negative impression of the tool, such as a nick or depression on a surface made by a screwdriver or a crowbar attempting to pry open items like a cash box, window, or door. It is possible to measure the impression to determine the size and shape of the tool that was used, as well as other characteristics of the tool.

ABRASION MARKS

Abrasion marks are made when surfaces slide across one another (Figure 16-3). Objects such as pliers, knives, axes, or a gun barrel make this type of mark. The harder surface leaves scratch marks or striations on the softer surface. Sometimes abrasion and indentation marks are made at the same time, such as a pry bar scratching one side of a doorjamb while leaving an impression on the wood on the other side.

CUTTING MARKS

Cutting marks are produced along the edge as a surface is cut (Figure 16-4). Saws, wire cutters, and other devices that cut through surfaces like wire, bolts, hinges, locks, and bones leave cutting marks.

Cutting Marks on Bones

The type of saw blade used to dismember a body can be determined by examining the cut surface of the bone. In his book entitled, *Dead Men Do Tell Tales,* Dr. William Maples describes a collection of cow bones belonging to the C. A. Pound Human Identification Lab in Florida. Cuts made by various saws were used to prepare a reference catalog of cut marks. Each type of saw leaves its own unique marks on the bone, as shown in Figure 16-5. Microscopic examination of the saw marks on the bone provides clues to what type of blade was used. At times, a body might have two different sets of saw marks when a criminal switches saws because the first one isn't working well or is too messy.

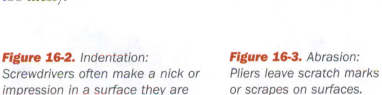

Digging Deeper
with Forensic Science e-Collection

Do more research on the work of forensic anthropologist William Maples using the Gale Forensic Sciences eCollection on school.cengage.com/forensicscience. William Maples wrote about his work identifying tool marks found on bones. Find out how Maples evaluated the marks, such as holes and fractures, on the bones he examined. How are tool marks found on bones similar to rifling marks found on bullets? Bones left outside for a period of time may be eaten by animals. Discuss how scientists try to distinguish between marks made from animal activity and those made from weapons.

Figure 16-2. Indentation: Screwdrivers often make a nick or impression in a surface they are used on.

©ACE STOCK LIMITED/Alamy

Figure 16-3. Abrasion: Pliers leave scratch marks or scrapes on surfaces.

©WoodyStock/Alamy

Figure 16-4. Cutting: Hack saws leave chopping marks on a surface of contact.

©AP IMAGES

Figure 16-5. Saw marks on bones.

Type of Saw	Cut Characteristics	Teeth Mark Pattern	Example
Stryker	Circular areas of short radius; some overlapping marks	Few teeth marks	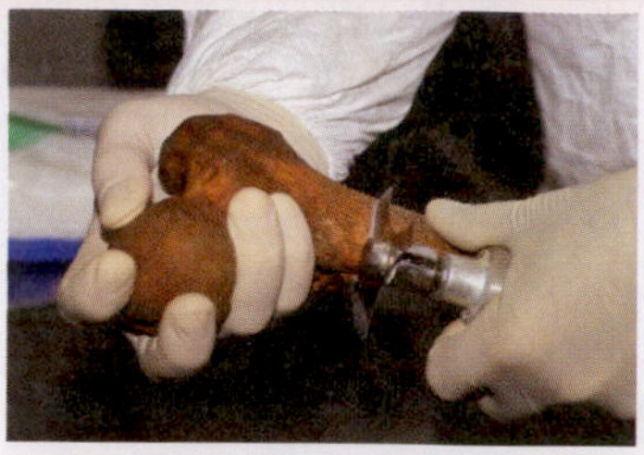 ©Steve Turner / Alamy
Band saw	Very smooth cut	Few teeth marks; straight fine cut; seldom overlapping marks	 ©David Young-Wolff/ PhotoEdit, Inc
Hack saw	Overlapping marks	Look like a tiny tic-tac-toe board with thousands of squares	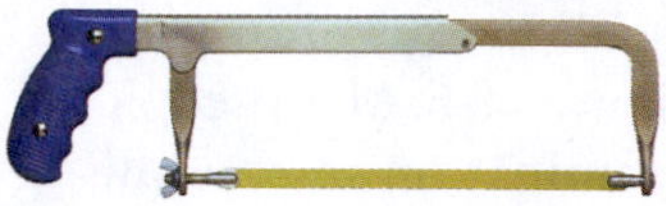 ©Image Farm Inc./ Alamy
Chain saw	Blade goes directly through bone; messy cut	Roughened edge	 ©Burke/Triolo/Brand X Pictures/Jupiterimages
Table saw	Parallel, curved striations	Ridge grooves	 ©Steve Skjold/Alamy
Handsaw	Rough cut with overlapping marks	Irregular cut	 ©Burke/Triolo/Brand X Pictures/Jupiterimages
Circular saw	Parallel curved striations	Ridged grooves	 ©Burke/Triolo/Brand X Pictures/Jupiterimages

TOOL SURFACE CHARACTERISTICS

Tools have unique characteristics resulting from manufacturing processes and from use over time. These characteristics help investigators differentiate one tool mark's impression from another. Tools change over time as they are used repeatedly, as shown in the three hammers in Figure 16-6. Nicks, ridge marks, and blemishes may develop on the striking surface of the tool. These unique characteristics from natural wear and tear can affect impressions made by the tools, making them unique as well.

Other factors that help make tool mark impressions more recognizable are oxidation or rusting of tools and uneven sharpening.

TOOL MARK EXAMINATION

All crime-scene investigators approach an investigation methodically and carefully. When a crime scene contains tool mark evidence, tool mark experts, as seen in Figure 16-7, are called in for evidence collection and preservation. Tool mark evidence includes the mark left at the scene as well as the tool. The tool may be recovered from the suspect or may have been left at the crime scene or discarded elsewhere. The expert may or may not have the benefit of finding a tool used during the crime. In that case, the investigator will search for indirect evidence of any tools used during the crime, such as tool mark impressions. Any evidence found is photographed and preserved to create a permanent record of the evidence in its original state of discovery. This record of evidence may be introduced at trial.

DOCUMENTING THE EVIDENCE

The best way to document tool and tool mark evidence is to use photography. All evidence is photographed with a measuring device to show the appropriate scale for reference.

Figure 16-6. *Unique markings such as nicks and ridge marks can be seen on the striking surface of three hammers.*

Figure 16-7. *A tool mark expert at work.*

©Cengage Learning

SAM YEH/AFP/Getty Images

Photographing Tools

When photographing a tool, experts focus on any scratches or gouges on the surface. Oblique lighting is preferred to direct lighting because it casts shadows and highlights details that are not easily visible under direct lighting. Magnesium smoking is often used on dark-colored tools. In this method, magnesium ribbon is burned, which produces a white, powdery film of magnesium oxide that coats the surface of the tool. Coating the tool highlights the detail during photography.

Photographing Tool Marks

While photographing and recording the tool mark evidence, the expert searches the surface of the tool mark for bits of foreign material using oblique lighting. If the surface is painted, the expert records if any paint was chipped away and provides a description. Photographing the tool mark and making an impression of it are common ways to further analyze the evidence. Impression marks are collected and preserved by the method of casting. Once the cast is ready, it is removed and smoked with a layer of magnesium oxide for photography. The expert also includes a ruler or other measuring device to indicate the scale of the tool mark.

CASTING IMPRESSIONS

If it is possible, tool mark evidence should be collected and preserved for analysis. The crime scene investigator may actually cut a part from the door or door jamb that contains tool mark evidence. If this is not possible, casting is a commonly used method of preserving impressions of the tool marks. The cast impression will retain the unique indentation marks made by a specific tool. A variety of silicone or rubber-based casting materials can be used to record impression marks, as shown in Figure 16-8. Different materials may work better according to the climate. The liquid casting material generally takes about 10 to 30 minutes to dry and solidify.

Figure 16-8. *Types of casting material.*

Material	Description
AccuTrans auto-mix casting system	Silicone base material applied by extruder gun
Mikrosil casting material	Putty that requires a separate catalyst to harden; applied by spatula
DuroCast	Compound that requires a separate catalyst to harden; applied with a spatula
Liquid silicone	Applied by extruder gun or from tube
Room-temperature silicone vulcanizing rubber	Silicone mold rubber; requires a separate catalyst to harden at room temperature

Surfaces should be examined before casting material is applied to determine if the area should be dusted for fingerprints. The use of magnetic dusting powder and silicone material allows for fingerprints to be lifted from the surface before impression marks are cast. In addition, the size of the impression should be measured and recorded. This provides a permanent record of the size and surface appearance.

COLLECTING AND PRESERVING A SAMPLE

All evidence is carefully transported to the forensic laboratory to protect it from damage. When tools are found, they are collected and packaged separately in containers or boxes and then taken to a laboratory for analysis. The object containing the tool mark or the actual impression cast is also packaged in a container or box and taken to the laboratory. Small objects may be wrapped with clean paper and placed in small containers or plastic bags, while larger objects can be packed in cartons or boxes.

All evidence must also be correctly labeled and processed. Important information that must be recorded includes case number, who collected the evidence, a signature, where it was found, when it was found, and why the evidence was collected. The proper chain of custody must be maintained.

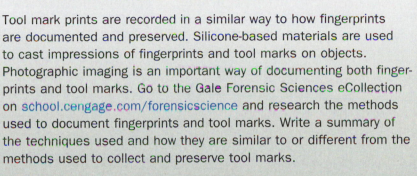

Digging Deeper
with Forensic Science e-Collection

Tool mark prints are recorded in a similar way to how fingerprints are documented and preserved. Silicone-based materials are used to cast impressions of fingerprints and tool marks on objects. Photographic imaging is an important way of documenting both fingerprints and tool marks. Go to the Gale Forensic Sciences eCollection on school.cengage.com/forensicscience and research the methods used to document fingerprints and tool marks. Write a summary of the techniques used and how they are similar to or different from the methods used to collect and preserve tool marks.

ANALYZING TOOL MARK EVIDENCE

The goal of a laboratory tool mark analysis is to identify major characteristics that define the type of tool used in the crime. The analysis also aims to identify unique characteristics, such as nicks and blemishes, that might distinguish between the same kinds of tools. If a tool mark's characteristics match a suspect tool, the mark will be examined further using advanced forensic comparison microscopes.

If tools are found in the suspect's possession, they may be used to create a new set of tool marks to be compared to those found at the crime scene. Crime scene tool marks and suspect tools are never fitted together to show a match. This would damage the integrity of the evidence. Instead, impressions from the crime scene and impressions made from the suspected tool are compared in the laboratory.

Another important feature of tools used in identification are serial numbers. A serial number is a unique number assigned to an object for identification purposes. Many tools are stamped or printed with these numbers. For quality control, the manufacturer commonly assigns them, making it easier to identify a batch of tools if a defect is found. Investigators use serial numbers to trace a tool to the manufacturer and even to the store where it was purchased. Criminals often try to remove serial numbers. Recovering the numbers has become an important technique for investigators.

NEW TECHNOLOGY IN TOOL MARK IDENTIFICATION

Forensics experts have traditionally used visual pattern recognition to compare marks found at crime scenes with those made with suspect tools. This

method was attacked in a famous 2000 Florida court case that established knife blade tool mark evidence to be inconclusive because of inadequate scientific testing. In response, research projects at the Department of Energy's Ames Laboratory at Iowa State University began developing new technology to scientifically determine the uniqueness of tool marks.

Research projects at the Ames Laboratory include building a tool mark image database and developing an algorithm to statistically analyze the images. A forensic comparison microscope acquires the images. The image database includes many tools, such as screwdrivers, pliers, wire cutters, bolt cutters, tin snips, wood chisels, and pry bars.

Another research project at the Ames Laboratory involves the use of three-dimensional characterization methods and statistical methods to distinguish tool marks. The researchers use a *profilometer,* a scanning tool that measures the depth or height of tool marks. It then uses the information to create a contour map of the marks from the scan to precisely identify a tool mark. This new technology allows forensic specialists to match unique marks on a tool to the marks made by a tool at the crime scene.

TOOL MARK EVIDENCE IN THE COURTROOM

After tool mark comparisons are scientifically analyzed, the expert tool mark witness prepares a written report to present to a jury. Original evidence, such as a tool or an actual tool mark, is presented in court whenever possible. However, casting and magnified images of tool mark comparisons are presented at a trial. Such evidence may be used to link a series of crimes. As tool mark analysis technology continues to advance in the future, forensic experts will be even better able to help the courts convict criminals.

SUMMARY

- Tools have major and minor surface differences that can help differentiate one tool's impression from another.

- Tool marks are categorized into one of three categories: indentation marks, abrasion marks, and cutting marks.

- The marks made by tools can link a tool to a crime scene and ultimately link the owner of the tool and potential suspect in a crime.

- Tool mark evidence should be photographed, documented, collected, or cast.

- Advancing technology for distinguishing tool marks is helping forensic experts help the courts fight crime and find criminals.

Richard Crafts (1986)

Pan Am stewardess Helle Crafts disappeared after she returned home from a trip to Germany. Her body could not be located, and her husband Richard Crafts, a potential suspect, passed a lie detector test. The police investigated and found that her husband had purchased a new freezer the week before Helle's disappearance and rented a wood chipper a few days before Helle's return from Germany. An eyewitness remembered seeing a man using a wood chipper along the banks of the Housatonic River early one morning. The date was November 19, two days after Helle's return from Europe.

©Susan Van Etten

The police systematically examined the area along the river and the river itself. They found more than 50 pieces of human bone, including parts of a finger, several thousand strands of hair similar to Helle's, two tooth caps, and several ounces of human tissue. A chain saw with its serial numbers filed off was found in the river. Identification of the body was based primarily on one of the recovered tooth caps. It was positively matched to dental records of Helle Crafts. Police theorized that Richard was facing a financially disastrous divorce and had decided to kill Helle upon her return from Germany. After killing her with a blunt object, Richard placed her body in the freezer. He then cut her body into small pieces and fed her remains through the wood chipper and into the river. Richard Crafts was convicted of murder and sentenced to a term of 50 years in prison.

William Maples's *Dead Men Do Tell Tales* (1995)

William Maples was one of America's greatest forensic anthropologists and used tool marks to help solve grisly murders. In the book *Dead Men Do Tell Tales*, Dr. Maples describes a case involving the discovery of a headless body. Bite marks on the remains of the body were identified as those belonging to sharks. Maples determined from fine saw marks on the victim's neck vertebra that the head had been sawed off before the sharks had attacked the body. From the pattern made on the bone, Maples was able to identify a hacksaw as the tool used to separate the head from the body.

Think Critically Explain how shoe prints and tool marks are similar types of evidence. Be sure to mention both as class and individual evidence.

Dr. David P. Baldwin and Colleagues, Forensic Scientists and Tool Mark Experts

Dr. David Baldwin has been the director of Ames Laboratory's Environmental and Protection Sciences Program since 1999. In addition, he has been the director of the Midwest Forensics Resource Center (MFRC) since 2002. This center assists midwestern state crime labs, universities, the Federal Bureau of Investigation, the Department of Justice, the Department of Energy, and the Bureau of Alcohol, Tobacco, Firearms, and Explosives. The MFRC is focused on forensic science training, education, and research. One of the MFRC's major goals is to develop new techniques and to expand technology for conducting forensic science work. By providing scientific equipment for analyzing crime-scene evidence, the MFRC hopes to help forensic investigators in crime laboratories and law enforcement agencies solve crimes. Baldwin believes that "Through our casework-assistance program, we make available to our partners experts and instrumentation that they don't have in their own crime labs."

David Baldwin has extensive training in analytical chemistry and instrumentation, and he has worked in several different areas of environmental chemistry research. His work, and that of his collaborators in forensic research, includes developing ways to analyze tool mark uniqueness and statistical tests to analyze tool mark individuality. One of the methods developed by Baldwin and Ames Lab senior chemist Sam Houk is a new type of mass spectroscopy to identify metals, ceramics, and other materials commonly found in tools.

Another Iowa State University professor of materials science and engineering, Scott Chumbley, is using three-dimensional characterization methods and statistical methods to distinguish tool marks to match crimes to criminals. Chumbley uses a scanning device called a profilometer to accurately measure the height or depth of tool marks and then develop a three-dimensional contour map and profile of each tool.

A technician prepares to analyze tool marks.

AP IMAGES

Todd Zdorkowski works alongside David Baldwin as the associate director of the MFRC. Work in forensics is not as glamorous as it appears on TV, says Zdorkowski. He says, "The backlogs are huge and always growing. Most analytical procedures must be done with a high degree of analytical precision and careful attention to every detail associated with handling the evidence and performed on deadline." He goes on to say, "Every examiner's work must be inspected and corroborated by another examiner before it can be released, and some examiners also find their work challenged in public."

A specialized degree in tool mark examination does not currently exist. Many investigators begin with a degree in the sciences, specifically chemistry and chemical engineering, for proper training in scientific analysis and using chemistry instrumentation. Skills and expertise in forensics and tool mark analysis are often acquired through graduate degrees, job experience, and certified training programs.

Learn More About It

To learn more about the work of a tool mark expert, go to school.cengage.com/forensicscience.

True or False

1. The cut edges of a board show abrasion marks.

2. Hammers can create indentation marks.

3. The cutting pattern of a used hacksaw blade is the same as a Stryker saw.

4. Rust, chips, and dents on a hammer can produce unique tool mark impressions.

5. Tool marks are considered to be circumstantial evidence.

6. If a tool mark impression is matched to a specific tool, then the owner of that tool must have been at the crime scene.

7. Pry bar and crowbar marks cannot be distinguished from each other.

8. New screwdrivers of the same size and brand always leave the same tool mark impression.

9. In court, a lawyer will fit the tool into the tool mark impressions to demonstrate a match.

10. Tool impressions can be cast using silicone-based materials.

Short Answer

11. Two different tools were used to strike a piece of board. Two impressions were made in the wood. Assuming the same person used the tools to make the impressions, list what characteristics could be used to help distinguish one tool from the other.

12. Distinguish between the following types of marks found on a surface: abrasions, cuts, and indentations.

13. List the steps taken when collecting and preserving tool mark impression evidence.

14. Summarize the importance of technology in tool mark examination and analysis for the courtroom. Include at least two different technologies in your summary.

Connections

Mathematics Forensic scientists are currently creating tool mark databases. How can they be analyzed using mathematical methods?

Further Study

Research and report on the significance of tool mark impressions in solving the Lindbergh kidnapping case.

Bibliography

Books and Journals

Cannon, Roger J. "Justice for All," *Security Management*, 34(3): 63, March 1990. From *Forensic Science Journals.*

Du Pasquier, E., J. Hebrad, P. Margot, and M. Ineichen. "Evaluation and comparison of casting materials in forensic sciences applications to tool marks and foot/shoe impressions," *Forensic Science International,* 82(1): 33–43, September 15, 1996. From *Forensic Science Journals.*

Evans, Colin. *The Casebook of Forensic Detection.* New York: Berkley Trade, 2007.

Fisher, B. *Techniques of Crime Scene Investigation.* Boca Raton, FL: CRC Press, 2000.

Maples, William, and Michael Browning. *Dead Men Do Tell Tales: The Strange and Fascinating Cases of a Forensic Anthropologist.* Jackson, TN: Main Street Books, 1995.

Sehgal, V. N., S. R. Singh, M. R. Kumar, C. K. Jain, K. K. Grover, and D. K. Dua. "Tool marks comparison of a wire cut ends by scanning electron microscopy—A forensic study," *Forensic Science International* 36(1–2): 21–29, January 1988. From *Forensic Science Journals.*

Web sites

Gale Forensic Sciences eCollection, school.cengage.com/forensicscience.

http://wi.essortment.com/forensicinvesti_rdxu.htm

http://www.firearmsid.com/Case%20Profiles/ToolmarkID/toolmark.htm

http://www.forensicmag.com/articles.asp?pid=52

http://www.channel4.com/science/microsites/S/science/society/forensic_marks.html

http://www.forensischinstituut.nl/NFI/en/Typen+onderzoek/Items/Examination+of+surface+marks+impressions+and+shapes.htm

http://www.fbi.gov/hq/lab/handbook/intro15.htm

http://www.crime-scene-investigator.net/otherimpressionevidence.html

http://www.ameslab.gov/final/News/2004rel/toolmarks.htm

http://www.abc.net.au/science/news/stories/s1089160.htm

http://www.external.ameslab.gov/forensics/Baldwin.htm

http://www.external.ameslab.gov/news/insider6-01forensics.htm

http://www.iastate.edu/Inside/2003/0314/forensics.shtml

ACTIVITY 16-1
TOOL MARKS: SCREWDRIVERS AND CHISELS

Scenario:

A break-in has occurred. Bud and Arthur are both suspects. A screwdriver was used in the course of a burglary, and pry marks were left behind. Can you identify which screwdriver was used in the crime?

Objectives:

By the end of this activity, you will be able to:
1. Analyze the photographs of the tools to detect any individualizing features.
2. Measure tool in the photographs accurately.
3. Determine which tool produced the tool mark impression.

Time Required to Complete Activity: 30 minutes

Materials:

(per group of two)
pictures of tool marks (this activity)
6" ruler or caliper
hand lenses

Safety Precautions:

None

Procedure:

1. Examine the photos of each of the tool marks pictured on page 482.
2. Note any special identifying characteristics that would make a screwdriver or chisel unique and easy to identify, such as size, shape, and unique markings.
3. Measure the blade of the screwdriver or chisel along its tip and at its widest point using a metric ruler or caliper. Record information on the table provided.

Questions:

1. List the characteristics that helped you distinguish the different types of tools.
2. The tool mark impression that matched the crime scene was number _____ owned by _____.
3. Would there be sufficient evidence for an arrest? Why or why not?

Crime scene tool marks

1

2

3 chisel

4 chisel

5

6

7

8

9

10

11

12

13

14

15

16

Art's tools

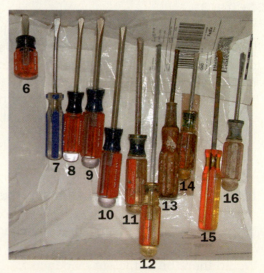

Bud's tools

4. Viewing photographs of the tools and tool mark impressions limits the analysis. If you had the actual tool and the tool impression or were able to view a three-dimensional casting of the impression, what other measurements and observations would you be able to make?

5. Provide an example of one type of technology that will provide a more extensive examination of tools and tool marks.

©Cengage Learning

Data Table: Screwdriver and Chisel Comparison

Screwdriver	Description of Tool Mark	Width at Tip (mm)	Width at Widest Point (mm)
1			
2			
3			
4			
5			
6			
7			
8			
9			
10			
11			
12			
13			
14			
15			
16			
Crime Scene			

Further Research:

Instead of working with photographs, use four different screwdrivers to prepare tool impressions. Select one screwdriver to be the crime-scene tool and prepare an impression on a piece of soft wood (like pine) to serve as crime-scene evidence. Take photographs and measurements, and prepare a cast of the tool impressions. Prepare your testimony as an expert witness on tool marks. Include measurement charts of four screwdrivers and the crime scene photographs along with casts to support your testimony. Be prepared to present evidence to the class and defend your findings.

ACTIVITY 16-2
TOOL MARKS: HAMMER MATCH

Scenario:
Hammers have been collected from five suspects. Can you match the hammer to its strike mark?

Objectives:
By the end of this activity, you will be able to:
1. Analyze the photographs of the tools to detect any individualizing features.
2. Record the measurements of the hammers' diameters by using the photographs provided.
3. Determine which tool produced the tool mark impression.

Time Required to Complete Activity: 15 minutes

Materials:
(groups of two students)
diagrams provided
hand lenses

Safety Precautions:
None

Procedure:

Part A: Measurement of Hammer
1. Using photos (Set A) on the next page, answer the following questions. Record the diameter of each hammerhead after reading the outside jaw measurement of each of the calipers. Record your answer to the nearest 0.1 mm. You may want to use a hand lens when reading the scale. Record your measurements in the Data Table.

Data Table: Hammer Width

Hammer	Width (mm)
A (Orange Ring Neck)	
B (Two Bands)	
C (Silver Neck)	
D (Baby Red Handle)	
E (Large Uncle Waldo)	
Casting from crime scene impression = 41 mm	

Read the calipers to measure each of the hammerhead widths

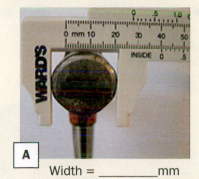

A
Width = _____mm

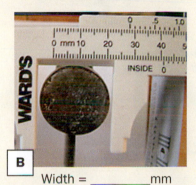

B
Width = _____mm

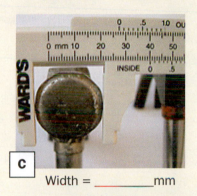

C
Width = _____mm

D
Width = _____mm

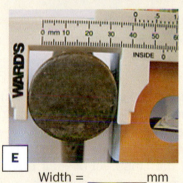

E
Width = _____mm

©Cengage Learning

2. It was determined that the murder was committed with a hammer whose head measured approximately 41 mm. Do any of the evidence hammers seem to match the impression found at the crime scene?

Part B: Hammer Photographs

1. Photographs of five hammers collected from five different suspects are shown below. Examine each photograph. They are not to scale.
2. Compare each photograph to the casting made of the crime-scene hammer impression.
3. Determine which hammer is a match and circle the areas that correspond between impression and suspect hammer.

Evidence impression made from a crime scene.

©Cengage Learning

Questions:

1. In Part A, which hammer is a crime-scene match? Explain your answer.
2. In Part A, a student made an error when measuring one of the hammers. Identify the photograph with the error and explain what mistake was made.
3. One student in the class got different measurements from the rest of the class. All of their measurements were between 35 mm and 54 mm. Can you describe the source of the error?
4. In Part B, which hammer photograph matches the cast impression? Justify your answer.
5. The head of the crime scene lab was not satisfied with the quality of the photographs of the hammerheads. What suggestions would you give to the photographer to improve his or her evidence collection techniques?
6. A hammer matched the impressions made at a crime scene. Why is this insufficient evidence for conviction?
7. The prosecutor tried to convince the jury that the defendant was guilty because he owned a hammer that produced an impression similar to that found at the crime scene. As the defendant's attorney, how would you argue that matching a tool to crime-scene impressions was insufficient evidence to convict?

ACTIVITY 16-3
HAMMER STRIKES ON WOOD

Scenario:

Police received a report of a construction site break-in. The specifics of the case suggested that it might be an inside job, involving one of the site's employees. The break-in involved a locked wooden toolbox that had clearly been broken with a hammer. The toolbox had contained cash. There was no hammer located at the site of the break-in, and it was assumed the thief used a personal hammer. Police collected all of the site's workers' hammers for testing. Your task is to produce cast impressions for comparison to the tool mark evidence found at the crime scene.

Objectives:

By the end of this activity, you will be able to:
1. Produce hammer impressions in wood from eight different suspects' hammers.
2. Compare the hammer impressions to the crime-scene impression.
3. Determine which hammer could have been used at the crime scene.

Time Required to Complete Activity: 45 minutes

Materials:

(groups of three to four students)
8 hammers of varying size (for all groups to share) labeled 1 through 8
1 strike board (per group)
labeling pen
calipers or metric ruler
crime-scene board with hammer impression
silicone casting material
digital camera (optional)
9 three-by-five-inch cards
1 roll of adhesive tape
9 evidence bags
9 evidence labels
small flashlight
forceps

Safety Precautions:

Wear safety goggles during this activity. Care must be taken when pounding with hammers to avoid contact with other students. Students should be warned that any unsafe act would lead to their removal from the activity.

Procedure:

Part A: Cast the Crime-Scene Impression

1. Using the silicone casting material, prepare a cast of the crime-scene impression. Apply sufficient silicone to completely fill the indentation.
2. Allow the silicone to harden thoroughly before removing (about 15 minutes).
3. Let the impression dry while you are making the strike impressions in Part B.
4. Note any distinctive markings in the silicone casting and measure the casting. Record your observations on the Data Table.

Part B: Make Strike Impressions

5. Obtain your strike board.
6. Divide and number the board into eight areas. Number the areas 1 through 8.
7. Obtain each of the eight different hammers and make a single hammer impression on the strike board in the appropriate place on the board. The same person should make all strike marks with the same amount of force.
8. (Optional) Photograph each impression with a caliper or metric ruler in place using a flashlight held at an oblique angle to enhance the shadowing, thereby improving the photograph.
9. Examine each of the strike marks with a flashlight held at an oblique angle. Use the adhesive tape to lift and secure trace evidence to the 3" x 5" card. Label each card with the number of the strike mark from which the evidence was recovered.
10. Record any trace evidence found, and using forceps, properly bag and label the evidence. (Optional) Photograph any trace evidence before removal.
11. Using your calipers, measure to the nearest 0.1 millimeter of each impression, and record in Data Table.

Part C: Comparisons of Crime Scene Casting to Hammer Mark Impressions

12. Compare the measurements from your casting to the measurements of your hammer strikes. Is there a match?
13. Prepare a report of your findings to a jury of your classmates. Support your conclusions using data (and photographs) or sketches of the hammer in question and the crime-scene castings made.

Data Table: Hammer Impressions

Hammer Impression	Diameter (mm)	Any Trace Evidence?	Other Observations
1			
2			
3			
4			
5			
6			
7			
8			
Crime scene			

Questions:

1. How could you improve the reliability of your results?
2. What characteristics of the hammer impression were used to match a specific hammer to the crime-scene impression?
3. What additional evidence could be collected that would strengthen the prosecutor's case?

Further Research:

Devise an experiment to determine the effect of a person's size or strength on the clarity of an impression made in a soft pine board. Include in your experiment the following:

a. question to be tested (what do I want to know)
b. hypothesis (what you think is the answer, as an "if, then" statement)
c. design of the experiment (how to do it)
d. what data will be collected and reported
e. analysis of the data collected (what does it mean)
f. conclusion (what have you found)

CHAPTER 17

Ballistics

THE WASHINGTON, D.C., SNIPER KILLINGS

Beginning in October 2002, a series of sniper murders terrorized residents of the Washington, D.C., area. Ten people were killed and three others wounded. At several different crime scenes, the police were able to collect shell casings and recover bullet fragments from the victims. Investigators determined that most of the shootings were related to the same 0.223-caliber firearm. Police apprehended two suspects—John Allen Muhammad, 41, and John Lee Malvo, 17—and discovered a rifle in the suspects' car. The recovered crime-scene evidence matched this rifle.

Technology can play a major role in helping police match firearms to ballistic evidence in cases such as this. A computerized firearms database, a collection of digital images of bullets and **shell casings** recovered nation-wide by police, can help investigators determine if any weapon has been involved in other crimes. Whenever police recover a bullet or shell casing, they photograph it and enter it into the database for matching.

This kind of ballistic evidence can be linked to a specific gun, but further evidence is needed to identify the shooter. In the case of the D.C. serial sniping, the police were able to find suspect fingerprints at two different crime scenes. In addition to the fingerprint evidence, small traces of DNA found in saliva left at the scenes helped identify the suspects. Finally, handwriting analysis of a letter and writing found in their car all pointed to Muhammad and Malvo as the snipers. Both men were found guilty—Muhammad was sentenced to death and Malvo was sentenced to life without parole.

Lee Malvo.

John Muhammad.

©AP IMAGES (both)

490

OBJECTIVES

By the end of this chapter you will be able to

✔ Discuss the differences between a handgun, a rifle, and a shotgun.

✔ Distinguish between a bullet and a cartridge.

✔ Discuss rifling on a gun barrel and how it affects the flight of the projectile.

✔ Explain the relationship between barrel size and caliber.

✔ Explain how bullets are test-fired and matched.

✔ Discuss the role of ballistics recovery and examination at the crime scene.

✔ Determine the position of the shooter based on bullet trajectory.

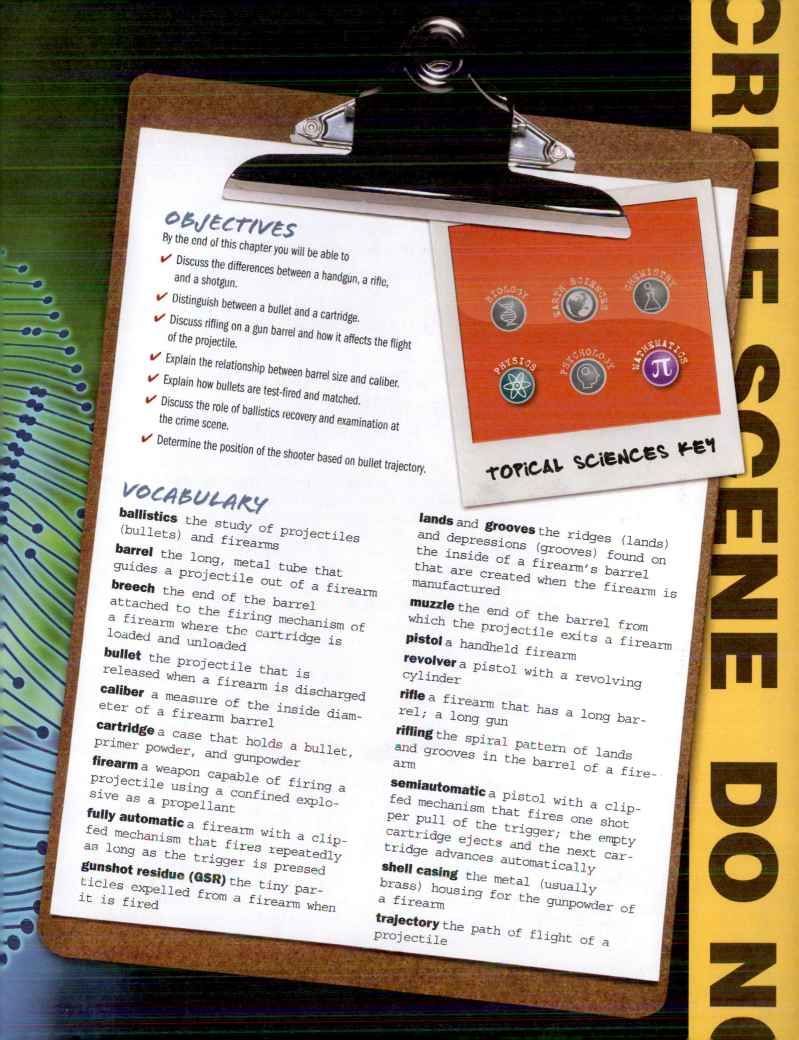

TOPICAL SCIENCES KEY

VOCABULARY

ballistics the study of projectiles (bullets) and firearms

barrel the long, metal tube that guides a projectile out of a firearm

breech the end of the barrel attached to the firing mechanism of a firearm where the cartridge is loaded and unloaded

bullet the projectile that is released when a firearm is discharged

caliber a measure of the inside diameter of a firearm barrel

cartridge a case that holds a bullet, primer powder, and gunpowder

firearm a weapon capable of firing a projectile using a confined explosive as a propellant

fully automatic a firearm with a clip-fed mechanism that fires repeatedly as long as the trigger is pressed

gunshot residue (GSR) the tiny particles expelled from a firearm when it is fired

lands and **grooves** the ridges (lands) and depressions (grooves) found on the inside of a firearm's barrel that are created when the firearm is manufactured

muzzle the end of the barrel from which the projectile exits a firearm

pistol a handheld firearm

revolver a pistol with a revolving cylinder

rifle a firearm that has a long barrel; a long gun

rifling the spiral pattern of lands and grooves in the barrel of a firearm

semiautomatic a pistol with a clip-fed mechanism that fires one shot per pull of the trigger; the empty cartridge ejects and the next cartridge advances automatically

shell casing the metal (usually brass) housing for the gunpowder of a firearm

trajectory the path of flight of a projectile

INTRODUCTION

Figure 17-1. *High-speed photograph captures the path of a bullet shot through an apple.*

©Harold & Esther Edgerton Foundation, 2007, courtesy of Palm Press, Inc.

Ballistics is the study of bullets and firearms. A **firearm** is a weapon, such as a gun, capable of firing a projectile (Figure 17-1) using a confined explosive. Ballistic evidence helps police answer many questions pertaining to a crime scene. These questions include:

- What type of firearm was used?
- What was the caliber of the bullet?
- How many bullets were fired?
- Where was the shooter standing?
- What was the angle of impact?
- Has this firearm been used in a previous crime?

HISTORY OF GUNPOWDER AND FIREARMS

More than one thousand years ago, the Chinese invented gunpowder. Gunpowder is potassium nitrate (saltpeter), charcoal, and sulfur. When ignited, it expands to six times its original size, causing a violent explosion. The Chinese used it to make fireworks and to shoot balls of flaming material at their enemies.

Years later, in 14th-century Europe, inventors learned they could direct the explosive force of gunpowder down a cylinder to move a deadly projectile, an object that is launched through the air. The projectiles launched from these early firearms were very effective in piercing suits of armor and wounding the enemy at a great distance.

Figure 17-2. *Matchlock weapons used wicks to ignite the gunpowder.*

Very old matchlock gun

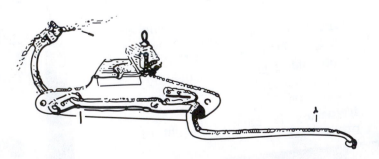

The inside of a matchlock

For a firearm to work reliably, it must effectively ignite the gunpowder. The earliest firearms, matchlock weapons, had wicks to carry a flame to the gunpowder (Figure 17-2). Later, matchlock weapons were replaced by flintlock weapons, which used sparks from a chip of flint instead of wicks to ignite the powder, allowing them to work even in damp weather. These weapons were muzzle-loaders, which meant that the user put the gunpowder and the projectile down the firearm's barrel (muzzle) and packed them into position.

Percussion firing replaced the flintlock method with the introduction of the **cartridge**, a case that holds a **bullet** (a pointed projectile), a small amount of primer powder, and the gunpowder (Figure 17-3). A hammer hit the primer powder, which exploded, igniting the gunpowder. From the late 1500s, firearms that used this method were more effective than flintlocks in shooting the bullet in the desired direction. Cartridges were loaded into the gun from the opposite end of the barrel—the **breech**. These breech-loading firearms could be loaded for firing more quickly than the older, muzzle-loading firearms.

LONG GUNS AND HANDGUNS

Modern firearms are divided into two basic types—long guns and handguns. Long guns, such as rifles and shotguns (Figure 17-4), require the use of two hands for accurate firing. **Rifles** fire bullets, whereas shotguns can fire either small round pellets (shot) or a single projectile called a slug.

Handguns fired with one hand are called **pistols**. American inventor Samuel Colt developed and patented a model in 1835 (Figure 17-5). Colt's weapon, unlike its predecessors, had a cylinder that could be loaded with several cartridges and fired in rapid succession. It was called a **revolver**, because the cylinder that held the cartridges turned as it fired.

Today, handguns can be further classified as revolvers or semiautomatic firearms. Revolvers hold six cartridges in the cylinder. The semiautomatic permits the loading of up to 10 cartridges into a magazine (clip), which is then locked into the grip of the firearm. **Semiautomatic** weapons, which fire only one bullet per pull of the trigger, differ from **fully automatic** weapons, which fire repeatedly as long as the trigger is pressed. In both, the empty cartridge ejects and the next cartridge advances automatically.

Figure 17-3. A percussion-cap firearm ignites the gunpowder using a primer that explodes upon impact by a small hammer.

©Pier Photography/Alamy

Did You Know?

Much of the credit for the modern firearm is given to French army officer and inventor Henri-Gustave Delvigne. In the 1820s and 1830s, he experimented with innovative rifles, bullets that expanded to fit rifled barrels, and other advances in firearms.

Figure 17-4. Shotguns, which are often used for hunting game birds, can fire many small pellets at once.

©Radlund & Associates

Figure 17-5. Colt .45s were one of the more popular early models of revolvers in the United States.

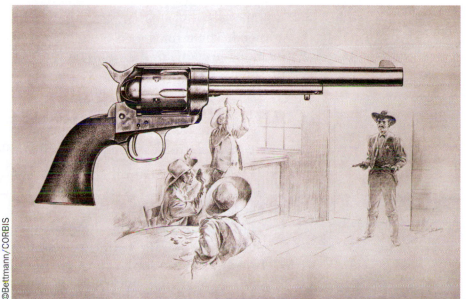

©Bettmann/CORBIS

FIREARMS AND RIFLING

An archer will hit a target with greater accuracy if there is a twist on the end of the arrow feathers. This same principle of twisting the projectile is part of the design of "rifled guns" or rifles. The word rifle originally referred to the **grooves**, or indentations, in the rifle's barrel. The ridges, or raised areas, that surrounded the grooves are called **lands**. Within the gun's barrel, lands and grooves cause a bullet to spiral when exiting the barrel of the gun, much in the same way a football spirals when thrown. This **rifling** pattern left on the bullet is specific to the firearm. Today, technology allows us to identify the patterns of land and grooves in pistols and rifles (Figure 17-6).

Ballistics experts realize that, even as guns of the same model are produced, the tools used to make the rifling within the gun's barrel wear the metal in such a way that each gun has a unique pattern. It is impossible to produce two identically rifled gun barrels. As a gun is fired, the barrel marks each bullet with its own unique pattern. Therefore, a bullet can be matched to the specific gun from which it was fired.

Figure 17-6. Bullets fired from a firearm show patterns of lands and grooves that match the barrel pattern of the firearm.

Base of bullet   Left twist   Right twist

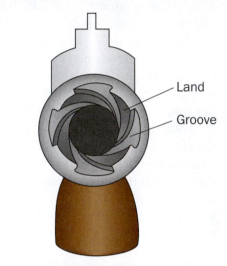

Land

Groove

BULLETS, CARTRIDGES, AND CALIBERS

A bullet is a projectile propelled from a firearm. Bullets are normally made of metal. The term *bullet* is often incorrectly applied to the cartridge, which includes primer powder, gunpowder, the bullet, and the casing material that holds them all together.

ANATOMY OF A CARTRIDGE

The typical cartridge is composed of the following parts (Figure 17-7):

1. The *bullet* (the projectile) can be composed of lead, copper, or combinations of various metals. It can be metal-jacketed, hollow-pointed, or even plastic-coated.

2. The *primer powder* mixture initiates the contained explosion that pushes the bullet down the barrel. The primer is struck by the firing pin of the firearm. The pressure causes the powder to ignite. The firearm's firing pin may strike the bottom of the cartridge *casing* in the centerfire cartridge, or it might strike anywhere on the rim of the rimfire cartridge.

3. The *anvil* and *flash hole* provide the mechanism of delivering the explosive charge from the primer powder to the *gunpowder*.

4. The *headstamp* on the bottom of the cartridge casing identifies the caliber and manufacturer (Figure 17-8).

Figure 17-7. *The typical center fire cartridge.*

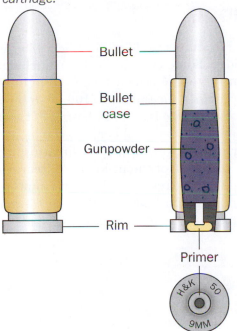

Figure 17-8. *Ammunition has a headstamp—either a logo or name identifying the caliber and manufacturer. The headstamp is located on the bottom of the cartridge casing. Shown is the headstamp of a Luger 9mm.*

©Cengage Learning

"Magnum" or "special" associated with the cartridge refers to the use of a higher-energy gunpowder or more gunpowder. "Rim fire" or "center fire" refers to where on the rear of the cartridge the firing pin strikes the cartridge casing—either along its rim or in the center. Smaller-caliber cartridges tend to be rim fire, while larger cartridges are usually center fire. Generally, center-fire cartridges are more powerful than rim-fire cartridges.

HOW A FIREARM WORKS

The sequence of events in the firing of a bullet is (Figure 17-9):

1. Pull the trigger and the firing pin of the firearm hits the base of the cartridge, igniting the primer powder mixture.

2. The tiny explosion—not much more than a spark—of the primer powder mixture on the anvil delivers a spark through the flash hole to the main gunpowder supply.

Figure 17-9. *The sequence of events in the firing of a bullet.*

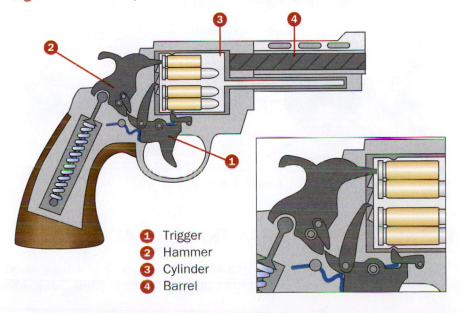

❶ Trigger
❷ Hammer
❸ Cylinder
❹ Barrel

Shotgun shells are measured in gauge. The 10-gauge, 12-gauge, 16-gauge, or 20-gauge refers to the number of round lead balls per pound of load in the shot. The larger the gauge number, the smaller the inside diameter of the barrel would be. A 12-gauge shotgun will propel fewer (but larger-sized) lead shot than a 20-gauge shotgun.

3. The main gunpowder supply ignites, and the pressure of the explosion pushes the bullet from the casing and into the barrel of the firearm. The amount of gunpowder and the mass of the projectile in a cartridge determines the speed of the bullet.

4. The bullet follows the lands and grooves pattern of the barrel and begins its spiral before it leaves the barrel.

CALIBER OF THE CARTRIDGE

Bullets (and their cartridges) are named by caliber and length. The **caliber** is a measure of the diameter of the cartridge. Some common calibers include: .22, .25, .357, .38, .44, and .45. These cartridges are usually measured in hundredths of an inch. Thus, a .45-caliber cartridge measures 45/100 of an inch in diameter (almost ½ an inch). The .357 cartridge is 35.7/100, or 357/1,000 of an inch. The European method of naming firearm caliber uses the metric system for measurement of cartridge diameter. Nine-millimeter firearms fire 9 mm bullets. Caliber also refers to the diameter of the inside of a firearm's barrel. Because the bullet moves through the barrel, the caliber of ammunition should match the firearm that shoots it. If a bullet is removed from a wound or crime scene, its caliber can link it to the weapon used to fire it.

THE STUDY OF BULLETS AND CARTRIDGE CASINGS

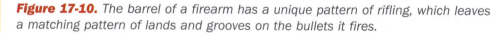

A significant part of ballistics involves examining used bullets and their spent cartridge casings for telltale markings left on them by the firearm that shot them. For example, as a gun is fired, the barrel marks each bullet with its own unique pattern of lands and grooves. By examining the lands and grooves, a bullet investigators can match to the gun from which it was fired (Figure 17-10). Investigators compare bullets and spent cartridge casings from a crime scene with bullets and spent cartridges shot from the suspected firearm. To get a known bullet for comparison, investigators test-fire the weapon into a water tank or gel block. This captures the bullet without damaging it. Then, they can compare the markings on known bullets with those on the suspect bullets.

Figure 17-10. The barrel of a firearm has a unique pattern of rifling, which leaves a matching pattern of lands and grooves on the bullets it fires.

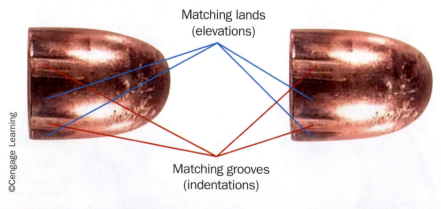

©Cengage Learning

MARKS ON SPENT CARTRIDGE CASINGS

Firing pin marks left on the spent cartridge casings can also be used to identify a firearm (Figure 17-11). Firing pin marks are impressions made on

Did You Know?

According to a recently released study, (November 2007) evidence matching the chemical composition of a crime scene bullet to that of a specific production lot has been discredited. It apparently had no scientific credibility, although it has been used as evidence in criminal proceedings for over forty years! It is expected that several convicted criminals may win appeals if their conviction was based on this faulty evidence.

the bottom of the cartridge by the firing pin as it strikes the bottom of the cartridge when the firearm is shot.

Breechblock markings are another kind of mark left on spent cartridge casings. When the firearm is shot, the explosive force pushes the bullet forward. At the same time, an equal and opposite force pushes the cartridge casing backward against the breechblock, which prevents the cartridge from shooting toward the user as it recoils. Breechblock marks are produced as the cartridge casing moves backward and strikes the breechblock. The markings are unique to the firearm and can be matched if the spent cartridge casings are found (Figure 17-12).

Figure 17-11. *Depending on the firearm and cartridge used, the fire pin mark appears on the rim or the center of the spent cartridge.*

Figure 17-12. *Breech marks come in various forms, sometimes as parallel lines, circular lines, or stippled at the bottom of the cartridge casings.*

Center-fire impression comparison.

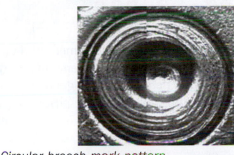

Circular breech mark pattern.

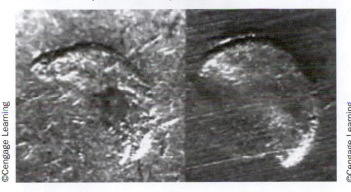

Rim-fire impression comparison.

Stippled breech mark pattern.

Other marks left on spent cartridge casings include extractor and ejector marks, which are minute scratches produced as the cartridge is placed in the firing chamber (by the extractor) and removed from the chamber after firing (by the ejector). These marks are produced only in semiautomatic and fully automatic weapons. In revolvers, cartridges are hand-fed into the revolving cylinder and have to be removed by hand as well.

Figure 17-13. *Gases are expelled from a firearm when it is fired and the gunpowder explodes.*

GUNSHOT RESIDUES

Because all firearms explode gunpowder, they produce **gunshot residues (GSR)** when fired (Figure 17-13). These residues are the traces of smoke and particles of unburned powder carried sideways from the firearm by the expansion of gases as the bullet is fired. Gunshot residues containing nitrates can stick to the person holding the firearm and leave evidence on the shooter.

The amount of GSR decreases as the distance between firearm and victim increases.

Investigators look for the presence of GSR when attempting to recreate a crime scene. If someone fired a gun, GSR could be found on his or her hands or clothing. GSR can be removed by washing, but chemical testing can often detect residue despite the attempted removal. The distance between the weapon and the victim can be determined by examining the GSR pattern on the body of a victim.

DATABASES

Digging Deeper
with Forensic Science e-Collection

Some people believe that the development of a ballistic fingerprint database that catalogs the rifling patterns of all registered guns could help solve future criminal investigations. Others worry that this is an unnecessary infringement on the privacy of gun owners. Go to the Gale Forensic Science eCollection on school.cengage.com/ forensicscience and research the topic of ballistic databases and write a report on your findings. How are these databases useful for investigators? What kind of information do they store? What controversies surround their implementation?

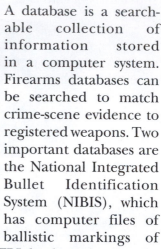

A database is a searchable collection of information stored in a computer system. Firearms databases can be searched to match crime-scene evidence to registered weapons. Two important databases are the National Integrated Bullet Identification System (NIBIS), which has computer files of ballistic markings of firearms used in previous crimes, and Drugfire, an FBI database that focuses on cartridge casings. In 2000, these databases were merged to form the National Integrated Ballistics Network (NIBIN).

TRAJECTORY

Did You Know?

Newton's third law of motion states that every force has an equal and opposite force that balances it. In the case of firearms, the explosive force that pushes the bullet out of the barrel is balanced by recoil. Recoil is the backward force that the bullet exerts on the gun when it is fired. The recoil affects both the spent cartridge casing and the shooter.

An important part of a ballistics investigation is determining where a shooter was located during a crime. Investigators look for clues at a crime scene to help them calculate a bullet's trajectory to figure out where a shooter discharged the firearm. **Trajectory** is the path of the propelled bullet. For example, if a trajectory angle is downward, the position of the shooter was above his or her target. From the angle of trajectory, the path back to the shooter can be traced.

CALCULATING TRAJECTORY

Trajectory can be calculated by finding two reference points along the flight path of the projectile. Ignoring gravity, if you assume that bullets travel in straight lines, two reference points will define a single line. An investigator can then assume that the shooter discharged the firearm somewhere along that line.

Reference points can be bullet holes in an object, such as a wall or a window, or can be a bullet wound on a victim. Less-specific reference points include GSR on objects or piles of spent cartridge casings. Even a single victim's body can have the two reference points needed to calculate where the shooter stood—an entry wound and an exit wound. In such cases, the investigator might position the corpse as it was at the time of impact and use a metal or wooden dowel to indicate the path of the bullet. Investigators can also use lasers to trace straight-line paths that can help them determine the position of the shooter or shooters.

A firearm crime is often complicated, and trajectory may be difficult to determine. Sometimes bullets ricochet, become damaged, and do not provide a direct path for measurement. Investigators may also have to use trace evidence, such as footprints, fingerprints, or DNA samples from hair or saliva, to determine a shooter's whereabouts during a crime.

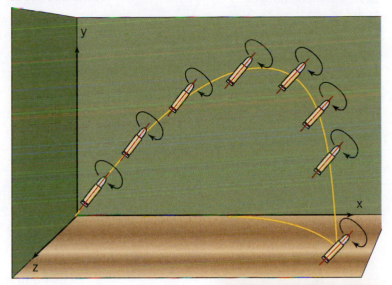

Figure 17-14. *A bullet's path (trajectory) is slightly curved, because as it moves forward toward the target, gravity also pulls it downward. The diagram below is highly exaggerated to demonstrate this effect.*

GRAVITY AND TRAJECTORY

Understanding the physics of trajectory can also help an investigator determine where a shooter was located during a crime. Two major forces are acting on a bullet once it is fired: the forward force of the gunshot and the downward force of gravity (Figure 17-14). A bullet begins to drop as it leaves the barrel of the firearm. If the shot is taken at a very distant object, the line of sight to the target must be adjusted to compensate for the effect of gravity on the bullet. If the target is closer, there would be less adjustment. Wind speed and direction are also factors affecting adjustments the shooter must make to hit the target.

DETERMINING THE LOCATION OF A SHOOTER

A bullet is found in the upholstery of a car's front driver's seat. The bullet first penetrated the car's front side window and then the seat. The bullet seemed to have come from an apartment building across the street. However, no one knows from which window or floor the shot was fired. The police recreated the crime scene and determined the path of the bullet using the hole in the car's window and the bullet hole in the seat as the reference points. Using a laser beam, they projected a line creating the approximate trajectory path of the bullet from the building to the car. They also measured the distance from the car to the apartment building as 60 feet or 720 inches . Figure 17-15, on the next page, shows the relationship of the objects at the scene.

To determine the position of a shooter, the distance between the shooter in the building and the bullet hole in the car seat must be determined. This requires at least two reference points from which to project a line back to the source of the shooter in the building. To determine the distance between the shooter and hole in the car seat, you set up a direct proportion using the two right triangles.

Did You Know?

Timeline of Advances of Ballistic Examination

1923 FBI Bureau of Forensic Ballistics established

1929 Weapons used in the St. Valentine's Day massacre identified by matching bullets

1930s Earliest gunshot residue (GSR) test performed

1968 Scanning microscope first used to examine GSR for comparison and match

1992 FBI established the Drugfire database, which compiles details on bullet and cartridge casing markings

1996 U.S. Bureau of Alcohol, Firearms, and Tobacco (AFT) establishes a database for spent ammunition

2000 FBI and AFT begin merging their databases to establish the National Integrated Ballistics Network (NIBIN)

Figure 17-15. *Notice the right triangle formed (in red).*

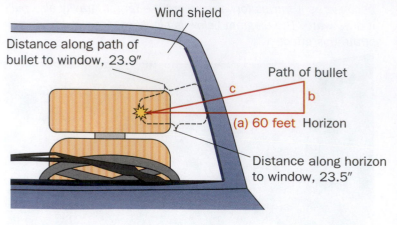

Wind shield

Distance along path of bullet to window, 23.9″

Path of bullet

c

b

(a) 60 feet   Horizon

Distance along horizon to window, 23.5″

EXAMPLE: CALCULATION USING DISTANCE VS. DROP

A car's seat has been shot through the car's window. The bullet hole is located four feet above the ground. The nearest building is 60 feet (or 720 inches) away along the horizon. If you measure the horizontal distance from the broken window to the bullet hole in the car seat horizontally and compare this distance to the diagonal length of the bullet path from the hole in the car's window to the hole in the seat, you can set up a ratio.

Here are the calculations:

$$\frac{\text{Distance to window}}{\text{Distance along horizon}} = \frac{\text{Distance to shooter } (c)}{\text{Distance to side of building}}$$

$$\frac{23.9 \text{ inches}}{23.5 \text{ inches}} = \frac{c}{720 \text{ inches}}$$

$$c = 732.3 \text{ inches}$$

Now you know two sides of the right triangle: the horizontal distance between the bullet hole in the car seat and the building and the diagonal distance between the bullet hole in the car seat and the shooter. You can use the Pythagorean theorem to determine the length of the third side of the triangle, which is the height of the shooter above the horizon (labeled *b* in Figure 17-15).

The Pythagorean theorem states that for a right triangle, the square of the length of the hypotenuse (longest side) is equal to the sum of the squares of the two other sides. Thus, in this example:

Hypotenuse = distance to the shooter

a = distance to the building

b = height of the shooter from the horizon, not from the ground

$\text{Hypotenuse}^2 = (a)^2 + (b)^2$

$(732.3 \text{ in})^2 = (720 \text{ in})^2 + b^2$

$536{,}300 \text{ in}^2 = 518{,}400 \text{ in}^2 + b^2$

$536{,}300 \text{ in}^2 - 518{,}400 \text{ in}^2 = b^2$

$17{,}900 \text{ in}^2 = b^2$

$b = 133 \text{ in} \sim 11 \text{ ft}$

Digging Deeper
with Forensic Science e-Collection

According to the Warren Report, which investigated the assassination of President John F. Kennedy, the path of a bullet took some turns as it struck bone. Go to the Gale Forensic Science eCollection on school.cengage.com/forensicscience and research the Kennedy assassination. Examine the trajectory patterns associated with the case. What path did the Warren Report describe? How did this help determine where the shooter was located? What kinds of controversies are associated with these findings? Make a poster that explains how the investigators used ballistics to determine how many shooters were involved and where the shooter was located.

If the building is 60 feet away, then the shooter was about 11 feet higher than the height of the bullet hole in the seat, which was at 4 feet. This height of about 15 feet off the ground (11 ft plus 4 feet) predicts that the shooter was on the second floor.

BULLET WOUNDS

Eyewitness accounts of a shooting are not always accurate. It is often helpful to examine bullet wounds on victims to confirm or dispute a witness's story. For example, if someone claims that a victim was running away from a shooter at the time of a shooting, an examination of the wounds on the body can confirm or dispute whether the victim was indeed shot in the back.

Determining which wound is the entrance wound and which is the exit wound is an important step in determining what happened at a crime scene. Generally, entrance wounds are smaller than exit wounds, because the skin is somewhat elastic, and it stretches when a bullet enters the body. Therefore, the size of the entry wound will be smaller than the bullet. Exit wounds are generally larger, because as the bullet moves through the body, it may collect and carry body tissue and bone with it.

Another way of determining entry and exit wounds is to look at clues on the body near each wound. For example, if the bullet penetrates clothing first, fibers may be embedded in the wound pointing in the direction of penetration. Also, investigators can examine the presence of GSR, which is usually found only around entrance wounds. Furthermore, if the bullet is fired when the muzzle is in contact with the skin, the hot gases released from the muzzle flash may burn the skin, leaving a telltale mark (Figure 17-16).

Bullets usually do not travel smoothly through a victim's body. Bones, organs, and other tissues bend their paths, causing a tumbling effect. The tumbling bullet may also tear a larger, more irregular exit wound than expected. A bullet may ricochet off bone and do considerable internal damage before exiting. A bullet may not exit the body at all.

Several factors influence whether a discharged bullet will pass through a victim or remain lodged somewhere in the body. For example, if the bullet has a high speed, it may have enough energy to pass directly through the body, leaving both an entrance wound and an exit wound. High-speed bullets are more likely to pass through a body than are low-velocity bullets. Small-caliber bullets, such as a .22 caliber, tend to lodge within the body, while larger-caliber bullets will pass through.

Figure 17-16. *Two GSR patterns from different distances. The darker pattern (left) was produced by a closer shot.*

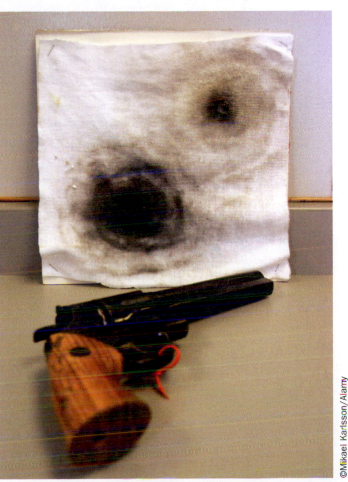

SUMMARY

- Ballistics is the study of bullets and firearms, which are weapons capable of firing a projectile using a confined explosive, such as gunpowder.

- Modern firearms are divided into two basic types—long guns and handguns—that require two hands or one, respectively, for accurate firing.

- Handguns can be further classified as revolvers or semiautomatic firearms, depending on the feeding mechanism.

- Bullets fired from a firearm show patterns of lands and grooves that match the rifling pattern in the barrel of the firearm.

- A cartridge consists of primer powder, gunpowder, a bullet (a pointed projectile), and the casing material that holds them all together.

- The caliber of a cartridge is a measure of its diameter and is identified along with the name of the manufacturer on the headstamp.

- In addition to examining lands and grooves on a bullet, investigators can examine firing pin marks, breechblock marks, and extractor and ejector marks on a spent cartridge casing to match evidence at a crime scene with a specific firearm.

- Gunshot residues found on victims, shooters, or nearby objects can help investigators recreate a crime scene.

- Investigators often use national databases to match crime-scene evidence to registered weapons.

- Using at least two reference points, an investigator can recreate a bullet's trajectory and determine where a shooter was located during a crime.

- Two major forces are acting on a bullet once it is fired: the forward force of the gunshot and the downward force of gravity.

- Examination of the wounds on a body can determine where a bullet entered and exited the victim.

CASE STUDIES

Sacco and Vanzetti (1920)

Three different types of shell casings were found at the scene of a payroll holdup. Two security guards were killed. When suspects Sacco and Vanzetti were arrested, they were both in possession of loaded guns of the same caliber and ammunition from the same three manufacturers as the casings found at the crime scene. Both were anarchists who openly advocated the violent overthrow of the government.

The trial opened in 1921. More than 140 witnesses were called to testify. The one fact that seemed incontrovertible was that the bullet that killed one of the security guards was so ancient in its manufacture that no similar ammunition could be found to test-fire Sacco's weapon except those equally ancient cartridges found in his pocket upon arrest.

The defendant's lawyer, Fred Moore, aggressively turned the trial from murder to politics. He accused the prosecution of trying the men as part of the Red Scare of 1919–1920. It soon turned into a worldwide spectacle of "patriots" versus "foreigners." Despite Moore's tactics, the defendants were found guilty and sentenced to death. In 1927, a committee was appointed to review the case. The comparison microscope had been recently invented and was used to conclusively link the murder bullet with Sacco's gun. Both men were executed in the electric chair in 1927.

Lee Harvey Oswald (1963)

©AP Photo

The Warren Commission concluded that Lee Harvey Oswald worked alone in his assassination of President John F. Kennedy on November 22, 1963. The bullets that were fired at President Kennedy were analyzed for chemical composition and left conflicting conclusions as to the number of bullets fired and their origin. However, many believe that the evidence seems to contradict the Warren Commission report on this point.

Three bullets were reported to have been fired. The path of one of the bullets went through President Kennedy and then through Governor Connally, who was also in the car. Of the three shots that Oswald is said to have fired, one missed the car completely. A second bullet struck Kennedy in the back and then proceeded through his body into Connally's back, out through his chest, striking his right wrist, and then proceeding through his left thigh. The last bullet struck Kennedy in the back of the head and was fatal.

Joseph Gerace (2004)

Albany Police Officer Joseph Gerace was trapped between parked cars in downtown Albany a few blocks from the State Capitol Building. Daniel Reed drove his car erratically toward the officer and Gerace, fearing for his life, fired at Reed. One of the shots accidentally killed David Scaringe as he crossed a nearby street. Ballistics testing confirmed the shot that killed Scaringe was fired from Officer Gerace's firearm. The officer left the scene of the incident while police tried to help Scaringe. Gerace was later charged with reckless endangerment.

Think Critically Although there are specific mathematical formulas and the opportunity to make precise measurements, the accuracy of trajectory calculations is still hard to "guarantee." Explain why.

Forensics Technician

Unlike a research scientist, a science technician applies scientific knowledge and laboratory methods toward specific, real-world problems. A forensic science technician helps investigators in all stages of a criminal investigation. He or she will gather evidence at a crime scene, analyze the evidence in a laboratory, and present reports to investigators as to the nature of the evidence. A forensic science technician may also be called to court to present findings as an expert witness during a trial. These technicians may specialize in a particular field of forensics, such as firearms or DNA analysis.

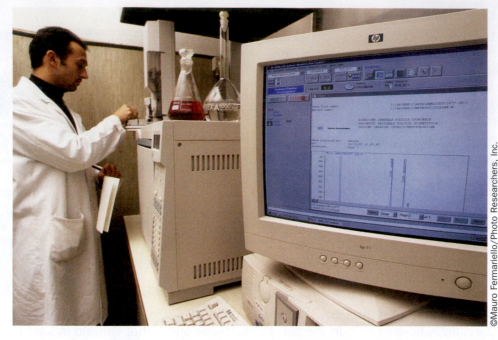

A technician analyzes chemical evidence using a gas chromatograph.

©Mauro Fermariello/Photo Researchers, Inc.

One of the most important parts of a forensic science technician's job is proper collection of evidence at a crime scene. After a crime scene has been fully mapped and documented, forensic science technicians may carefully collect and label ballistics evidence, such as bullets, cartridge casings, or weapons left at the crime scene, for further analysis at the laboratory. They may collect hair, blood, saliva, or skin samples from a victim or from a weapon for DNA analysis to match that found on a suspect. Such tissue samples must also be properly labeled and cannot be contaminated with tissue brought in from outside the crime scene. Technicians also collect non-organic materials, such as paint flakes or glass fragments, to match with that of a suspect's car. They also collect fibers from clothing, rope, and other fabrics to compare with that found on the victim or suspect. Other chemical evidence, such as poisons or gunshot residues, are collected and analyzed for their concentrations and chemical compositions.

Like a research scientist, forensic scientists must use precision, remain objective, and carefully record all steps of their laboratory analyses. They gather evidence, carry out tests, analyze the results, and present their findings in reports that may be used during an investigation or a trial. Introducing personal feelings into an investigation or performing sloppy work can lead to unreliable conclusions, which may incorrectly free a suspect or incriminate an innocent person. Also, improper handling or labeling of evidence during any stage of the investigation can lead to the destruction of evidence that might never again be recovered. This position requires a two- to four-year education that includes courses in chemistry, physics, math, and criminal justice.

Learn More About It
To learn more about the work of a forensic ballistics expert, go to school.cengage.com/forensicscience.

CHAPTER 17 REVIEW

True or False

1. The lands and grooves of a barrel's rifling improve the accuracy of a bullet.

2. The caliber of a cartridge is always measured in one-hundredths of an inch.

3. Firing pin marks are found on the back of the bullet.

4. Lands and grooves help match a crime-scene bullet with its shell casing.

5. The amount of gunshot residue on a victim is usually proportional to the distance between the victim and the shooter.

Multiple Choice

6. Shotguns are examples of
   a) handguns
   b) long guns
   c) revolvers
   d) semiautomatic weapons

7. Which of the following best describes the trajectory of a projectile?
   a) the height of the shooter
   b) the path of the flight of a bullet
   c) the housing for the bullet's gunpowder
   d) the pattern of lands and grooves on the projectile

8. Which of the following is not part of a cartridge?
   a) barrel
   b) bullet
   c) gunpowder
   d) primer powder

9. The caliber of a bullet is related to its
   a) diameter
   b) length
   c) speed
   d) weight

10. Semiautomatic pistols store cartridges in a
    a) magazine (clip)
    b) cylinder
    c) firing pin
    d) muzzle

Short Answer

11. What kind of information can be learned from gunshot residue (GSR) examination?

_____

12. Describe how investigators match a bullet to a firearm.

_____

13. Why are bullets fired into a gel tank in a forensics ballistic lab?

_____

14. Compare and contrast an entry wound and an exit wound produced from a bullet.

_____

15. Using the following terms, explain how the different parts of a gun and cartridge enable the bullet to be fired from a gun:
    a. trigger
    b. hammer
    c. firing pin
    d. gunpowder
    e. energy
    f. bullet
    g. lands and grooves

16. What is NIBIS and how is it used to help solve crimes?

_____

Bibliography

Books and Journals

David, A. J., and L. Exline. *Current Methods in Forensic Gunshot Residue Analysis*. Boca Raton, FL: CRC Press, 2000.

Di Maio, Vincent J. *Gunshot Wounds: Practical Aspects of Firearms, Ballistics, and Forensic Techniques*. Boca Raton, FL: CRC Press, 1999.

Evans, Colin. *The Casebook of Forensic Detection: How Science Solved 100 of the World's Most Baffling Crimes*. New York: Berkley Trade, 2007.

Heard, Brian J. *Handbook on Firearms and Ballistics: Examining and Interpreting Forensic Evidence*. Boca Raton, FL: CRC Press, 1997.

Pejsa, Arthur J. *Modern Practical Ballistics*. Minneapolis, MN: Kenwood Publishers, 2001.

Platt, Richard. *Crime Scene*. Englewood Cliffs, NJ: Prentice Hall, 2006.

Web sites

Gale Forensic Science eCollection. school.cengage.com/forensicscience.

Eng, Paul. "Uncovering Convincing Evidence," www.abc.news.com, Oct. 30, 2003.

http://en.wikipedia.org/wiki/Bullet

http://www.firearmsid.com/Bullets/bullet1.htm

http://library.med.utah.edu/WebPath/TUTORIAL/GUNS/GUNBLST.html

http://www.firearms.iD.com

http://www.bls.gov/oco/ocos115.htm

ACTIVITY 17-1
BULLET TRAJECTORY

Objective:

By the end of this activity, you will be able to:
1. Analyze three different crime scenes.
2. Determine information about the shooter's position.

Time Required to Complete Activity: 40 minutes (teams of two students)

Introduction:

The device pictured below is a model representing three different bullet trajectories that will be used in the three scenarios in this activity.

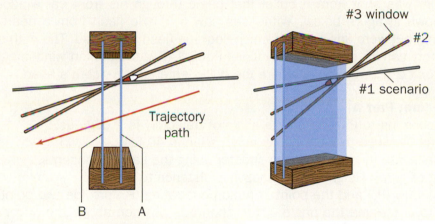

The angle of impact (the bullet's path) is the angle created by the pathway of the bullet and the horizon. To determine this angle, at least two points along the trajectory path must be identified. These two points could be an entry wound (A) and exit wound (B) or possibly a windshield penetration (A) before entering the body (B). By identifying the location of the shooter, we may be able to collect additional evidence to help identify the firearm.

Recall the Law of Tangents. This law states that:

Tan of an angle of elevation = $\dfrac{\text{Opposite side (B)}}{\text{Adjacent side (A)}}$

If the distance from the wound to side A can be measured, side B can be calculated.

If the angle of elevation and side A can be measured, then it is possible to calculate side B.

Materials:

ruler
calculators with sine function or tangent table

Safety Precautions:

None

Scenario 1:

A victim was shot from a bullet that came through his front car window as shown in figure below. Witnesses saw a muzzle flash from a nearby building, but were unsure from which floor the flash originated. The path of trajectory can be determined by using Point 1 (P_1), the broken windshield, and Point 2 (P_2), the point where the bullet entered the victim's head.

Procedure: Part A

1. According to the sketch the angle of trajectory is 2 degrees.
2. The distance to the building in question is 40 feet.
3. Calculate the height of the shooter using the Law of Tangents.

Height of gun = Height above horizon + distance to ground

The window (P_1) and the point of head penetration (P_2) provide two points used to determine the angle of the shooter's position above the driver's location.

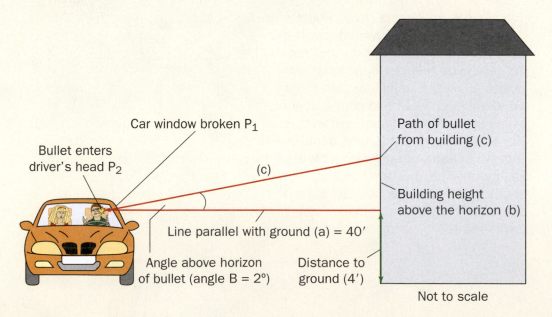

Not to scale

Victim's head   Car window   Path of bullet

P₂   P₁

moved in a downward direction about 2 degrees from the gun through the window to the victim from right to left and downward.

A second method of determining the height of the shooter uses a laser pointer as pictured on the right. A laser pointer can be projected toward the location of the shooter and may assist in determining the shooter's position. Neither method is perfectly accurate, but both will give a fair approximation.

Calculation Position Using the Law of Tangents
    Distance to building ≈ 40 feet
    angle of elevation (−) = 2

Solving using the Law of Tangents:
    Tan of angle of elevation = opposite/adjacent

$$\text{Tan of } 2° = \frac{b \text{ (height of building above wound or above the horizon)}}{a \text{ Distance to the building (40 ft)}}$$

$$.0349 = \frac{b}{40 \text{ feet}}$$

$$.0349 \, (40 \text{ ft}) = b$$

$$1.34 \text{ ft} = b$$

Shooter height = 1.34 ft + 4 ft = 5.34 ft

The shooter was located on the first floor.

Scenario 2:

Refer to the figure at the beginning of the activity.

Witnesses saw a victim fall while riding his bike. He had been struck in the head by a bullet. When the crime-scene investigators arrived, they calculated the angle of elevation of the shooter to be about 6.5 degrees. The distance to the building from which the bullet was fired was 152 feet, and the height of the entry wound on the victim while on his bike measured 6 feet above the ground.

Solve for height using the tangent method. Show your work.

Height = _____ feet

Scenario 3:

Part A:

A man is shot from a hotel window while sitting on the hood of his car. Use the following information to determine from which window the shot came.

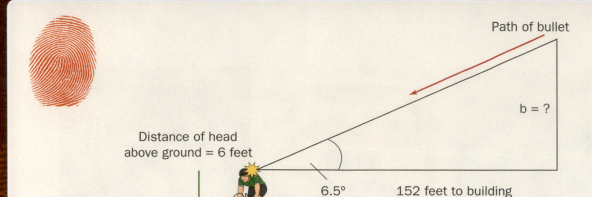

Path of bullet

b = ?

Distance of head
above ground = 6 feet

6.5°   152 feet to building

Solve for height using the tangent method. Show your work.
Height = _____feet

Scenario 3:

Part A:

A man is shot from a hotel window while sitting on the hood of his car. Use the following information to determine from which window the shot came. The trajectory angle is 25 degrees.

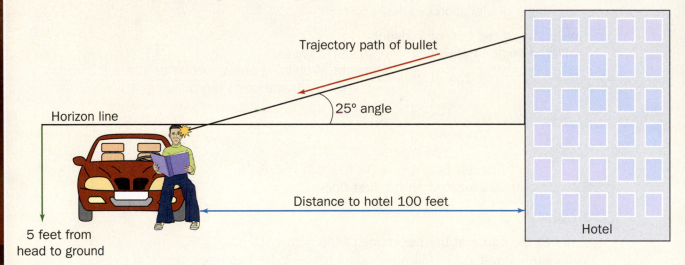

Trajectory path of bullet

25° angle

Horizon line

Distance to hotel 100 feet

Hotel

5 feet from
head to ground

At what distance above the ground was the shot fired? Show your work. This will help locate the correct floor.

Part B:

Using the diagram at the top of the next page, determine the correct window.

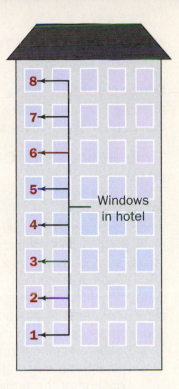

Windows in hotel

Head of victim

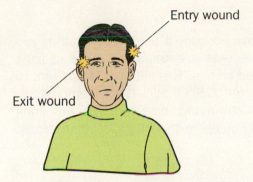

Entry wound

Exit wound

The bullet was fired from which window?
Draw lines illustrating how you arrived at your conclusion.

Final Analysis:

1. List problems that might interfere with the accuracy of your results.
   a.
   b.
   c.

2. What problems would be encountered if we couldn't accurately determine the trajectory angle?
   a.
   b.

Solve the following:

3. Angle of entry (trajectory) = 15° and the distance to the building is 700 feet

   Height of shooter ~_____ feet (above the horizon since the person could be sitting or standing and not be at ground level)

4. Angle of entry (trajectory) = 27° and the distance to the building is 60 feet

   Height of shooter ~_____ feet above the horizon

5. Angle of entry (trajectory) = 35° and the distance to the building is 85 feet

   Height of shooter ~_____ feet above the horizon

ACTIVITY 17-2
FIRING PIN MATCH

Objectives:
By the end of this activity, you will be able to:
Compare firing pin impressions from different sources.

Time Required to Complete Activity:
20 minutes

Introduction:
When cartridge shell casings are recovered from a crime scene, they are photographed and compared to NIBIS records to determine if these casings match any found at previously committed crimes. This allows investigators to link a series of crimes to the same perpetrator. Shell casings can demonstrate certain identifying markings, such as ejector marks, breech marks, and firing pin impressions. In this activity, you will compare the firing pin impressions. Your comparison should include:
- Caliber of the cartridge
- Headstamp marking of the manufacturer
- Location of the firing pin strike
- Description of the unique firing pin characteristics

Materials:
(per student working in pairs)
pencil
lab sheet of firing pin photographs
hand lens
stereomicroscope (optional)

Safety Precautions:
None

Scenario:
Three suspects were apprehended and accused of robbery. Empty shell casings were found at two different crime scenes during the past month. They are labeled A through L. Police test-fired firearms belonging to the suspects and compared firing pin impression marks made by those found on crime-scene casings.

Procedure:

1. View each shell casing with a hand lens or stereomicroscope to determine each of the following: the caliber, headstamp, location of firing pin strike (center or rim), and description of firing pin marks. Record your data on Data Table 1.

2. Using a pen or pencil, circle or mark the unique patterns on each casing.

3. Using the three cartridge casings from the three suspects and your information from Data Table 1, determine if any of the crime-scene casings match casings from the suspects.

Cartridge shell casings made from test firing guns from the three suspects.

Suspect 1

Suspect 2

Suspect 3

Data Table: Comparison of casings

	Suspect 1	Suspect 2	Suspect 3
Caliber			
Headstamp marking			
Firing pin strike (center or rim)			
Description of mark			

Evidence cartridge shell casings recovered from previous robberies.

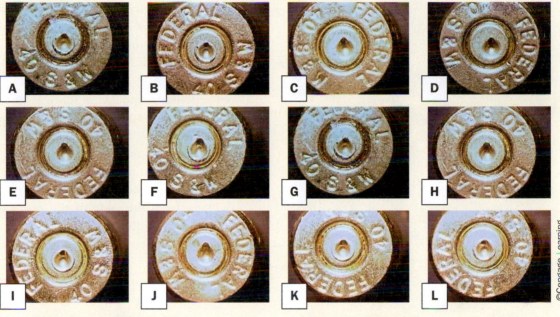

A   B   C   D

E   F   G   H

I   J   K   L

©Cengage Learning

Final Analysis:

1. Of the three suspects, which one(s) could you link to the crimes?
2. Based on the shell-casing matches, which of the three suspects could *not* be linked?
3. Describe specific (unique) characteristics that linked one of the suspect's casings to the crime-scene casings.
4. If you were a prosecuting attorney, what argument could you provide to the defense's claim that "if a suspect's cartridge shell casings were *not* found at a crime scene, he must be innocent"?
5. Crime labs today are better able to compare and analyze ballistic evidence. Describe two advances in technology that have enabled a better use of ballistics evidence in solving crime.

Further Study:

Research the Washington, D.C. sniper case mentioned at the beginning of the chapter. Explain how ballistics evidence was used to link the two suspects to the serial killings.

Glossary

Investigators reconstruct the crime.

©Bill Pugliano/Getty Images

A

abrasion mark a mark produced when a surface slides across another surface

agglutination the clumping of molecules or cells caused by an antigen–antibody reaction

algor mortis the cooling of the body after death

allele an alternate form of a gene; for example, a gene for human hair color may have alleles that cause red or brown hair

amorphous (Ch. 4) without a defined shape; fibers composed of a loose arrangement of polymers that are soft, elastic, and absorbing (for example, cotton)

amorphous (Ch. 14) without shape or form; applied to glass, it refers to having particles that are arranged randomly instead of in a definite pattern

anabolic steroids synthetic compounds which enhance muscle growth

analytical skills the ability to identify a concept or problem, to isolate its component parts, to organize information for decision making, to establish criteria for evaluation, and to draw appropriate conclusions

angiosperm a flowering plant that produces seeds within a fruit

anthropology the scientific study of the origins and behavior as well as the physical, social, and cultural development of humans.

antibodies proteins secreted by white blood cells that attach to antigens

antigen–antibody response a reaction in which antibodies attach to specific antigens to bind foreign substances

antigens any foreign substance or cell in the body that reacts with antibodies

arch a fingerprint pattern in which the ridge pattern originates from one side of the print and leaves from the other side

autolysis the spontaneous breakdown of cells as they self-digest

autopsy a post mortem examination of a corpse

B

ballistics the study of projectiles (bullets) and firearms

barrel the long, metal tube that guides a projectile out of a firearm

Becke Line line created as refracted light becomes concentrated around the edges of the glass fragment

bifurcation the branching of a fingerprint ridge

breech the end of the barrel attached to the firing mechanism of a firearm where the cartridge is loaded and unloaded

buccal swab swabbing the mouth with a sterile cotton swab for the purpose of obtaining a DNA sample

bullet the projectile that is released when a firearm is discharged

C

caliber   a measure of the inside diameter of a firearm barrel

cartridge   a case that holds a bullet, primer powder, and gunpowder

cause of death   the immediate reason for a person's death (such as heart attack, kidney failure)

cell-surface protein   proteins embedded in the cell membrane

chain of custody   the documented and unbroken transfer of evidence

chromosome   a cell structure that contains genetic information along strands of DNA

circumstantial evidence   (indirect evidence) evidence used to imply a fact but not prove it directly

class evidence   material that connects an individual or thing to a certain group *(see individual evidence)*

clay   the smallest type of soil particles that have the capacity to absorb and hold water

CODIS (Combined DNA Index System)   a database of DNA information that allows the exchange of information between law enforcement agencies

comparison microscope   a compound microscope that allows the side-by-side comparison of samples, such as of hair or fibers

controlled substance   a drug or other chemical compound whose manufacture, distribution, possession, and use is regulated by the legal system

core   a center of a loop or whorl

cortex   the region of a hair located outside of the medulla containing granules of pigment

concentric fracture pattern   a circular fracture pattern in glass

counterfeiting   the production of an imitation of currency, works of art, documents, and name-brand look-alikes for the purpose of deception

crime-scene investigation   a multidisciplinary approach in which scientific and legal professionals work together to solve a crime

crime-scene reconstruction   a hypothesis of the sequence of events from before the crime was committed through its commission

crystalline   regularly shaped; fibers composed of polymers packed side by side, which make it stiff and strong (for example, flax)

cuticle   the tough outer covering of a hair composed of overlapping scales

cutting mark   a mark produced along the edge as a surface is cut

Genes determine our physical traits.

© Myrleen Ferguson Cate/PhotoEdit, Inc

D

death   the cessation, or end, of life

decomposition   the process of rotting and breaking down

deductive reasoning   deriving the consequences from the facts using a series of logical steps

delta   a triangular ridge pattern with ridges that go in different directions above and below a triangle

density   the ratio of the mass of an object to its volume, expressed by the equation: density = mass/volume

direct evidence   evidence that (if true) proves an alleged fact, such as an eyewitness account of a crime

direct transfer   the passing of evidence, such as a fiber, from victim to suspect or vice versa

DNA fingerprint   pattern of DNA fragments obtained by examining a person's unique sequence of DNA base pairs (also called DNA profiling)

DNA probe   a molecule labeled with a radioactive isotope, dye, or enzyme that is used to locate a particular sequence or gene on a DNA molecule

document analysis   the examination of questioned documents with known material for a variety of analyses, such as authenticity, alterations, erasures, and obliterations

One task of a coroner is determining manner of death.

©Robert Voets/CBS Photo Archive via Getty Images

document expert a person who scientifically analyzes handwriting

drug a chemical substance that affects the processes of the mind or body; a substance used in the diagnosis, treatment, or prevention of a disease; a substance used recreationally for its effects on the mind or body, such as a narcotic or hallucinogen

E

electrophoresis a method of separating molecules, such as DNA, according to their size and electrical charge using an electric current passed through a gel containing the samples

enzyme a chemical which can speed up chemical reactions

epiphysis the presence of a visible line that marks the place where cartilage is being replaced by bone

exemplar a standard document of known origin and authorship used in handwriting analysis

exine outer layer of the wall of a pollen grain or spore

eyewitness a person who has seen someone or something and can communicate these facts

F

fact a statement or assertion of information that can be verified

fiber the smallest indivisible unit of a textile, it must be at least 100 times longer than wide

fingerprint an impression left on any surface that consists of patterns made by the ridges on a finger

firearm a weapon capable of firing a projectile using a confined explosive as a propellant

first responder the first police officer to arrive at a crime scene

forensic anthropology the study of physical anthropology as it applies to human skeletal remains in a legal setting

forensic entomology the study of insects as they pertain to legal issues

forensic palynology the use of pollen and spore evidence to help solve criminal cases

forensic relating to the application of scientific knowledge to legal questions

forgery the making, adapting, or falsifying of documents or other objects with the intention of deceiving someone

fraudulence when a financial gain accompanies a forgery

Fry and Daubert Rulings federal court decisions allowing judges to use their discretion in admitting expert witness testimony into court proceedings

fully automatic a firearm with a clip-fed mechanism that fires repeatedly as long as the trigger is pressed

G

gene segment of DNA in a chromosome that contains information used to produce a protein or an RNA molecule

geology the study of soil and rocks

glass a hard, amorphous, transparent material made by heating a mixture of sand and other additives

gunshot residue (GSR) the tiny particles expelled from a firearm when it is fired

gymnosperm a plant with naked seeds that are not enclosed in a protective chamber (fruit), such as an evergreen

H

hair follicle the actively growing root or base of a hair containing DNA and living cells

humus material in the upper layers of soil made up of the decaying remains of plants and animals

I

indentation mark a mark or impression made by a tool on a softer surface

individual evidence a kind of evidence that identifies a particular person or thing

instar one of the three larval stages of insect development

J

joints locations where bones meet

K

keratin a type of fibrous protein that makes up the majority of the cortex of a hair

lands and **grooves** the ridges (lands) and depressions (grooves) found on the inside of a firearm's barrel that are created when the firearm is manufactured

L

larva (plural **larvae**) immature form of an animal that undergoes metamorphosis (for example, a maggot)

latent fingerprint a hidden fingerprint made visible through the use of powders or other techniques

Pharmacists provide controlled substances by prescription, so accidental overdoses do not occur.

©ThinkStock Images//Jupiter Images

latent impressions hidden impressions requiring special techniques to be visualized

leaching the removal of minerals and clay as water drips through the soil

leaded glass glass containing lead oxide

lines of convergence a two-dimensional view of the intersection of lines formed by drawing a line through the main axis of at least two drops of blood that indicates the general area of the source of the blood spatter

livor mortis the pooling of the blood in tissues after death resulting in a reddish color to the skin

Locard's Exchange Principle contact between individuals and locations that leads to an exchange of trace evidence

logic the process of forming conclusions from assumptions and known facts

Technicians use sophisticated tools to find trace evidence.

logical   conclusions drawn from assumptions and know facts

loop   a fingerprint pattern in which the ridge pattern flows inward and returns in the direction of the origin

luminol   a chemical which glows in the presence of blood

M

manner of death   one of four means by which someone dies (i.e., natural, accidental, suicidal, or homicidal)

mechanism of death   the specific body failure that leads to death

medulla   the central core of a hair fiber

melanin granules   bits of pigment found in the cortex of a hair

metabolism   the rate at which the body burns food and liberates energy

mineral   a naturally occurring, crystalline solid formed over time on Earth

mineral fiber   a collection of mineral crystals formed into a recognizable pattern

minutiae   the combination of details in the shapes and positions of ridges in fingerprints that make each unique; also called ridge characteristics

mitochondrial DNA   DNA found only in the mitochondria that is inherited only through mothers

monomer   a small molecule that may bond to other monomers to become a polymer

muzzle   the end of the barrel where the projectile exits a firearm

N

narcotic   an addictive drug, such as opium, that relieves pain, alters mood and behavior, and causes sleep or feelings of mental numbness

natural fiber   a fiber produced naturally and harvested from animal, plant, or mineral sources

National Integrated Ballistics Information Network (NIBN)   stores ballistics information

neutron activation analysis   a method of analysis that determines composition of elements in a sample

Ninhydrin   a chemical which reacts to the amino acids given off by the skin and can be used to produce a fingerprint

nitrogenous bases   the building blocks of DNA and RNA—guanine, cytosine, thymine adenine, and uracil

normal line   a line drawn perpendicular to the interface surface of two different media

nucleotide   a molecule made up of a sugar, phosphate group, and a nitrogenous base

nuclear DNA   DNA found in the nucleus of the cell

O

observation   what a person perceives using his or her senses

obsidian   volcanic glass

opinion   personal belief founded on judgment rather than on direct experience or knowledge

ossification   the process that replaces soft cartilage with hard bone by the deposition of minerals

osteobiography the physical record of a person's life as told by his or her bones

osteoblast a type of cell capable of migrating and depositing new bone

osteoclast a bone cell involved in the breaking down of bone and the removal of wastes

osteocyte an osteoblast that becomes trapped in the construction of bone; also known as a living bone cell

osteoporosis weakening of bone, which may happen if there is not enough calcium in the diet

P

palynology the study of pollen and spore evidence to help solve crime cases

paper bindle a folded paper used to hold trace evidence

patent fingerprint a visible fingerprint that happens when fingers with blood, ink, or some other substance on them touch a surface and transfer the pattern of their fingerprint to that surface

patent impressions two-dimensional impressions that are already visible

PCR (polymerase chain reaction) a method used to rapidly make multiple copies of a specific segment of DNA; can be used to make millions of copies of DNA from a very small amount of DNA

perception interpreting information received from the senses

pH scale a scale used to identify if a substance is acidic, basic, or neutral

pistil the female reproductive part of a flower where eggs are produced

pistol a handheld firearm

plastic fingerprint a three-dimensional fingerprint made in soft material such as clay, soap, or putty

plasma the liquid portion of the blood

plastic impressions three-dimensional impressions cast in soft materials, such as soil and snow or blood

point of origin a three-dimensional view formed using lines of convergence and angles of impact of at least two different drops of blood to identify the source and location of blood spatter

poison a naturally occurring or manufactured substance that can cause severe harm or death if ingested, inhaled, or absorbed through the skin

Forensic palynologists examine soil, dirt, and dust for pollen and spores.

©Joe Raedle/Getty Images

pollen "fingerprint", also called a **pollen profile**, the number and type of pollen grains found in a geographic area at a particular time of year

pollen grain a reproductive structure that contains the male gametes of seed plants

pollination the transfer of pollen from the male part to the female part of a seed plant

polymer a substance composed of long chains of repeating units

primary crime scene the location where the crime took place

pupa (plural **pupae**) the stage in an insect's life cycle when the larva forms a capsule around itself and changes into its adult form

Q

questioned document any signature, handwriting, typewriting, or other written mark whose source or authenticity is in dispute or uncertain

Medical examiners remove a body from a crime scene after forensic investigators have done their initial investigation.

R

radial fracture pattern   a fracture pattern in glass which appears as lines radiating from the point of impact

radioactive probe   a molecule used to tag a region of DNA for the purpose of making it visible in a DNA profile

red blood cells   donut-shaped cells that carry oxygen throughout the body

refraction   the change in the direction of light as it changes speed when moving from one substance into another

refractive index   a measure of how light bends as it passes from one substance to another

restriction enzyme   a molecule that cuts a DNA molecule at a specific, base sequence

revolver   a pistol with a revolving cylinder

ridge pattern   the recognizable pattern of the ridges found in the end joints of fingers that form lines on the surfaces of objects in a fingerprint, they fall into three categories: arches, loops, and whorls

rifle   a firearm that has a long barrel; a long gun

rifling   the spiral pattern of lands and grooves in the barrel of a firearm

rigor mortis   the stiffening of the skeletal muscles after death

rock   a hard substance made up of minerals

S

sand   granules of fine rock particles

satellite drop of blood   secondary drops formed when some blood breaks free from the main contact drop of blood

secondary crime scene   a location other than the primary crime scene, but that is in some way related to the crime, where evidence is found

secondary transfer   the transfer of evidence such as a fiber from a source (for example, a carpet) to a person (suspect), and then to another person (victim)

semen   fluid produced by the body containing sperm cells

semiautomatic   a pistol with a clip-fed mechanism that fires one shot per pull of the trigger; the empty cartridge ejects and the next cartridge advances automatically

serology   the study of the chemistry of the blood

shell casing   the metal (usually brass) housing for the gunpowder of a firearm

silicon dioxide (SiO_2)   the chemical name for silica

silt   a type of soil whose particles are larger than clay and smaller than sand

skeletal trauma analysis   the investigation of bones and the marks on them to uncover a potential cause of death

soil   a mixture of minerals, water, gases, and the remains of dead organisms that covers Earth's surface

soil profile   a cross section of horizontal layers, or horizons, in the soil that have distinct compositions and properties

sole (outsole)   the pattern on the bottom of a piece of footwear

spore   an asexual reproductive structure that can develop into an adult found in certain protists (algae), plants, and fungi

stereomicroscope   a binocular (two eyepiece) microscope used to study trace evidence or compare samples

stamen   the male reproductive part of a flower consisting of the anther and filament where pollen is produced

STR (short tandem repeat)   tandem (next to each other) repeats of short DNA sequences (two to five base pairs) with varying numbers of repeats found among individuals

synthetic fiber   a fiber made from a man-made substance such as plastic

T

tempered glass   treated glass which does not shatter

ten card   a form used to record and preserve a person's fingerprints

textile   a flexible, flat material made by interlacing yarns (or "threads")

tire groove   a depression in the tread pattern

tire rib   a ridge of tread running down the tread area and around the circumference of the tire

tool mark   any impression, scratch, or abrasion made when contact occurs between a tool and another object

toxicity   the degree to which a substance is poisonous or can cause injury

toxin   a poisonous substance naturally produced by certain plants, animals, and bacteria that is capable of causing disease or death in humans; a subgroup of poisons

trace evidence   small but measurable amounts of physical or biological material found at a crime scene

track width   the distance from the center of the tread pattern on the left tire to the center of the tread pattern on the corresponding right tire

trajectory   the path of flight of a projectile

tread pattern   the unique design of a tire's surface

turning diameter   a measure of how tight a circle can be driven by a vehicle

V

VNTR (variable number of tandem repeats)   tandem (next to each other) repeats of a short DNA sequence (9 to 80 base pairs) with varying numbers of repeats among individuals

W

weathering   formation of soil through the action of wind and water on rock

wheelbase   the distance from the center of the front axle on a vehicle to the center of the rear axle

white blood cells   cells that police the body destroying foreign materials

whorl   a fingerprint pattern that resembles a bull's-eye

Y

yarn   fibers that have been spun together

Appendix A

Table of *Sin(e)* (Angle)

Angle	Sin (a)	Angle	Sin (a)	Angle	Sin (a)	Angle	Sin (a)
0	0.0000	26	0.4383	52	0.7880	78	0.9781
1	0.0174	27	0.4539	53	0.7986	79	0.9816
2	0.0348	28	0.4694	54	0.8090	80	0.9848
3	0.0523	29	0.4848	55	0.8191	81	0.9876
4	0.0697	30	0.5000	56	0.8290	82	0.9902
5	0.0871	31	0.5150	57	0.8386	83	0.9925
6	0.1045	32	0.5299	58	0.8480	84	0.9945
7	0.1218	33	0.5446	59	0.8571	85	0.9961
8	0.1391	34	0.5591	60	0.8660	86	0.9975
9	0.1564	35	0.5735	61	0.8746	87	0.9986
10	0.1736	36	0.5877	62	0.8829	88	0.9993
11	0.1908	37	0.6018	63	0.8910	89	0.9998
12	0.2079	38	0.6156	64	0.8987	90	1.0000
13	0.2249	39	0.6293	65	0.9063		
14	0.2419	40	0.6427	66	0.9135		
15	0.2588	41	0.6560	67	0.9205		
16	0.2756	42	0.6691	68	0.9271		
17	0.2923	43	0.6819	69	0.9335		
18	0.3090	44	0.6946	70	0.9396		
19	0.3255	45	0.7071	71	0.9455		
20	0.3420	46	0.7193	72	0.9510		
21	0.3583	47	0.7313	73	0.9563		
22	0.3746	48	0.7431	74	0.9612		
23	0.3907	49	0.7547	75	0.9659		
24	0.4067	50	0.7660	76	0.9702		
25	0.4226	51	0.7771	77	0.9743		

Appendix B

Table of Tangents

Angle	Tangent	Angle	Tangent	Angle	Tangent
0	0.0000	30	0.5773	60	1.7317
1	0.0175	31	0.6008	61	1.8037
2	0.0349	32	0.6248	62	1.8804
3	0.0524	33	0.6493	63	1.9622
4	0.0699	34	0.6744	64	2.0499
5	0.0875	35	0.7001	65	2.1440
6	0.1051	36	0.7265	66	2.2455
7	0.1228	37	0.7535	67	2.3553
8	0.1405	38	0.7812	68	2.4745
9	0.1584	39	0.8097	69	2.6044
10	0.1763	40	0.8390	70	2.7467
11	0.1944	41	0.8692	71	2.9033
12	0.2125	42	0.9003	72	3.0767
13	0.2309	43	0.9324	73	3.2698
14	0.2493	44	0.9656	74	3.4862
15	0.2679	45	1.0000	75	3.7306
16	0.2867	46	1.0354	76	4.0091
17	0.3057	47	1.0722	77	4.3295
18	0.3249	48	1.1105	78	4.7023
19	0.3443	49	1.1502	79	5.1418
20	0.3639	50	1.1916	80	5.6679
21	0.3838	51	1.2347	81	6.3095
22	0.4040	52	1.2798	82	7.1099
23	0.4244	53	1.3269	83	8.1372
24	0.4452	54	1.3762	84	9.5045
25	0.4663	55	1.4279	85	11.4157
26	0.4877	56	1.4823	86	14.2780
27	0.5095	57	1.5396	87	19.0404
28	0.5317	58	1.6001	88	28.5437
29	0.5543	59	1.6640	89	56.9168

Appendix C

Bone Length Charts

All measurements are in centimeters. (2.54 cm = 1 inch)

American Caucasian Males

Factor x Bone Length	Plus	Accuracy
Stature (cm) = 2.89 × humerus	+ 78.10 cm	± 4.57
Stature(cm) = 3.79 × radius	+ 79.42 cm	± 4.66
Stature(cm) = 3.76 × ulna	+ 75.55 cm	± 4.72
Stature (cm) = 2.32 × femur	+ 65.53 cm	± 3.94
Stature (cm) = 2.60 × fibula	+ 75.50 cm	± 3.86
Stature (cm) = 1.82 × (humerus + radius)	+ 67.97 cm	± 4.31
Stature (cm) = 1.78 × (humerus + ulna)	+ 66.98 cm	± 4.37
Stature (cm) = 1.31 × (femur + fibula)	+ 63.05 cm	± 3.62

American Caucasian Females

Factor x Bone Length	Plus	Accuracy
Stature (cm)=3.36 × humerus	+ 57.97 cm	± 4.45
Stature (cm)=4.74 × radius	+ 54.93 cm	± 4.24
Stature (cm)=4.27 × ulna	+ 57.76 cm	± 4.30
Stature (cm)=2.47 × femur	+ 54.10 cm	± 3.72
Stature (cm)=2.93 × fibula	+ 59.61 cm	± 3.57

Caucasian, Both Sexes

Factor x Bone Length	Plus	Accuracy
Stature = 4.74 × humerus	+ 15.26 cm	± 4.94
Stature = 4.03 × radius	+ 69.96 cm	± 4.98
Stature = 4.65 × ulna	+ 47.96 cm	± 4.96
Stature = 3.10 × femur	+ 28.82 cm	± 3.85
Stature = 3.02 × tibia	+ 58.94 cm	± 4.11
Stature = 3.78 × fibula	+ 30.15 cm	± 4.06

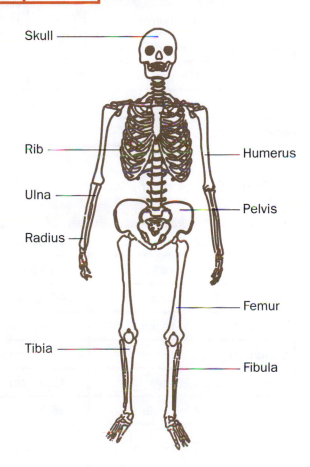

Skull

Rib

Ulna

Radius

Tibia

Humerus

Pelvis

Femur

Fibula

African-American and African Males

Factor x Bone Length	Plus	Accuracy
Stature = 2.88 × humerus	+ 75.48 cm	± 4.23
Stature = 3.32 × radius	+ 85.43 cm	± 4.57
Stature = 3.20 × ulna	+ 82.77 cm	± 4.74
Stature = 2.10 × femur	+ 72.22 cm	± 3.91
Stature = 2.34 × fibula	+ 80.07 cm	± 4.02
Stature = 1.66 × (humerus + radius)	+ 73.08 cm	± 4.18
Stature = 1.65 × (humerus + ulna)	+ 70.67 cm	± 4.23
Stature = 1.20 × (femur + fibula)	+ 67.77 cm	± 3.63

African-American and African Females

Factor x Bone Length	Plus	Accuracy
Stature = 3.08 × humerus	+ 64.67 cm	± 4.25
Stature = 3.67 × radius	+ 71.79 cm	± 4.59
Stature = 3.31 × ulna	+ 75.38 cm	± 4.83
Stature = 2.28 × femur	+ 59.76 cm	± 3.41
Stature = 2.49 × fibula	+ 70.90 cm	± 3.80

All Ethnic Groups or, if Ethnicity is Unknown, Both Sexes

Factor x Bone Length	Plus	Accuracy
Stature = 4.62 × humerus	+ 19.00 cm	± 4.89
Stature = 3.78 × radius	+ 74.70 cm	± 5.01
Stature = 4.61 × ulna	+ 46.83 cm	± 4.97
Stature = 2.71 × femur	+ 45.86 cm	± 4.49
Stature = 3.01 × femur	+ 32.52 cm	± 3.96
Stature = 3.29 × tibia	+ 47.34 cm	± 4.15
Stature = 3.59 × fibula	+ 36.31 cm	± 4.10

Appendix D

Celsius-Fahrenheit Conversion Table

°Celsius	°Fahrenheit	°Celsius	°Fahrenheit	°Celsius	°Fahrenheit	°Celsius	°Fahrenheit	°Celsius	°Fahrenheit
– 40	– 40	– 10	14.0	20	68.0	50	122.0	80	176.0
– 39	– 38.2	– 9	15.8	21	69.8	51	123.8	81	177.8
– 38	– 36.4	– 8	17.6	22	71.6	52	125.6	82	179.6
– 37	– 34.6	– 7	19.4	23	73.4	53	127.4	83	181.4
– 36	– 32.8	– 6	21.2	24	75.2	54	129.2	84	183.2
– 35	– 31.0	– 5	23.0	25	77.0	55	131.0	85	185.0
– 34	– 29.2	– 4	24.8	26	78.8	56	132.8	86	186.8
– 33	– 27.4	– 3	26.6	27	80.6	57	134.6	87	188.6
– 32	– 25.6	– 2	28.4	28	82.4	58	136.4	88	190.4
– 31	– 23.8	– 1	30.2	29	84.2	59	138.2	89	192.2
– 30	– 22.0	0	32.0	30	86.0	60	140.0	90	194.0
– 29	– 20.2	+ 1	33.8	31	87.8	61	141.8	91	195.8
– 28	– 18.4	2	35.6	32	89.6	62	143.6	92	197.6
– 27	– 16.6	3	37.4	33	91.4	63	145.4	93	199.4
– 26	– 14.8	4	39.2	34	93.2	64	147.2	94	201.2
– 25	– 13.0	5	41.0	35	95.0	65	149.0	95	203.0
– 24	– 11.2	6	42.8	36	96.8	66	150.8	96	204.8
– 23	– 9.4	7	44.6	37	98.6	67	152.6	97	206.6
– 22	– 7.6	8	46.4	38	100.4	68	156.2	98	208.4
– 21	– 5.8	9	48.2	39	102.2	69	156.2	99	210.2
– 20	– 4.0	10	50.0	40	104.0	70	158.0	100	212.0
– 19	– 2.2	11	51.8	41	105.8	71	159.8		
– 18	– 0.4	12	53.6	42	107.6	72	161.6		
– 17	+ 1.4	13	55.4	43	109.4	73	163.4		
– 16	3.2	14	57.2	44	111.2	74	165.2		
– 15	5.0	15	59.0	45	113.0	75	167.0		
– 14	6.8	16	60.8	46	114.8	76	168.8		
– 13	8.6	17	62.6	47	116.6	77	170.6		
– 12	10.4	18	64.4	48	118.4	78	172.4		
– 11	12.2	19	66.2	49	120.2	79	174.2		

Formulas: $T_c = \dfrac{5}{9}(T_r - 32)$ $T_r = \dfrac{9}{5}T_c + 32$

Index

livor mortis, 312–13
Locard's exchange principle, 22, 56
logical conclusions, 7
long bones, 369
long guns, 493
loop, 137
LSD, 254
Luminol, 207
lying, 12

M

Madrid train bombings, 140
maggots, 320, 322
magnesium smoking, 474
Magnuson, John, 290
Malvo, John Lee, 490
manila, 82
man-made fibers, 83–84
manner of death, 311
Maples, William, 471, 477
marijuana, 254
Markov, Georgi, 261
Marquis, Frank, 60
Marsh, George, 87
Mayer, Johann Christoph Andreas, 134
Mayfield, Brandon, 140
McCourt, Helen, 171
McCurdy, Elmer, 375
McDonald, Pete, 445
MDMA (Ecstasy), 254
mechanism of death, 311–12
medical examiners, 24, 210, 446
medulla, 52, 53, 55
melanin, 52
memory, 8
mercury, 258
mescaline, 254
metals, 257–58
methadone, 255
methamphetamines, 255
methanol, 256
microscopic techniques
   hair analysis, 51, 56–57
   polarizing light, 79
   pollen and spore analysis, 116

Midwest Forensics Resource Center (MFRC), 478
Milne, Lynne, 120
mineral fibers, 83
minerals, 341, 344–45
minutiae, 138, 139
mitochondrial DNA, 161, 373
modified natural fibers, 83
monomers, 83
morphine, 255
M proteins, 201
Muhammad, John Allen, 490
Mullis, Kary, 164, 172
muzzle, 492

N

NAA (neuron activation analysis), 48, 57, 411
Napoleon Bonaparte, 60
narcotics, 253, 255
National Integrated Ballistics Network (NIBIN), 498
National Integrated Bullet Identification System (NIBIS), 498
natural death, 311
natural fibers, 81–83, 84
neuron activation analysis (NAA), 48, 57, 411
9/11/01, 6
ninhydrin, 141
nitrogenous bases, 161
Noguchi, Thomas, 446
nonseed plants, 108. See also spores
normal line, 399
N proteins, 201
Nutt, Susan, 410
nylon, 83–84

O

observations
   case studies, 10–11
   definition of, 4–5
   in forensics, 9
   tips for improving, 7–8
   by witnesses, 5–7

obsidian, 396
Oetting, Lillian, 31
olefins, 84
oolites, 345
opinion, 7
opium, 255
Orfila, Mathieu, 253
ossification, 363
osteobiography, 365
osteoblasts, 363
osteoclasts, 363–64
osteocytes, 363
osteoporosis, 364, 365
Oswald, Lee Harvey, 503
ovoid bodies, 54
oxycodone, 255

P

Packer, Alfred, 375
palynology, 108, 117, 120. See also pollen; spores
paper bindles, 26, 27–28
papilla, 51
patent fingerprints, 138
patent impressions, 432
paternity, 168
Payne, Roger, 87–88
PCP, 254
PCR (polymerase chain reaction), 164, 166, 172
pelvis, male vs. female, 367–68
perception, 4–5
pesticides, 257–58
pH, of soil, 342–43
photographs
   of shoe impressions, 434–35
   of tool marks, 473–74
physical evidence, 23
Pilobus fungi, 113
pistil, 110
pistols, 493
Pitchfork, Colin, 170
plain weave, 85, 86
plant fibers, 81
plasma, 197
Plaster of Paris, 436
plastic fingerprints, 138